LOST IN SPACE

A star-forming section of the Cat's Paw Nebula as captured by the James Webb Space Telescope

LOST IN SPACE

5,000 FACTS FOR NAVIGATING THE UNIVERSE

NEIL DEGRASSE TYSON
and LINDSEY NYX WALKER

A STARTALK BOOK

NATIONAL GEOGRAPHIC
Washington, D.C.

A lunar halo, also known as a Moon ring or winter halo, curves around Chile's La Silla Observatory.

CONTENTS

The Atlas V rocket launches the Lucy spacecraft from Cape Canaveral, Florida, in 2021.

INTRODUCTION

We live in a universe of wonder and mystery. We gaze into the night sky, dazzled by immensity. Any time we look up, our curiosity outweighs our knowledge, and we keep asking, What? When? Where? Why? Especially Why?

Why do we ask "Why?" Because we are human, and our brains work that way. We crave understanding. We observe and measure and calculate so that we can come closer to the cosmos, or at least find delight in understanding.

Science—the pursuit of all that is objectively true in the universe—has long driven us to explore, fueled by curiosity. Curiosity emerges in the youngest of us. The toddler pokes at dirt, tastes it, to get to know the world better. And that thread is at the heart of scientific pursuit through human history. Ancient astronomers sought rules by which the Sun, Moon, planets, and stars behave and interact.

The astrophysicists of today design and operate exquisite technologies. They not only probe the tiniest particles of matter that manifest the very forces governing our existence but also explore beyond what our eyes can behold—and, on occasion, beyond what our minds can comprehend. And that's just what we'll explore in these pages. So join us as we get lost in space together.

This adventurous accumulation of knowledge brings us to the spirit and intention of this book. Each of the 5,000 facts collected here represents a step—some small, some large—along the path that humans have traveled. That path is partly physical, but mostly intellectual as we ask ever more questions. The path is not straight; there are forks in the road, even U-turns, detours, and adjustments to our orientation. But every step—every fact—builds on what came before.

So, let's navigate from the beginning, where our history and earliest understanding originate: the home base we call planet Earth. What can we see? What can we glean? What can we figure out? Then, as we reach out to space, via our balloons and planes, our jets and rockets, our satellites and telescopes, we become ever more enlightened. Thanks to the methods and tools of science, we've come to know more about the Sun and Moon, as well as the planets with moons of their own—even coming to view our very own Earth as just another planet in this complex system.

Our search for facts takes us farther and farther out as we come to know the stars and then the galaxies they inhabit, including our own Milky Way—just a blip in the cosmos. Ultimately, the human quest for knowledge arrives at the most remarkable, nearly unthinkable, but objectively real concepts of space and time. Fact by fact, we have navigated our way, only to find even more unanswered questions.

You can choose to start this joyride at the beginning and follow all the way through, or you can dip in wherever, whenever, discovering a chance to laugh at the way we have juxtaposed these majestic facts with our favorite fictions.

As always, don't forget to go outside every now and then and look up into the universe—where, as you'll discover, the sky is not the limit.

A protostellar jet in the star-forming region Sh2-284, at the edge of the Milky Way, is eight light-years across, almost double the distance from the Sun to neighboring Alpha Centauri.

1

Before you can ascend toward the stellar splendors beyond, before you can get lost among the sprawling distant galaxies laced with strange new worlds, you must first understand where you are: the wondrous world below your feet. Welcome to planet Earth, the only home we've ever known.

Astrophotographer Rogelio Bernal Andreo took this self-portrait at Tenaya Lake in Yosemite National Park, California, with the Milky Way as his backdrop.

HOME

BASE

CURRENT FACT COUNT 0035

HISTORY OF THE PLANET

Earth was birthed in the chaotic tumble of a newly forming star system, where it survived countless violent bombardments to become the only known haven to life.

- If the entire 14-billion-year timeline of cosmic history, from the Big Bang to today, were placed upon a calendar year, the solar system would form around the beginning of September.
- Earliest life would form in mid-September.
- Multicelled organisms would form in mid-December.
- The entire history of civilization, including all knowledge of human events captured from the earliest writing to present day, would unfold during the final fifteen seconds of December 31.

Rock of ages: Australia's Pilbara region is 3.6 billion years old, predating atmospheric oxygen.

IN THE BEGINNING

- **ABOUT 4.6 BILLION YEARS AGO, A CLOUD OF INTERSTELLAR GAS AND DUST COLLAPSED, THE FIRST STEP IN CREATING OUR SOLAR SYSTEM.**
- The cloud's collapse may have been triggered by the explosion of a nearby star, called a supernova.
- The cloud flattened as it swirled, and matter began to clump together, heat up, and collapse into a singular, massive object at its center.
- That object became our Sun.
- The remaining bits of debris coagulated into dozens of protoplanets, including Earth, that swirled around our young star.
- Space rocks bombarded our newborn world in collisions that helped the planet grow larger and larger.
- The early solar system formed from dust in less than 200,000 years. That's like a human pregnancy lasting just 12 hours.
- **THE KANT NEBULAR HYPOTHESIS, PROPOSED IN THE MID-18TH CENTURY BY PHILOSOPHER IMMANUEL KANT, EXPLAINS THAT THE SOLAR SYSTEM FORMED FROM THE COLLAPSE OF A SWIRLING GAS CLOUD.**
- Kant postulated the hypothesis before we understood the true age of the universe or even the planet itself. It remains the most plausible and accepted explanation.
- French intellectual Pierre-Simon Laplace proposed a similar nebular hypothesis in the early 19th century, suggesting that the planets formed while a vast cloud collapsed due to gravity.
- **EARLY EARTH WAS A SCORCHING HELL-SCAPE, PLAGUED BY FREQUENT VOLCANIC ERUPTIONS AND CONSTANT BOMBARD-MENTS BY SPACE ROCKS OF EVERY SIZE.**
- Most of young Earth's surface was so hot that nearly its entire surface was a molten ocean of liquid rock.
- We call that period (4.6 to 4.0 billion years ago) the Hadean, named for Hades, Greek god of the underworld.
- As Earth cooled, the rocks settled according to their density, forming layers like a cake.
- The densest components—mostly iron and nickel—sank and collected in the planet's center: Earth's core today.
- The lightest rocks—mainly silicate with quartz and feldspar minerals—floated to the surface to form Earth's upper crust.
- **FEW ROCKS FROM THE HADEAN PERIOD ON EARTH SURVIVE TODAY.**
- Earth's oldest known rock, dated to 4.011 billion years ago, wasn't found on Earth but as an Earth meteorite on the Moon.

An illustration depicts the formation of a planetary system around its host star.

- Its true origin remains contested, but if it did derive from Earth rather than the Moon, it was possibly ejected by an ancient impact.
- Earth's interior is still sweltering, heated by the leftover energy of the planet's formation and the decay of radioactive elements.
- The intense volcanic eruptions on young Earth spewed gases like carbon dioxide and water vapor, creating a thick atmosphere.
- **A FEW HUNDRED MILLION YEARS AFTER ITS FORMATION, EARTH'S SURFACE COOLED ENOUGH TO SUSTAIN LIQUID WATER.**
- Earth's mantle (the hot layer between its crust and core) might hold more water than all the oceans combined.
- Earth's surface water might have burst out from inside the planet, with volcanic eruptions unleashing the water vapor locked in deep rocks.
- Once early Earth cooled below water's boiling point, liquid rain fell and filled the empty basins to form a primeval ocean.
- How did Earth get its water in the first place? The sources are still debated, but one popular hypothesis posits that water-rich asteroids collided with our young planet.
- **THREE TO FOUR BILLION YEARS AGO, EARTH'S SURFACE CRACKED LIKE THE SHELL OF A HARD-BOILED EGG, CREATING A UNIQUE GLOBAL SYSTEM OF SHIFTING PLATES.**
- Earth's plates constantly jockey for position, creating new land and destroying old.
- So far, Earth is the only known planet with this feature of plate tectonics, though some planets and moons show evidence of tectonic activity in their ancient pasts.
- **ANCIENT MICROBES APPEARED ON EARTH AT LEAST 3.5 BILLION YEARS AGO: THE AGE OF THE OLDEST ACCEPTED FOSSILS OF LIFE.**
- Previous relatives of the earliest known microbes, though not preserved as fossils themselves, left evidence of their existence as banded iron deposits called stromatolites: mushroom-shaped mounds off Australia's western coast.

CURRENT FACT COUNT 0146

THE PERIODIC TABLE OF ELEMENTS

Most elements were created during the deaths of massive stars. A few were born of the universe itself, and some are only known to exist in the labs of Earthling scientists.

- **TODAY'S PERIODIC TABLE CURRENTLY CONTAINS 118 ELEMENTS, WITH MORE YET TO BE DISCOVERED.**
- The periodic table was first conceived by Russian scientist Dmitri Mendeleev in 1869, 10 years after Charles Darwin's publication of *On the Origin of Species*.
- Elements in the periodic table are defined by their atomic number: the number of protons in the atom's nucleus.
- The chemical properties of an element derive from its atomic structure.
- Each column of the periodic table represents groups of elements that chemically bond with other elements in the same way.
- Mendeleev's original periodic table, published in 1869, had 63 elements.

Unlike what its name suggests, the lead in pencils is actually made from the mineral graphite.

- Using his periodic table, Mendeleev correctly predicted the existence of three elements that would later be found in his lifetime: gallium, scandium, and germanium.
- **MODERN PHYSICS HAS IDENTIFIED EVERY ELEMENT IN THE NATURAL UNIVERSE, AND PHYSICISTS HAVE CREATED MORE IN LABORATORIES THAT THE UNIVERSE HAS NEVER SEEN.**
- Of the 118 known elements, 92 were forged in the Big Bang and in the cores of high-mass stars. The rest we make in laboratories.
- Many elements on Earth can take multiple forms. For example, carbon is the element that makes up both the active ingredient in pencils and diamonds.
- Observations reveal that the same elements we observe on Earth occur throughout the universe. No element is unique to Earth.

✦ TOP 10 MOST PREVALENT ELEMENTS IN EARTH'S CRUST

- Oxygen (46.6%)
- Silicon (27.7%)
- Aluminum (8.1%)
- Iron (5%)
- Calcium (3.6%)
- Sodium (2.8%)
- Potassium (2.6%)
- Magnesium (2.1%)
- Titanium (0.4%)
- Hydrogen (0.1%)

OPPOSITE: Mendeleev's first periodic system, with space for elements yet to be discovered

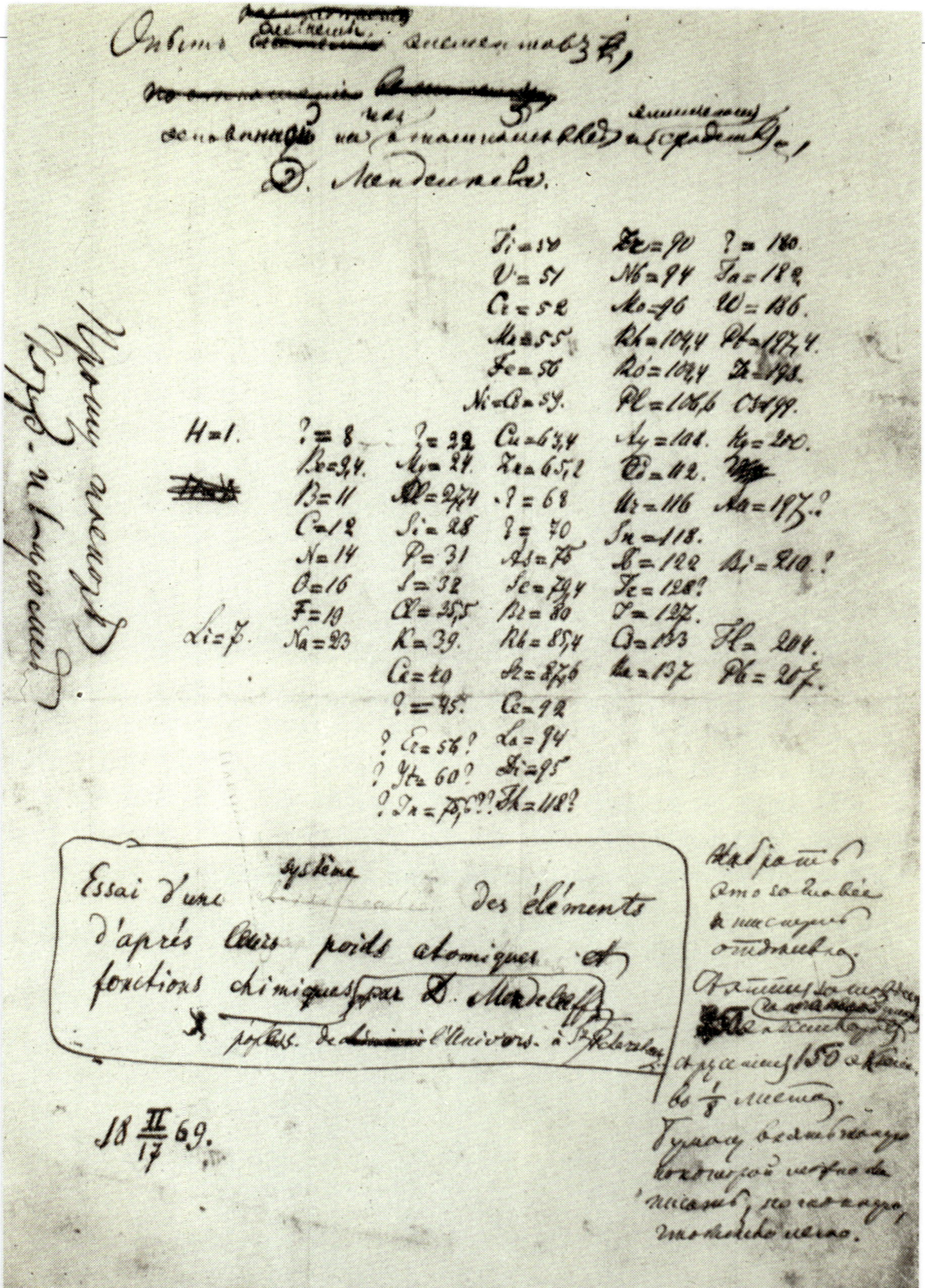

Опыт системы элементов, основанной на их атомном весе и химическом сходстве.

Д. Менделеева.

			Ti=50	Zr=90	?=180.
			V=51	Nb=94	Ta=182
			Cr=52	Mo=96	W=186.
			Mn=55	Rh=104,4	Pt=197,4.
			Fe=56	Ru=104,4	Ir=198.
			Ni=Co=59.	Pd=106,6	Os=199.
H=1.	?=8	?=22	Cu=63,4	Ag=108.	Hg=200.
[illegible]	Be=9,4.	Mg=24.	Zn=65,2	Cd=112.	[illegible]
	B=11	Al=27,4	?=68	Ur=116	Au=197?.
	C=12	Si=28	?=70	Sn=118.	
	N=14	P=31	As=75	Sb=122	Bi=210.?
	O=16	S=32	Se=79,4	Te=128?	
	F=19	Cl=35,5	Br=80	J=127.	
Li=7.	Na=23	K=39.	Rb=85,4	Cs=133	Tl=204.
		Ca=40	Sr=87,6	Ba=137	Pb=207.
		?=45.	Ce=92		
		?Er=56?	La=94		
		?Yt=60?	Di=95		
		?In=75,6??	Th=118?		

Essai d'une système des éléments d'après leurs poids atomiques et fonctions chimiques par D. Mendeleeff

profess. de l'Univers. à St Pétersbourg.

18 II/17 69.

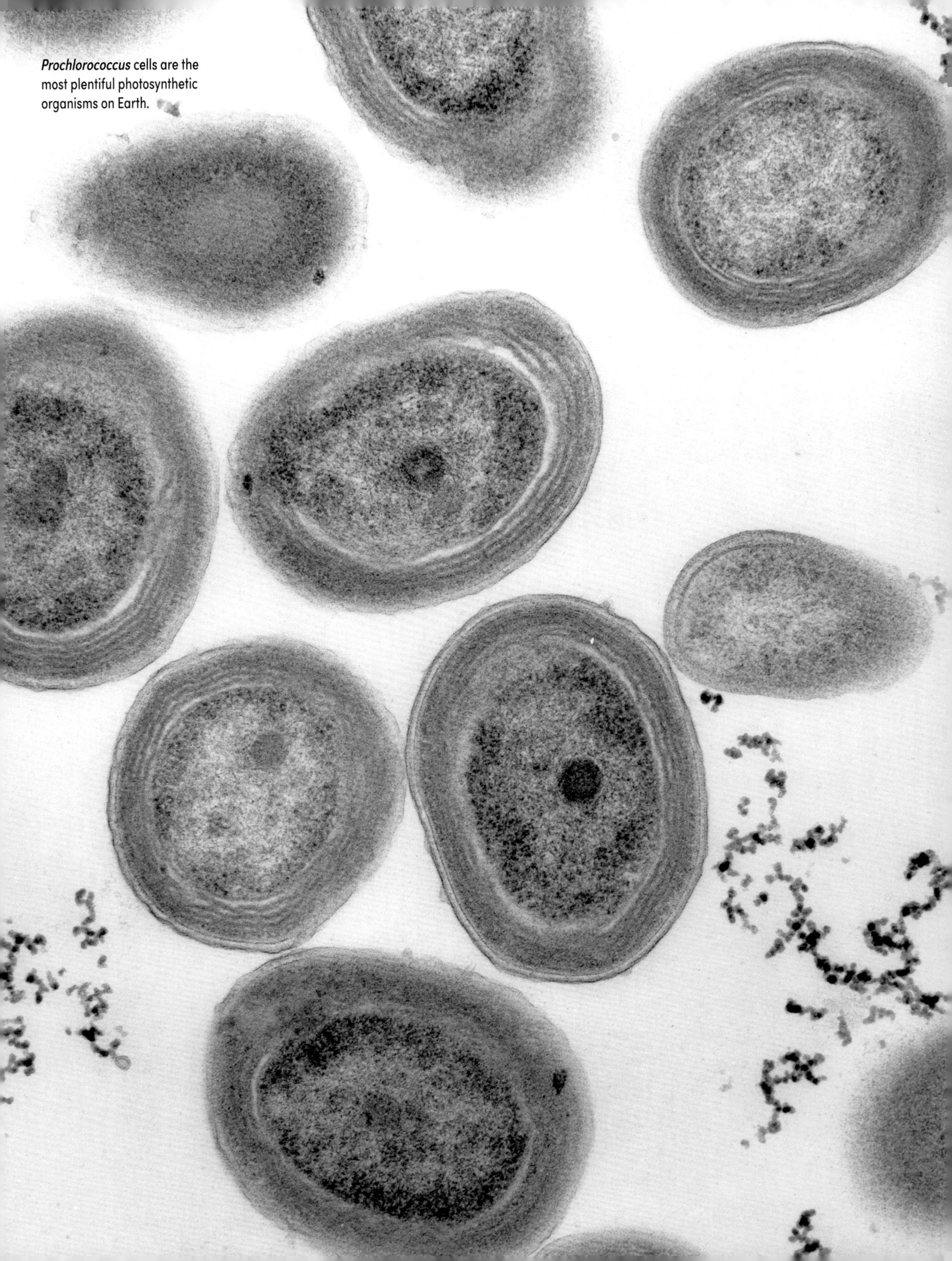

Prochlorococcus cells are the most plentiful photosynthetic organisms on Earth.

✦ WHEN THE UNIVERSE FIRST FORMED ...

- there were only three elements: hydrogen,
- then helium,
- followed by lithium.
- **STARS CREATE ELEMENTS THROUGH NUCLEAR FUSION BY COMBINING LIGHTER ELEMENTS INTO HEAVIER ONES.**
- The birth, life, and death of stars create and disperse elements throughout the universe.
- Star life cycles form the elements that make up our bodies. As Carl Sagan famously said, "We are made of star-stuff."
- Many heavy elements form when stars die and explode in a supernova or collapse to create a neutron star or black hole.
- The gold in your jewelry comes from the collisions of neutron stars billions of years ago.
- **THE MOST ABUNDANT ELEMENT IN THE UNIVERSE IS HYDROGEN.**
- Hydrogen makes up 75 percent of the mass of the universe.
- But it is found only in trace amounts on Earth.
- Earth's earliest atmosphere was likely composed of hydrogen and helium, like many of the atmospheres of the solar system's outer planets.
- Earth lost most of its atmospheric hydrogen and helium because it doesn't have strong enough gravity to hang on to such light gases.
- All of early Earth's oxygen was locked tight in its rocks.
- Earth's oldest gaseous oxygen likely came from photosynthesis of a tiny blue-green cyanobacterium.
- Since oxygen readily reacts with silicon, aluminum, iron, and other common ingredients of Earth's crust, it took a long time for oxygen to increase in Earth's atmosphere.
- Around the world is an ancient layer of red-and-black rock known as a banded iron formation that contains the records of Earth's earliest gasps of oxygen.

A cross section of a banded iron formation in Karijini National Park, Australia

1980s actor Mr. T decked in the element gold

- Banded iron formations were created when blooms of cyanobacteria released oxygen, which reacted with dissolved iron in the oceans, creating iron particles that settled to the seafloor.
- Cyanobacteria helped create most of the iron mined today, which is sourced largely from these banded iron formations.
- **OXYGEN ALSO INTERACTS AND BONDS EASILY WITH OTHER ATOMS.**
- Earth's abundance of free oxygen is a by-product of photosynthesis.
- Thank the oceans for the oxygen you breathe: More than half of it comes from photosynthetic algae and bacteria.
- About 20 percent of Earth's atmospheric oxygen comes from just one group of marine bacteria called *Prochlorococcus*.
- There are three octillion *Prochlorococcus* in the ocean.
- That many *Prochlorococcus* weigh more than 200 million Volkswagen Beetles.

✦ ELEMENTAL

- Aluminum is the third most common element found in Earth's crust and the most abundant metal, yet it was not discovered until 1825.
- **ALUMINUM WAS ONCE CONSIDERED AS PRECIOUS AS SILVER.**
- Because of aluminum's atomic urge to combine with other elements, it's nearly impossible to find in pure form. It must be chemically extracted from other mined materials. For decades, that process was costly, time-consuming, and yielded only small amounts of the partially pure element.
- The Washington Monument was clad with a six-pound pyramid of aluminum in 1884. Two years later, the cost of the "precious metal" plummeted when a recently graduated chemist derived a highly efficient method to extract aluminum at an industrial scale. Now we buy rolls of it at the grocery store and discard it after a single use.
- Like their corresponding planet's position in the solar system, uranium (element 92), neptunium (element 93), and plutonium (element 94) lie in consecutive order on the periodic table.
    - Uranus: Discovered in 1781. Uranium: Discovered in 1789
    - Neptune: Discovered in 1846. Neptunium: Discovered in 1940
    - Pluto: Discovered in 1930. Plutonium: Discovered in 1940
- **IN 1940, WHEN PLUTONIUM WAS DISCOVERED, PLUTO WAS STILL CONSIDERED A PLANET. THE ELEMENT NAME DIDN'T CHANGE 66 YEARS LATER, WHEN PLUTO'S CLASSIFICATION OFFICIALLY DID.**
- Oxygen is not itself flammable, but it does promote the combustion of other flammable materials. If oxygen were flammable, simply lighting a match would ignite Earth's atmosphere.
- The densest naturally occurring element is osmium. One cubic foot of it weighs nearly a ton.
- **THE HEAVIEST KNOWN ELEMENT, OGANESSON, IS SYNTHETIC.**
- You can pack significantly more osmium atoms into a given space than oganesson atoms, but a single oganesson atom weighs more than a single osmium atom.
- Oganesson is so unstable, it starts breaking down in less than a millisecond.
- Some superdense asteroids may contain elements much heavier than known elements on the periodic table.
- 33 Polyhymnia is one such superdense asteroid that may contain yet undiscovered superheavy elements.
- Polyhymnia is likely made from an element with 164 protons—that's 46 more than oganesson.
- Iridium is common on asteroids but rare on Earth because it bonds to iron, most of which sank to Earth's core as it formed.
- Earth's crust has a thin layer of iridium, key evidence of the Chicxulub asteroid collision that killed off the dinosaurs.

"I'M SURE THAT IN 1985, PLUTONIUM IS AVAILABLE IN EVERY CORNER DRUGSTORE, BUT IN 1955, IT'S A LITTLE HARD TO COME BY."

—DR. EMMETT BROWN, *BACK TO THE FUTURE*

✦ THE NOBLEST OF GASES

- The noble gases (helium, neon, argon, krypton, xenon, and radon) rarely react with other elements on Earth.
- Most noble gases were discovered or co-discovered by the British in the late 1800s.
- Upon discovering they don't chemically interact with other elements on the table, these elements were declared to be "noble," thus imbuing upon the sacred table features of Britain's own prejudiced class system.
- Oganesson is technically considered a noble gas as it resides in the same column as the other seven. But its noble traits end there. In fact, it may not be a gas at all but rather a solid.
- The inertness of noble gases makes them useful for many aeronautical applications, such as buoying blimps and pressurizing rocket fuel tanks.
- The harsh conditions of outer space might force usually inert noble gases to react, which created a rare molecule of argon and hydrogen in the Crab Nebula.
- **ASTEROIDS CONTAIN PLENTY OF USEFUL METALS, LIKE PLATINUM AND COBALT, THAT PRIVATE COMPANIES MAY MINE IN THE FUTURE TO SUPPLEMENT EARTH'S SUPPLIES.**
- Nebulium: the element that never was. In 1864, the Cat's Eye Nebula emitted a strange spectra inconsistent with any yet discovered. Sixty years later, the phenomenon was attributed to a state of the oxygen atom only possible in rarefied gaseous environs—like what you find inside nebulae.
- **DEEP UNDER EARTH'S SURFACE LURKS HELIUM THAT SCIENTISTS THINK DATES FROM THE PLANET'S FORMATION.**

- Vibranium is the fictional, nearly indestructible yet extremely lightweight metal element found in Black Panther's home, Wakanda.

- According to legend, vibranium was carried to Earth millions of years ago when a meteorite made entirely of the element crash-landed.

- **WATER IS SIMILARLY HYPOTHESIZED TO HAVE FIRST ARRIVED ON EARTH FROM A CRASH-LANDED SPACE ROCK, PERHAPS THE VERY SAME IMPACT EVENT THAT FORMED OUR MOON.**

- Our best hope for a vibranium-like material in the real world rests upon graphene, the strongest known material in the universe (about 200 times stronger than steel), as well as the thinnest. It's also the most conductive material on Earth.

- Because of quantum physics, many conductors become superconductors as the temperature drops near absolute zero.

✦ ELEMENTS NAMED AFTER ...

PEOPLE

- Lawrencium (103): named for Ernest Lawrence, who invented the cyclotron particle accelerator used in the element's discovery. It has no known existence outside the laboratory.
- Curium (96): named for Marie Skłodowska-Curie and her husband and research partner, Pierre Curie. The element's discovery in 1944 was kept classified until the end of World War II due to its ties to the Manhattan Project.
- Seaborgium (106): named for Glenn T. Seaborg, who helped discover at least 10 new elements and more than 100 isotopes
- Roentgenium (111): named for Wilhelm Röntgen, who discovered x-rays
- In Germany, x-rays are called Röntgen rays.
- Nobelium (102): named for Alfred Nobel, the inventor of dynamite
- Nobel used the wealth derived from inventing dynamite to fund the Nobel Prize in perpetuity.
- In 1901, the first ever Nobel Prize in Physics was awarded to Wilhelm Röntgen for his discovery of x-rays.
- Einsteinium (99): named for theoretical physicist Albert Einstein
- Einsteinium was discovered in the dusty radioactive cloud from the first test of the hydrogen bomb and later synthesized in a laboratory.

MYTHOLOGY

- Tantalum: named for King Tantalus of Greek mythology, who was condemned to an eternity of thirst and hunger while fruit and water lay just beyond his reach
- Niobium: named for the daughter of Tantalus, Niobe. Niobium's properties are so similar to tantalum's that they were believed to be one and the same for about five decades.
- Thorium: named after Thor, the Norse god of war. It was so named by its Swedish discoverer, Jöns Jakob Berzelius.
- Selenium: Also discovered by Jöns Jakob Berzelius, selenium is named after Selene, the Greek goddess of the Moon.
- Titanium: named for the immortal giants of Greek mythology due to its exceptional strength
- Helium (Sun): named for Helios, the Greek Titan of the Sun
- Iridium: This extremely rare element is found in unusually high abundance within some impact craters, as well as a specific, thin layer of Earth's crust that formed around 65 million years ago—evidence that it derives mainly from asteroids, including the one that ended the reign of the dinosaurs.

PLACES

- Ytterbium: named for Ytterby, Sweden

- Three other elements derived from the same mineral sample were subsequently named for the town:
    - Terbium
    - Erbium
    - Yttrium
- Scandium: extracted from a Yttrium sample and named for the Latin word (Scandia) for Scandinavia
- Americium: named for the Americas and sitting just below the following elements:
- Europium: named after Europe
- Francium: named for France
- Californium: named for the state of California
- Berkelium: named for Berkeley, California
- Polonium: discovered by Marie Skłodowska-Curie and named for her home country of Poland despite being discovered in Paris, France
- Moscovium: named for Moscow Oblast in Russia

THINGS

- Radium: named by Marie Skłodowska-Curie and Pierre Curie for the Latin word for "ray." Marie also coined the word "radioactivity."
- Technetium: named for the Greek word for "artificial." It's the first artificial element ever created.

CURRENT FACT COUNT 0313

THE ATMOSPHERE

Earth's atmosphere is that layered blanket of gas and particles that separates Earthlings nestled below from the void of outer space above.

- In thickness, Earth's atmosphere is to Earth what an apple's skin is to the apple.
- If Earth had no atmosphere, no one would benefit from the buoyant forces of air and everyone would weigh a few ounces more.
- **LAYERS OF THE ATMOSPHERE WITH INCREASING ALTITUDE:**
  - Troposphere (up to 12 miles above Earth's surface)
  - Stratosphere (around four to 31 miles up)
  - Mesosphere (31 to 53 miles up)
  - Thermosphere (53 to 375 miles up)
  - Exosphere (375 to 6,200 miles up)
- A so-called "pause" separates each layer of Earth's atmosphere where characteristics like temperature, chemistry, movement, and density rapidly change.

✦ TROPOSPHERE

- The height of the troposphere varies depending on which part of the planet it covers: It is highest at the Equator, lowest at the poles.
- **THE TROPOSPHERE HOLDS ALL THE AIR THAT PLANTS AND ALGAE USE FOR PHOTOSYNTHESIS.**
- It also contains all the oxygen we breathe.
- In the troposphere, temperatures drop with increasing altitude, but in the stratosphere, they warm with altitude.
- About 99 percent of the atmosphere's water vapor and aerosols are in the troposphere.
- Jet streams are windy belts blowing west to east near the boundary between the troposphere and stratosphere.
- Jet streams affect the pressure in the atmosphere, driving weather on the surface and influencing aircraft above.
- Earth's rotation drives the winds of the jet streams because spinning creates rotational momentum that propels the warm air rising near the Equator.
- The speed of Earth's rotation, along with the atmosphere and everyone and everything below and within it, moves fastest at the Equator (approximately 1,000 miles an hour) and gradually slows toward the North and South Poles, which remain nearly stationary.
- **THIS DIFFERENCE IN ROTATIONAL SPEED AFFECTS THE TRAJECTORY OF OBJECTS MOVING WITHIN THE ATMOSPHERE. THE PHENOMENON IS KNOWN AS THE CORIOLIS EFFECT.**
- Mount Everest's summit is exposed to the winds of the jet stream, which can flow in bursts so intense they drop oxygen levels, putting climbers at risk.
- The rotation of Earth can affect the outcome of football games. If a field is oriented north-south, then the Coriolis effect will deflect a 50-yard kick nearly a half inch to the right.
- In 2015, the Cincinnati Bengals barely clinched a close win with an overtime field goal kick, thanks in part to the Coriolis effect at work on the airborne football. The ball ricocheted off the upper left post and through the goal. The same force that sends hurricanes into a clockwise rotation (in the Northern Hemisphere) provided that game-winning fraction of an inch.

✦ STRATOSPHERE

- The stratosphere hosts the ozone layer, which is created when ultraviolet light reacts with molecular oxygen (O_2) to create ozone (O_3).
- Commercial planes fly in the lower stratosphere, where conditions are usually calmer than the troposphere below.
- Climate change is affecting the jet stream, intensifying storms and worsening airplane turbulence.

✦ MESOSPHERE

- The coldest temperature in Earth's atmosphere is about minus 130°F in the upper mesosphere.
- Most meteors burn up in the mesosphere.
- The top of the mesosphere has scarce water vapor that collects in glowing noctilucent clouds, the highest clouds in the atmosphere.

✦ THERMOSPHERE

- The thermosphere absorbs the Sun's high-energy x-rays and ultraviolet radiation, which causes a sharp increase in temperature.
- Temperatures increase with altitude in the thermosphere, from minus 185°F near the bottom of the layer to some 3600°F near the top.

- Bundle up: Thermosphere temperatures can reach thousands of degrees, but it would still feel cold to us because there are so few molecules to heat our skin—that is, if we could survive without a space suit.
- Heat is the measure of the vibration of molecules within a substance.
- When the Sun is highly active, the thermosphere absorbs the extra energy, which causes the layer to "puff up" and reach higher altitudes.
- **MOST OF OUR SATELLITES ORBIT IN THE THIN AIR OF THE THERMOSPHERE.**
- Auroras dance within the thermosphere.
- The International Space Station (ISS) orbits in the thermosphere at nearly 18,000 miles an hour.
- Continued collisions with the sparse air molecules at those high speeds create drag on the ISS.
- The ISS loses 1.2 miles in altitude each month. Occasional speed boosts via rockets keep it in orbit.

✦ IONOSPHERE

- The ionosphere is the layer where Earth's upper atmosphere and space meet.
- Here, solar radiation knocks electrons from atoms and molecules, forming a sea of negatively charged electrons and the now positively charged particles.
- Atoms and molecules that carry a positive or negative charge are called ions.
- The ionosphere increases on the day side of Earth, where solar radiation beats down onto the atmosphere.
- The ionosphere decreases on Earth's night side as the ions collide and recombine back into neutral atoms and molecules.
- The ionosphere is divided into three layers. The lowest layer disappears with every sunset and reemerges with every sunrise.
- Radio operators bounce radio waves off the electrically charged ionosphere, helping them extend the range of their transmissions.
- During daytime, the lowest layer of the ionosphere absorbs some AM radio waves, reducing the efficiency and distance of transmissions.
- This absorption can prove beneficial to reduce interference of competing radio signals.
- At night, when the ionosphere is at its thinnest, AM radio waves travel farther. That means you can hear more distant AM radio stations at night than you can during the day.
- But it also means radio signals transmitted at night are at higher risk of interfering with one another. That's why many AM channels go silent after dark.
- GPS satellites rely on radio waves to communicate with receivers on Earth. They must account and correct for the constant ionospheric fluctuations.

✦ EXOSPHERE

- The uppermost layer of Earth's atmosphere, called the exosphere, has no clearly defined boundary. Here, Earth's atmospheric particles and outer space blur together.
- Some scientists don't consider the exosphere part of the atmosphere at all.
- The exosphere is so rarefied that individual gas molecules almost never collide with one another.
- Even so, the scarce particles floating around the exosphere are enough to cause drag on orbiting satellites and eventually tug them to the ground without the occasional boost.

✦ THE KÁRMÁN LINE

- Just where does our atmosphere end and outer space begin? Simplest answer: at the Kármán line.
- The Kármán line is defined as "a physical boundary where aerodynamics stops and astronautics begins."
- It is named for Theodore von Kármán, the Hungarian physicist who first defined the boundary in the early 20th century. He was also head of the rocket science lab that became NASA's Jet Propulsion Laboratory.
- Kármán pinned the line at about 50 miles (80 km) above Earth's surface based on the available aeronautic technology of the day.
- Decades later, the Fédération Aéronautique Internationale (FAI) favored a rounder number and declared that space begins at 100 kilometers (62 mi) above sea level. They named it the Kármán line.
- The United States' Federal Aviation Administration (FAA) adopted a definition of space that begins at 50 miles (about 80 km), a closer reflection of Kármán's prediction.
- But NASA Mission Control remains true to Kármán's original definition and considers outer space to begin 76 miles (122 km) up, because that is approximately where airplanes require rocket boosters to remain aloft.
- Defining outer space via the Kármán line makes no sense on any other celestial object.
- Above the Kármán line, you can see stars in the daytime.
- If Earth's atmosphere were half its current pressure, the Kármán line would be 31 miles up (50 km). One-tenth its current pressure: 6.2 miles (10 km) up.

✦ FROM AIRGLOW TO DIRTY STORMS

- The molecules in Earth's atmosphere scatter incoming sunlight.
- The atmosphere deflects short wavelengths of blue and violet more readily, spreading them over a greater area than lower-frequency colors, giving the sky its azure hue.
- The sky is actually violet. The atmosphere scatters violet wavelengths even more than blue, just beyond what the human eyeball can see. Ask a spider, or another creature with superior eyesight, and the sky is assuredly violet.
- Near the horizon, the sky's blue color fades to white. That's because you're looking through a much thicker layer of atmosphere where light scatters along a longer path.
- The night sky is never completely dark. It always carries a faint glow caused by sunlight striking Earth's upper atmosphere, generating what's called airglow.
- Airglow can barely be seen from Earth's surface, but it can be seen from the International Space Station.
- Airglow can be many colors, creating layers of red, green, purple, and yellow visible light.
- **IF EARTH WERE SHRUNK DOWN TO THE SIZE OF A SCHOOLROOM GLOBE, THEN:**
  - The Moon would be the size of a softball 30 feet away.
  - Mars would sit one mile away.
  - The International Space Station would orbit half an inch from the globe's surface.
  - The Blue Origin and Virgin Galactic launches would reach two dimes' thickness from the surface.
  - Felix Baumgartner's 2012 daredevil jump would have begun one dime's thickness from the surface.
- **THE ERUPTION OF INDONESIA'S MOUNT TAMBORA IN 1815 UNLEASHED A TOWERING PLUME OF VOLCANIC ASH AND GAS 35.4 MILES INTO THE ATMOSPHERE WITH LONG-LASTING WORLDWIDE EFFECTS.**
- Tambora released 36 cubic miles of volcanic ash.
- Tambora also unleashed nearly 60 megatons of sulfur dioxide into the atmosphere, which reacts with water vapor in the stratosphere, creating Sun-blocking aerosols.
- Tambora's voluminous gaseous and ashy emissions blotted out the Sun.
- The ensuing cold and dark conditions earned 1816 the nickname "the year without summer."
- During the year without summer, cold snaps and unexpected storms caused crop failures and widespread starvation.
- The initial eruption killed 11,000 people in Indonesia, and the subsequent famine may have killed another 100,000 around the world.
- **SNOW FELL IN JUNE 1816 IN ALBANY, NEW YORK, AND DENNYSVILLE, MAINE.**
- It was during that gloomy summer of 1816 that Mary Shelley began to write *Frankenstein.*
- After Tambora's 1815 eruption, global temperatures took about 10 years to completely return to normal conditions.
- A reason to thank volcanic eruptions: The same aerosols released into the atmosphere can create dazzling sunsets and sunrises.
- Hunga Tonga-Hunga Ha'apai volcano's 2022 eruption in Tonga sent aerosols into the stratosphere that caused vibrant sunsets and sunrises in New Zealand, almost 1,500 miles away.
- **A VOLCANIC ERUPTION CAN PRODUCE LIGHTNING WITHIN ITS OWN PLUME.**
- Sometimes called "dirty thunderstorms," volcanic lightning can happen when particles of spewing ash collide and generate a static charge.
- The charge is then dispelled as sinister and stunning displays of lightning.

Hunga Tonga-Hunga Ha'apai's eruption shot ash 35 miles into the sky.

A sensitive digital camera can capture what the the human eye can't, including the airglow phenomenon, shown here above Canada's Dinosaur Provincial Park.

A long-exposure image catches a lightning strike over Colorado in 2019.

✦ LIGHTNING

- Lightning storms are nature's particle accelerators.
- Some highly energetic storms boot out electrons at near light speed. The phenomenon is dubbed "dark lightning."
- Dark lightning occurs when the emitted electrons collide with air molecules, emitting gamma rays in bursts—the highest energy on the electromagnetic spectrum, usually found around cataclysmic cosmic events and black holes.
- What causes lightning? We don't actually know.
- We know what it is and what conditions precede it, but we can't yet explain what triggers those conditions.
- One hundred bolts of lightning strike Earth every second.
- There are about 2,000 lightning storms occurring around the world at any given moment.
- When warm, moist air rises to meet the cool air above, warmer water droplets and frozen ice crystals collide. Those collisions strip electrons to form negatively and positively charged particles.
- Heavy, negatively charged particles and lighter, positively charged particles separate within the cloud, mimicking a battery.
- Eventually, the swell of negative charges in lower regions of the cloud overcomes the insulation properties of the surrounding air and discharges a surge of energy to the positive charges on the ground, on buildings, or, if you're the extremely unlucky sort, on you.
- Sometimes, negatively charged particles connect with positively charged particles within their own cloud or an adjacent one.
- A single bolt of lightning can superheat the air around it to about 50,000°F, five times hotter than the surface of the Sun.
- Thunder is the soundtrack of a lightning bolt's sonic boom. When lightning slashes through air, the superheated molecules rapidly expand, faster than the speed of sound. That's why you see lightning before you can hear it.
- The sound of thunder travels one mile in about five seconds, so you can approximate the distance to the lightning storm with a simple calculation: Count the number of seconds between the lightning bolt and sound of thunder, then divide by five (many people tend to forget the last step).
- Benjamin Franklin famously proved that lightning bolts were electrical discharges when he flew his kite during a Philadelphia storm in 1752 (don't try this at home).
- Contrary to popular lore, Franklin did not discover electricity. Rather, he demonstrated the "sameness of the electric matter with that of lightning," as he wrote in the *Pennsylvania Gazette*.
- **WITH THIS EXPERIMENT, FRANKLIN INVENTED THE FIRST LIGHTNING RODS—STILL IN USE TODAY—TO PROTECT HOMES AND BUILDINGS FROM LIGHTNING STRIKES AND ENSUING FIRES.**
- The Catholic Church renounced Franklin's invention for thwarting the will of God, who they believed wielded lighting to punish sinners. Ironically, their tall spires made churches highly susceptible to lightning strikes.

"THUNDER IS IMPRESSIVE; BUT IT IS THE LIGHTNING THAT DOES THE WORK."

—MARK TWAIN

✦ THE OZONE LAYER

- Ozone (O_3) is a molecule made of three oxygen atoms.
- Ozone forms naturally in the stratosphere from a series of reactions between ultraviolet radiation from the Sun and oxygen molecules in the atmosphere.
- **OZONE SUCCESSFULLY BLOCKS ULTRAVIOLET RADIATION, NOT BY REFLECTING IT, BUT BY TRAPPING IT INTO A CONSTANT, INESCAPABLE DANCE OF DO-SI-DO.**
- Move one: Ultraviolet light hits a molecule of O_3, breaking it apart into O_2 and O.
- Move two: That same ultraviolet light also breaks O_2 molecules into single O atoms.
- Move three: O atoms and O_2 molecules recombine into O_3, and the cycle begins anew.
- Ozone molecules work like a natural sunscreen, limiting harmful UV radiation from reaching Earth's surface.
- The term "ozone" is from the Greek word *ozein*—to smell—because of the gas's pungent odor.

✦ A HOLE IN THE OZONE?

- The ozone hole isn't actually a hole; it's a seasonally appearing swath of stratosphere above Antarctica with especially thin ozone.
- **GLOBAL POLLUTION OF HUMAN-MADE CHEMICALS PRIMARILY ACCOUNTS FOR THE REDUCTION OF OZONE.**
- These ozone-depleting substances typically break down into chlorine or bromine atoms when exposed to sunlight, which rapidly destroy ozone molecules.
- A single chlorine atom in the stratosphere can destroy more than 100,000 ozone molecules.
- Chlorine or bromine that can't reach the stratosphere—like the chlorine in your pool—doesn't affect the ozone layer.
- Common ozone-depleting compounds include chlorofluorocarbons (CFCs), hydrochlorofluorocarbons (HCFCs), carbon tetrachloride, methyl chloroform, halons, and methyl bromide.
- Before they were banned, these were commonly used in air conditioners, aerosols, refrigerators, fire extinguishers, and more.
- **WHY IS THE HOLE ABOVE ANTARCTICA, WHERE THE FEWEST HUMANS AND LEAST AMOUNTS OF POLLUTION ARE FOUND? AND WHY ONLY SEASONALLY?**
- First, Antarctica's uniquely cold winter temperatures help form colorful, high-altitude polar stratospheric clouds, which provide a perfect surface to facilitate reactions that unleash ozone-destroying compounds.
- Second, springtime sunlight breaks down these compounds to release chlorine and bromine atoms.
- Third, the polar vortex circulates strong winds around the South Pole, containing atmospheric molecules above the continent and prolonging the aforementioned ozone-depleting reactions.
- As the polar vortex breaks up in late spring, the thinned ozone mixes into the rest of the Southern Hemisphere ozone, making the "hole" temporarily vanish.
- The prettiest clouds you'll ever see are polar stratospheric clouds—or nacreous clouds—that shimmer in iridescent hues.
- Since many ozone-depleting compounds have been banned, the ozone hole has slowly healed—and may completely heal by 2070.
- Humans accidentally make ozone because of emissions from things like cars, power plants, and chemical plants.
- Anthropogenic ozone doesn't help repair the ozone hole—instead, it creates a harmful haze.

✦ THE MAGNETOSPHERE

- Earth's magnetic field dominates in a region called the magnetosphere.
- Churning molten iron in Earth's core generates our magnetic field.
- **EARTH HAS THE STRONGEST MAGNETOSPHERE OF ALL OUR SOLAR SYSTEM'S INNER ROCKY PLANETS.**
- Earth's magnetosphere protects us from the constant bombardment of charged particles and radiation from the Sun.
- Without our protective magnetosphere, the Sun's radiation would eventually strip away the molecules in our atmosphere.
- Mars may have looked a lot like Earth today if it hadn't lost its magnetosphere.

- Earth's magnetosphere traps the harmful solar particles in a pair of doughnut-shaped zones called Van Allen belts.
- The magnetosphere is shaped like a tadpole. The Sun-facing side is compressed by solar bombardment, and the night-facing side extends like a tail.
- Intense solar storms can break through the magnetosphere, wreaking havoc on electronics and electronic-dependent societies on Earth.
- **EARTH'S MAGNETIC FIELD FORMED AT LEAST 3.7 BILLION YEARS AGO—WHEN LIFE WAS JUST TAKING HOLD.**
- The blossoming of multicellular life on Earth some 600 million years ago may be due to the near collapse of the magnetosphere, during which hydrogen could escape to space and the larger oxygen atoms could increase in the atmosphere. This is called the Great Oxidation Event.
- Magnetic north and south on Earth have swapped places 183 times in the last 83 million years.
- If Earth's magnetic poles reversed, your compass would point to Antarctica rather than the North Pole.
- While the flip would pose no physical harm, it could throw society into chaos as electronics and GPS-dependent systems fail.
- On average, Earth's poles flip every 300,000 years, but the last one was much longer ago—some 780,000 years.

✦ AURORAS

- Thanks to the magnetosphere, we get to experience the gorgeous aurora borealis and the aurora australis, nicknamed the northern and southern lights.
- They dance across the night sky when charged particles from the Sun rake Earth's magnetic field and excite various gases.
- A Cro-Magnon cave painting in France may represent the earliest depiction of an aurora, dating to 30,000 B.C.
- The earliest known written record of an aurora dates to 2600 B.C. in China.
- **SOME IN NORTHERN CHINA SAW THEM AS THE FIERY EXHALATIONS OF DRAGONS.**
- Vikings imagined them as reflections from the battle shields of the Valkyries.
- In Philip Pullman's novel *The Golden Compass*, the aurora borealis separates parallel worlds.
- In the early 17th century, Italian astronomer Galileo Galilei named the phenomenon the aurora borealis. Aurora means "dawn" in Latin; borealis comes from the Greek word for "north."
- **GALILEO BELIEVED THE AURORA WAS SUNLIGHT REFLECTED OFF THE HIGH ATMOSPHERE—A MISTAKE THAT TOOK NEARLY FOUR CENTURIES TO CORRECT.**
- Galileo mentioned the aurora in his arguments for heliocentrism when he incorrectly surmised that it derives from sunshine glinting off the northern atmosphere. That misconception perpetuates to this day.
- Auroras only occur near the poles, where magnetic field lines funnel charged particles.
- Aurora colors come from different excited gases, like the glow of neon lights.
- The most common aurora color is green, caused by excited oxygen molecules.
- The blue color of auroras is caused by excited nitrogen molecules.
- The rarest aurora color is crimson, caused by excited oxygen atoms at higher altitudes.
- Auroras often look more exciting in photos because human eyes mainly see black and white in such low light.
- **AURORAS MAKE SOUND, BUT IN A RANGE KNOWN AS INFRASOUND, A FREQUENCY TOO LOW FOR HUMANS TO HEAR.**
- In May 2024, a coronal mass ejection on the Sun sparked an intense geomagnetic storm, sending aurora spirals across the sky for three nights.

✦ THE STREAK CALLED STEVE

- An aurora-like streak called STEVE (Strong Thermal Emission Velocity Enhancement) sometimes appears in the night sky. Like a celestial picket fence, it glows as a hazy purple arch with bright green stripes.
- STEVE was found (and named) by citizen scientists, making global headlines. Though it is associated with auroras, its exact cause is still a mystery.

CURRENT FACT COUNT 0425

GRAVITY: A BINDING FORCE

Gravity is the fundamental force that shapes and contours spacetime itself. Everything with mass—as well as all energy, no matter how small—exerts a gravitational force on every other object in the universe.

- Gravity transformed the universe from a hot, homogeneous primordial soup into a cosmic web of galaxy clusters, filaments, and giant voids.
- **GRAVITY INCREASES WITH THE MASS OF AN OBJECT.**
- Gravity's influence decreases by the square of the distance between two objects.
- For example: If the Moon were twice as far from Earth, its gravitational bond would weaken to one-fourth its current strength.
- **ISAAC NEWTON WAS THE FIRST TO RECOGNIZE THAT THE FORCE PULLING THINGS TO THE GROUND, LIKE APPLES FROM TREES, IS THE SAME FORCE THAT KEEPS THE MOON CIRCLING EARTH.**
- Newton determined that gravity also keeps planets, asteroids, and comets orbiting the Sun.

A 1993 German postage stamp depicts Sir Isaac Newton.

- Gravitational forces differ around Earth, depending on the amount of mass underground.
- Since different planets have different masses, their forces of gravity are different, too. A large, gaseous planet could have less gravity than a smaller, superdense planet that packs more mass.
- Our bodies have evolved perfectly to with-stand exactly 1 g—or normal gravity—the acceleration of an object falling to Earth.
- Earth gravity equals 32.2 feet a second squared. In other words, a falling object *accelerates* by 32.2 feet a second, *each passing second.*
- If you could maintain that acceleration on the way to the Moon, you'd experience 1 g the entire ride.
- You'd also get to the Moon in 2.5 hours. Right now, it takes about three days.
- If gravity on Earth were more than 1 g, we'd struggle to move.
- If gravity on Earth were less than 1 g, our muscles would begin to wither.
- During rocket launches, mission pilots throttle the main engine to maintain a 3-g force on the body. Any stronger and our internal organs face dire consequences, including circulatory failure and blindness.
- Astronauts lie down during launch to reduce the risk of brain damage from the sustained high g-force of launch. Our hearts and arteries can pump blood more easily to the brain and other organs in this position.
- **FIGHTER PILOTS ENDURE MUCH HIGHER DOSES OF GRAVITY—UP TO 9 G—BUT ONLY FOR A SECOND OR TWO.**
- If you wish to accelerate, you have two options: Change your direction or change your speed. Either will create a g-force.
- Anything or anyone moving at a constant speed cannot feel the motion of that speed unless either the speed or direction changes.
- In the 1978 film, Superman saves Lois Lane from a plummet by snatching her out of midair. But Lois had been in free fall for seven seconds, accelerating at the speed of gravity. At that speed, she's not much better off hitting the pavement than she is slamming into the Man of Steel.
- **TO SAVE THE DAY, SUPERMAN SHOULD FIRST MIRROR LOIS'S SPEED AND DIRECTION, GRAB HER, THEN SLOWLY DECELERATE.**
- The film *Ad Astra* could have saved significant margins on their effects budget by following the laws of physics. The astronauts were seen floating weightlessly in their spaceships at all times. But when a space vehicle is firing its engine, as was often the case throughout the film, it is accelerating, thus creating artificial gravity within.
- In the terrifying yet unlikely event of detachment from their spacecraft, an astronaut faces the real threat of getting lost in space.
- **IN THE ANIMATED FILM *WALL-E*, THE ADORABLE ROBOT USES A FIRE EXTINGUISHER TO NAVIGATE OPEN SPACE. THAT WOULD REALLY WORK.**

- In 1965, Ed White became the first American to walk in space. Though tethered, he did so with a handheld oxygen-jet gun.
- Today, space suits come equipped with a mini-jet pack called SAFER (Simplified Aid For EVA Rescue). The tiny jets of liquid nitrogen from SAFER's thrusters allow astronauts to maneuver back to safety.
- NASA astronaut Bruce McCandless II performed the first ever untethered spacewalk in 1984. With a pair of joysticks, he controlled the direction of nozzles spewing liquid nitrogen to self-propel through empty space.
- About the size of a mini-fridge, the jet pack offered enough thrust to traverse safely to and from the space shuttle *Challenger,* with no safety tether in sight.
- You might think of g-force as operating only downward, but it can manifest in any direction. When a race car driver makes a turn, the speed change and directional change combine to create a strong g-force that manifests not downward through their spine, but sideways. That's why they wear special padding and gear designed to restrain their heads and keep their shoulders affixed to their seats while they regularly experience 6 g's around tight turns.
- **BY THE WAY, RACE CAR SPELLED BACKWARD IS RACE CAR.**

Bruce McCandless II's record-setting untethered spacewalk in 1984

✦ NEED FOR SPEED

- *Top Gun*'s thrill-seeking fighter pilots Goose and Maverick "feel the need for speed." More accurately, they feel the need for acceleration—but that doesn't have the same ring to it.
- Mach 1: the speed of sound; about 760 miles an hour (variable by the temperature and density of the surrounding air molecules)
- Mach 2: twice the speed of sound; the speed of a Concorde commercial jetliner
- **MACH 5: ANYTHING ABOVE THIS SPEED IS CONSIDERED "HYPERSONIC"**
- Mach 25: the speed at which the Space Shuttle reenters Earth's atmosphere
- 1,000 miles an hour: the speed of Earth's rotation at the Equator
- 18 miles a second: the speed of Earth's orbit around the Sun
- 135 miles a second: the speed of the solar system around the Milky Way's center
- 200 miles a second: the speed at which the Milky Way galaxy and our neighboring Andromeda galaxy fall toward the Virgo cluster
- In the 2022 film *Top Gun: Maverick,* Tom Cruise's Maverick ejects from a fighter jet tearing through the sky at Mach 10.4—that's about 7,000 miles an hour. He imparted 400 million joules of kinetic energy into the surrounding air molecules—the explosive power of 220 pounds of TNT. Unless he ejected via a specialized protective capsule complete with aerobrakes, that move would render our hero very, very dead.
- 120 miles an hour: the terminal velocity of a human body in free fall due to air resistance
- **TO ACHIEVE A VELOCITY HIGHER THAN 120 MILES AN HOUR WOULD REQUIRE EITHER A BOOST FROM A JET PACK OR, IN THE CASE OF SKYDIVING DAREDEVIL FELIX BAUMGARTNER, AN ASCENT TO MORE RAREFIED ATMOSPHERE.**
- Baumgartner reached a supersonic 800 miles an hour during his famous jump from the stratosphere in 2012.

A team assists Chinese astronaut Jiang Xinlin upon his return from space in 2024.

✦ BIG G, LITTLE G

- The gravitational constant, G, is a mathematical formula used to calculate the proportional strength of attraction between two objects.
- **IT STEMS FROM ISAAC NEWTON'S LAW OF UNIVERSAL GRAVITATION (1687), WHICH UNIFIED THE CELESTIAL AND TERRESTRIAL REALMS UNDER A SINGLE PRINCIPLE.**
- The gravitational constant, sometimes called Big G, is different from acceleration due to gravity, the figure most of us know as "1 g" (little g).
- Newton estimated the value of Big G more than 100 years before scientists first measured it in a laboratory.
- **IN 1797–98, HENRY CAVENDISH AND HIS COLLEAGUES MEASURED BIG G TO AN ACCURACY OF ABOUT ONE PERCENT.**
- Its precise value remains difficult to pin down, despite numerous experiments.
- Newton developed his theories of gravity and planetary motion in the 17th century. Then came Einstein's general theory of relativity in the 20th century, which explains gravitational attraction as the result of curvatures in spacetime.
- If the gravitational constant differed by even one-tenth of one percent, the universe would be vastly different.
- So different that life may have never existed.
- If the gravitational constant were a smidge higher, the Sun would be far too hot for life to evolve on Earth.
- If the gravitational constant were a smidge lower, the Sun would be far too cool for Earth's surface to be habitable.

✦ GRAVITY SUCKS

- Astronauts are weightless in orbit because they are in free fall toward Earth. There's nothing magical about zero g. Anything in free fall exhibits weightlessness.
- Anything in orbit is in free fall. It just happens to free-fall toward Earth at the exact rate that Earth's surface curves away from it.
- The sci-fi TV series *The Expanse* explores the effect of varying gravity levels on the human body, as well as how we might adapt over many generations.
- In *The Expanse,* the belters (those who harvest resources from the asteroid belt and live in low gravity) have more lanky, lithe, and frail bodies than Earthlings or Martians.
- A belter's chest would collapse in Earth gravity, so belters visiting Earth must first adjust in a suspension tank.
- **SUPERMAN'S SUPERSTRENGTH IS A RESULT OF GROWING UP ON KRYPTON, A PLANET WITH SIGNIFICANTLY HIGHER GRAVITY THAN EARTH.**
- Astronauts aboard the International Space Station (ISS) prove our bodies can quickly adapt to low gravity.
- Over time, ISS astronauts' bones, muscles, and hearts begin to shrink.
- In zero g, astronauts will lose 1 to 2 percent bone density per month.
- **TO COMBAT BONE DENSITY LOSS IN ZERO G, ASTRONAUTS EXERCISE TWO HOURS EVERY SINGLE DAY.**
- Astronauts can grow up to two inches during their extended stay on the ISS. But they immediately shrink once reintroduced to gravity.
- Upon returning to Earth, ISS astronauts feel heavy and disoriented.
- They have such trouble walking, they must be carried from the capsule.
- **FOR RETURNING ISS ASTRONAUTS, EVEN TALKING OR HOLDING A PEN CAN FEEL FOREIGN AND DIFFICULT.**
- All these effects of mere weeks or months in zero g on the ISS prove that if we are to become a truly spacefaring species, spending years or lifetimes or multiple generations aboard a spaceship, we're going to need artificial gravity.

✦ ARTIFICIAL GRAVITY

- To experience a steady 1 g (the gravity equivalent to Earth), your spaceship should accelerate at exactly 32 feet a second each second.
- **THE BEST METHOD TO CREATE ARTIFICIAL GRAVITY IS TO EXPLOIT THE CENTRIFUGAL FORCE.**

"EVERYTHING EVERY DAY HERE ON EARTH IS BASED ON GRAVITY, AND YOU DON'T REALIZE IT UNTIL YOU DON'T HAVE IT ANYMORE."

—NASA ASTRONAUT PEGGY WHITSON

Experience centrifugal force in action on a carnival swing ride.

✦ CENTRIPETAL AND CENTRIFUGAL

- Centripetal and centrifugal forces describe an object moving around another, like Earth orbiting the Sun.
- Centripetal (literally "center-seeking") force represents the gravitational pull keeping the orbiting object at the same distance from the one it orbits.
- Centrifugal (literally "center-fleeing") force represents the tendency for the orbiting object to fly out from the orbit.
- **CENTRIPETAL FORCE IS AN ACTUAL FORCE; CENTRIFUGAL FORCE IS AN APPARENT OR IMAGINED FORCE.**
- Centripetal force changes the direction of the object at each point of its path, so it keeps moving in a circle.
- When a car makes a sudden sharp turn, a passenger feels the centrifugal force as they're flung in the opposite direction.
- The carnival ride called the Gravitron exploits centrifugal force. Riders stand against the walls of a spinning room, and so long as the room spins fast enough, centrifugal force keeps them pinned in place, even as the room tilts at an angle.
- The centrifugal force is strongest at Earth's Equator and weakest at the poles.
- **WHEN SANTA CLAUS LEAVES THE NORTH POLE WEIGHING IN AT 400 POUNDS, HE ARRIVES AT THE EQUATOR WEIGHING ONLY 399.**
- Stanley Kubrick's 1968 film *2001: A Space Odyssey* portrayed how we might exploit centrifugal force to attain artificial gravity. The living quarters of the *Discovery One* spaceship were built around the perimeter of a rotating ring.
- The ring must spin at a precise speed to enact a centrifugal force equivalent to the feeling of 1 g.
- The closer one moves from the edge toward the center of the spaceship, the lower the centrifugal force and thus the lower the gravity.
- **UNFORTUNATELY, THE RING WHIRLS ABOUT THREE TIMES TOO FAST TO CREATE 1 G. ANY ASTRONAUTS WANDERING NEAR THE OUTER EDGE WOULD LIKELY FIND THEMSELVES HELPLESSLY AFFIXED TO THE FLOOR, LIKE RIDERS ABOARD THE GRAVITRON AT A CARNIVAL.**
- In Kubrick's *2001,* based on the novel by Arthur C. Clarke, humanity built that spaceship, along with lunar bases, by 1994. They later colonized Jupiter's moons by 2061. Since humanity is still yet to achieve a station of such caliber or even a viable lunar habitat, it's looking unlikely we'll achieve that third triumph any time soon.

✦ EINSTEIN'S HAPPIEST THOUGHT

- The equivalence principle states that gravity is indistinguishable from acceleration.
- **IN 1907, ALBERT EINSTEIN ENVISIONED SOMEONE FALLING IN AN ELEVATOR SHAFT AFTER THE CABLE BROKE AND REALIZED THAT SOMEONE IN SUCH A FREE FALL EXPERIENCES COMPLETE WEIGHTLESSNESS.**
- Everything within that elevator shaft would free-fall at the exact same rate.
- A person could pull items from their pocket and watch them hover in midair.
- Conversely, if the elevator suddenly shot upward, the person would be pinned to the floor.
- Gravity, Einstein reasoned, is indistinguishable from acceleration.
- **HE DESCRIBED THESE IDEAS AS THE HAPPIEST THOUGHT OF HIS LIFE.**

CURRENT FACT COUNT 0425

BY THE NUMBERS: GRAVITY

0 G

the g-force experienced by astronauts aboard the International Space Station—or anything else in free fall

1 G

the g-force experienced by Earthlings at rest

16,000 G

about how much g-force the extremophile known as the tardigrade can endure. This near-unkillable microorganism is affectionately referred to as the water bear or moss piglet.

214 G

the highest g-force on record ever survived by a human, felt for a split second by Swedish professional race car driver Kenny Bräck during a high-speed crash and abrupt deceleration. He fractured several bones but lived to tell the tale.

400 G

about how much g-force you deliver to an unassuming insect when you smack it with a flyswatter

12 G

the highest g-force pulled by a roller coaster. Coney Island's Flip Flap Railway was one of the first coasters to execute the loop-the-loop maneuver. Reports of severe neck injuries, whiplash, and loss of consciousness by its thrill-seeking passengers expedited its decommission in 1902.

25,000

miles an hour: the escape velocity of anything from Earth's surface

100,000

miles a second: the escape velocity from the surface of a typical neutron star

10,000,000,000,000

pounds: the weight at the surface of a neutron star of an object that weighs 100 pounds on Earth

90

minutes: the time for any object to orbit within a few hundred miles of Earth's surface

25,900

feet per second: your peak speed (assuming no friction) if you fell through Earth's center

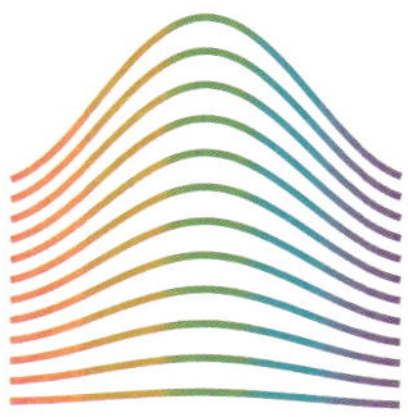

2

minutes: the time it would take an object to fall from chest height to its surface if dropped on Halley's comet

94,200

miles an hour: the escape velocity from the solar system at Earth's distance from the Sun

2,800

pounds: the weight at the Sun's surface of an object that weighs 100 pounds on Earth

250

pounds: the weight on planet Jupiter of an object that weighs 100 pounds on Earth

91

pounds: the weight on the planet Venus of an object that weighs 100 pounds on Earth

ZERO

pounds: the weight in the water of any freely swimming fish in the ocean

ZERO

pounds: the weight of the International Space Station while in orbit around Earth

17

pounds: the weight on the Moon of an object that weighs 100 pounds on Earth

ZERO

pounds: the weight of any object in orbit around any other object

CURRENT FACT COUNT 0449

THE SUN, OUR STAR

From brilliant sunsets to scorching summer temps, our solar system's largest star profoundly affects our planet.

✦ THE TURN OF THE SEASONS

- What causes the seasons? Not Earth's position in orbit around the Sun, but rather Earth's 23.5-degree tilt on its own axis.
- In fact, Earth's closest approach to the Sun occurs during the Northern Hemisphere's peak winter: early January. Earth's farthest distance from the Sun, its aphelion, occurs in early July.
- The intensity of light and heat reaching Earth's surface varies by latitude and the orientation of the tilting planet toward the Sun.
- The more a hemisphere tilts toward the Sun, the more light and heat it receives; the more it tilts away, the less.
- That's why the Northern and Southern Hemispheres experience opposite seasons. When it's summer in Canada, it's winter in Australia, and vice versa.
- The North Pole points toward the Sun the most in June; the South Pole in December.
- **THE SUMMER SOLSTICE MARKS THE POINT OF MAXIMUM TILT TOWARD THE SUN; THE WINTER SOLSTICE MARKS THE POINT OF MAXIMUM TILT AWAY FROM THE SUN.**
- The Northern and Southern Hemispheres share these two dates, but what one hemisphere calls the summer solstice,

Stonehenge frames the summer solstice in June 2018.

the opposite hemisphere calls the winter solstice.

- The summer solstice marks the longest day of the year—the day with the longest period of sunshine.
- The winter solstice marks the shortest day, or the day with the shortest duration of sunshine.
- The ancients were keen observers of the Sun, creating monuments and events to mark these moments.
- **STONEHENGE, COMPLETED AROUND 2500 B.C., ALIGNS WITH THE SUMMER AND WINTER SOLSTICES.**
- The Sun rises on the summer solstice and sets on the winter solstice directly over Stonehenge's Heel Stone.
- **INTI RAYMI, AN INCA FESTIVAL OF THE SUN, IS STILL CELEBRATED IN CUSCO, PERU, ON THE WINTER SOLSTICE IN JUNE.**
- The Antikythera mechanism, an artifact exhumed from an ancient Greek shipwreck site in 1901, is called the first analog computer. Though its true purpose and origin remain mysterious, most scholars understand it to be a manual orrery used to track the motion of the stars, Sun, and Moon and even predict astronomical events like eclipses.

The Inca festival of Inti Raymi at the ruins of the Sacsayhuamán fortress

The fascinating Antikythera mechanism

- The number of sunspots at any given moment is what tracks the 11-year cycle of the Sun's activity.
- **EVERY 11 YEARS THE POLARITY OF THE SUN FLIPS, SO A FULL CYCLING OF THE SUN IS ACTUALLY 22 YEARS.**
- Though dark against the Sun's surface, if removed and placed on the sky, sunspots would visibly glow.
- Sunspots generally come in pairs, like the positive and negative poles of a magnet.
- The average density of the Sun—from its core to surface—is about the same as water.
- The Sun's equator rotates once every 24.5 Earth days; the polar regions once every 34.3 Earth days.
- It takes approximately 50,000 years for photons created in the Sun's core to migrate to the surface, as they scatter hither and yon through all the gas particles within the Sun itself.
- The Sun has been fusing hydrogen to helium for five billion years and will continue for another five billion before it dies.
- **MORE THAN 90 PERCENT OF ALL THE ATOMS IN THE SUN ARE HYDROGEN.**

"THE GLORIOUS SUN, / STAYS IN HIS COURSE AND PLAYS THE ALCHEMIST, / TURNING WITH SPLENDOR OF HIS PRECIOUS EYE / THE MEAGRE CLODDY EARTH TO GLITTERING GOLD."

—WILLIAM SHAKESPEARE, *KING JOHN*

Sunset as seen from the DARC Observatory in Fresno County, California

CURRENT FACT COUNT 0533

THE MOON, OUR SATELLITE

Our early ancestors relied on the Moon's changing patterns to track time, tides, and life cycles. In modern times, we admired it enough to pay it a visit.

- Though luminous and bright, the Moon has no internal light source: It only beams back reflected sunlight.
- **THE MOON LOOKS WHITE BECAUSE OF AN ABUNDANCE OF THE HIGHLY REFLECTIVE ROCK-FORMING MINERAL ANORTHITE.**
- Although anorthite is abundant on the Moon, it's rare on Earth.

An illustration from Galileo Galilei's *Sidereus Nuncius* (1610)

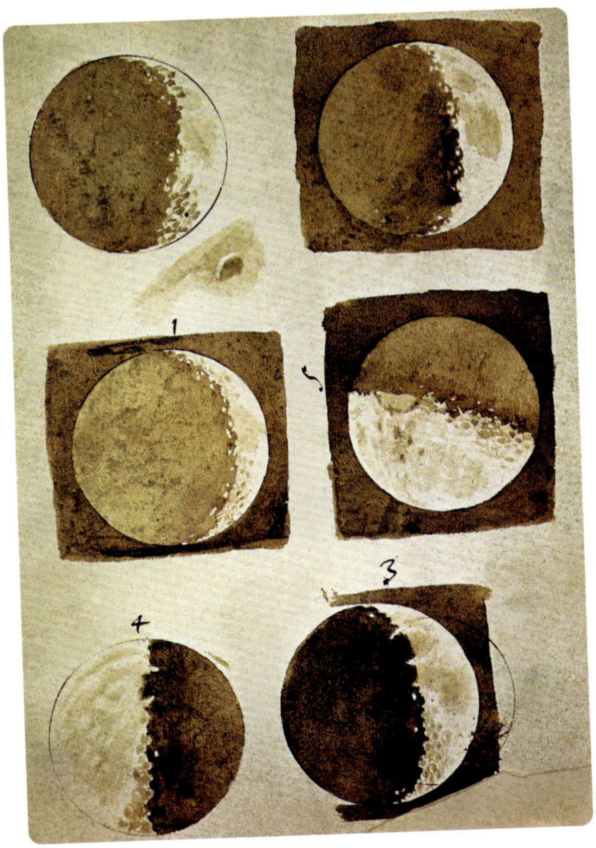

✦ MOON PHASES

- On Earth, we see phases of the Moon because the portion visibly lit by the Sun changes with the ever shifting spatial relationship of the three bodies.
- It takes about 28 days to cycle from a full moon to a new moon and back.
- The changes in between are called waxing (growing) and waning (slimming).
- The official names for all the phases of the Moon are:
  - New moon
  - Waxing crescent moon
  - First quarter moon
  - Waxing gibbous moon
  - Full moon
  - Waning gibbous moon
  - Last quarter moon
  - Waning crescent moon
- A half moon is actually a quarter moon—we see half of the Earth-facing half of the Moon.
- You've heard of rainbows; how about moonbows? They occur when water droplets in the atmosphere refract moon-light—like a rainbow at night.
- The Moon can show up in the daytime sky when it sits above the horizon at an angle where the Sun's reflected light is particularly bright.
- Earth reflects the Sun's light, too, in a phenomenon known as earthshine.
- **YOU CAN SEE EARTHSHINE JUST BEFORE OR AFTER A NEW MOON, WHEN THE DIMLY LIT FULL MOON APPEARS WITH THE THIN, BRIGHT CRESCENT.**
- The Moon absorbs most of the sunlight that strikes its surface and reflects only about 11 percent. Earth, however, is about 40 percent more reflective than the Moon, with some of that light bouncing back to the Moon's surface.
- That means Earth exhibits phases to a lunar resident (let's call them Lunarians) just as the Moon exhibits phases to us Earthlings. Lunarians would see a "new Earth" when Earthlings see a full moon, and a "full Earth" while Earthlings see a new moon.
- **ONE OF THE MOST ICONIC PHOTOGRAPHS EVER TAKEN IN OR OF SPACE (OR ANYWHERE) IS NAMED "EARTHRISE."**
- The title "Earthrise" is a misnomer. Lunarians would never witness Earth rise or set because the Moon remains tidally locked, showing only one face to Earth.
- Earth appeared to "rise" to the astronaut snapping the photograph because the Apollo command module was orbiting Moon's surface.
- **IN 1959, SOVIET SPACECRAFT LUNA 3 CAPTURED THE FIRST IMAGES OF THE MOON'S BACKSIDE. NO HIDDEN LUNARIAN CIVILIZATIONS WERE TO BE FOUND.**
- The stark line that separates the Moon's night and day—the line that defines its phase from our view on Earth—is dramatically named the terminator (not to be confused with Arnold Schwarzenegger's dramatic depiction of a dystopian cyborg-infested future).
- The best time to Moon-gaze isn't when it is full but during a waxing or gibbous crescent,

A lunar rainbow caused by full moonrise at Lower Yosemite Falls, Yosemite National Park, California

when the terminator casts deep shadows across the craggy terrain. A simple pair of binoculars is all you need to witness the striking contrasts.

- **THE GRAVITY ON THE MOON IS ONE-SIXTH THE GRAVITY ON EARTH, NOT NEARLY STRONG ENOUGH TO CONTAIN AIR MOLECULES REQUIRED FOR ANY MEANINGFUL ATMOSPHERE.**

- Shadows on the Moon appear a starker, deeper black than shadows cast across Earth's surface. That's because the Moon doesn't have an atmosphere to scatter light particles into otherwise unlit areas, as is the case on Earth.

✦ THE VIEW FROM EARTH

- Earth and the Moon are locked in synchronous rotation.

- That means the Moon accomplishes one rotation on its axis in the same amount of time it completes one orbit around Earth.

- **BECAUSE OF SYNCHRONOUS ROTATION, WE ALWAYS SEE THE SAME SIDE OF THE MOON.**

- Sorry, Pink Floyd. There's no dark side of the Moon. That's the far side—the side constantly turned away from Earth.

- And hey, Autobots, you might want to correct your 2011 film's title to *Transformers: **Far Side** of the Moon.*

✦ THE MARIA

- The Moon's dark patches were once thought to be water, and therefore were named *maria,* Latin for "seas."
- We now know them to be plains of ancient volcanic lava.
- In the West, we're used to seeing a face—the "man in the Moon"—in the patches of light and dark.
- **IN EAST ASIAN COUNTRIES (CHINA, JAPAN, KOREA) AND IN MESOAMERICAN MYTHS, THE MOON'S SPLOTCHES ARE PERCEIVED AS A RABBIT.**
- In India, the pattern we see on the Moon is envisioned as a set of hands.
- The Salish people of the Pacific Northwest explain the shapes on the lunar surface with a story about a toad that leaped onto the Moon to escape the clutches of a wolf.

✦ SPECIAL MOONS

- A blue moon is the second full moon in a single calendar month.
- Blue moons occur on average every two or three years, so whenever you say "once in a blue moon," make sure the thing happens at least once every two or three years.
- The average time between consecutive full moons is 29.5 days.
- So even in a leap year, February, with 29 days, will never be home to a blue moon.
- **THE PHRASE "BLUE MOON" MIGHT AS WELL HAVE BEEN INSTEAD DUBBED "GOBBLEDYGOOK MOON."**
- The phrase as we now understand it emerged in late 1940s America after a series of misinformed articles propagated a sentence from an *Old Farmer's Almanac* article that was steeped in false information.
- From there, the idea that "blue moon" means "second moon in a month" and a metaphor for an exceedingly rare event was perpetuated through other media until nobody could remember where it had originated.
- Sometimes, the Moon can appear blue, when atmospheric smoke or dust particles scatter high-frequency light, as was reported by some eyewitnesses after the infamous 1883 eruption of Krakatoa in Indonesia. But phenomena such as these are not timed to our calendars.
- The 1968 disaster film *Krakatoa, East of Java* chronicles that eruption—ignoring with its title the fact that Krakatoa happens to be *west* of Java.

"The Rabbit in the Moon" in the 16th-century *History of the Aztecs* by Bernardino de Sahagun

✦ TOTAL ECLIPSE OF THE ...

- Total lunar eclipses can happen only during full moons, when the Moon's orbit slowly enters Earth's conical shadow in space.
- During a lunar eclipse, the Moon will occasionally take on a deep, dark red color, depending on Earth's atmospheric conditions, as sunlight disperses through the air en route to the lunar surface.
- **FOR REASONS MYSTERIOUS, INSTEAD OF CALLING IT A ROSE MOON OR A RED DELICIOUS APPLE MOON OR A BING CHERRY MOON, PEOPLE HAVE PREFERRED "BLOOD MOON," LEADING TO ALL MANNER OF ASTROLOGICAL AND THEOLOGICAL PORTENTS.**
- The Moon's orbit around Earth is not a perfect circle. Its elliptical path brings it sometimes closer and sometimes farther from Earth, varying by +/- 5 percent across the month.
- A full moon that happens to occur during the Moon's closest approach to Earth has been called by sensationalists a super moon.

- Super moons are not rare, occurring three or four times a year.

- Super moons appear 5 percent larger than the average full moon.

- For reference, a 15.75-inch pizza is 5 percent larger than a 15-inch pizza, hardly deserving the label "super."

- Trifecta! A super blue moon—full, close, and the second in the month—is rare but can happen.

- **FULL MOONS ALWAYS LOOK UNNATURALLY LARGE ON THE HORIZON.**

- This famous moon-on-horizon illusion is enhanced by buildings, mountains, and trees on your horizon.

- The human brain's depth cues get easily confused when an object at a great distance is juxtaposed with objects just miles away in the foreground.

Phases of the "blood moon"

✦ MOONS BY THE MONTH

- Over the centuries, many agrarian cultures have given names to each full moon, month by month.

- **PEOPLE WHO GROW CROPS AND SPEND LOTS OF TIME OUTDOORS TEND TO CARE MORE ABOUT MOONLIGHT THAN CITY DWELLERS DO.**

- In 1850, nearly 7 in 10 people living in the U.S. were farmers.

- By 2020, fewer than 2 in 100 people were farmers.

- **SOME LONG-FORGOTTEN FULL MOON NAMES FROM NORTHERN HEMISPHERE CULTURES AROUND THE WORLD:**
  - January's full moon has been called the Wolf Moon, the Quiet Moon, and the Severe Moon.
  - February's full moon has been called the Snow Moon, the Hungry Moon, and the Bear Moon.
  - March's full moon has been called the Worm Moon, the Crow Moon, and the Sap and Sugar Moon.
  - April's full moon has been called the Pink Moon, the Breaking Ice Moon, and the Awakening Moon.
  - April's full moon is used by Christians to determine the date of Easter: the first Sunday following the first full moon on or after the spring equinox.
  - May's full moon has been called the Flower Moon, the Egg Laying Moon, the Planting Moon, and the Grass Moon.
  - June's full moon has been called the Strawberry Moon and the Hot Moon.
  - July's full moon has been called the Buck Moon, the Salmon Moon, and the Hay Moon.
  - August's full moon has been called the Sturgeon Moon, the Lynx Moon, and the Lightning Moon.
  - September's full moon has been called the Harvest Moon, the Corn Moon, and the Song Moon.
  - October's full moon has been called the Hunter's Moon, the Drying Rice Moon, and the Falling Leaves Moon.
  - November's full moon has been called the Beaver Moon, the Frost Moon, and the Mourning Moon.
  - December's full moon has been called the Cold Moon, the Long Night Moon, or the Oak Moon.

- If modern humans cared about the Moon as much as our ancestors did, we might come up with new names. In the U.S., we might say:
  - A full moon in April, if before the 15th: Tax Return Moon
  - A full moon in April if after the 15th: Late-Fee Moon

- The full moon in June in the Northern Hemisphere crosses low in the sky, retaining an amber tinge the whole way.

- **THE AMBER-TINGED FULL MOON IN JUNE IS OTHERWISE KNOWN AS THE HONEY MOON.**

- The full moon that lands closest to the September equinox, either preceding or following it, is famously called the Harvest Moon.

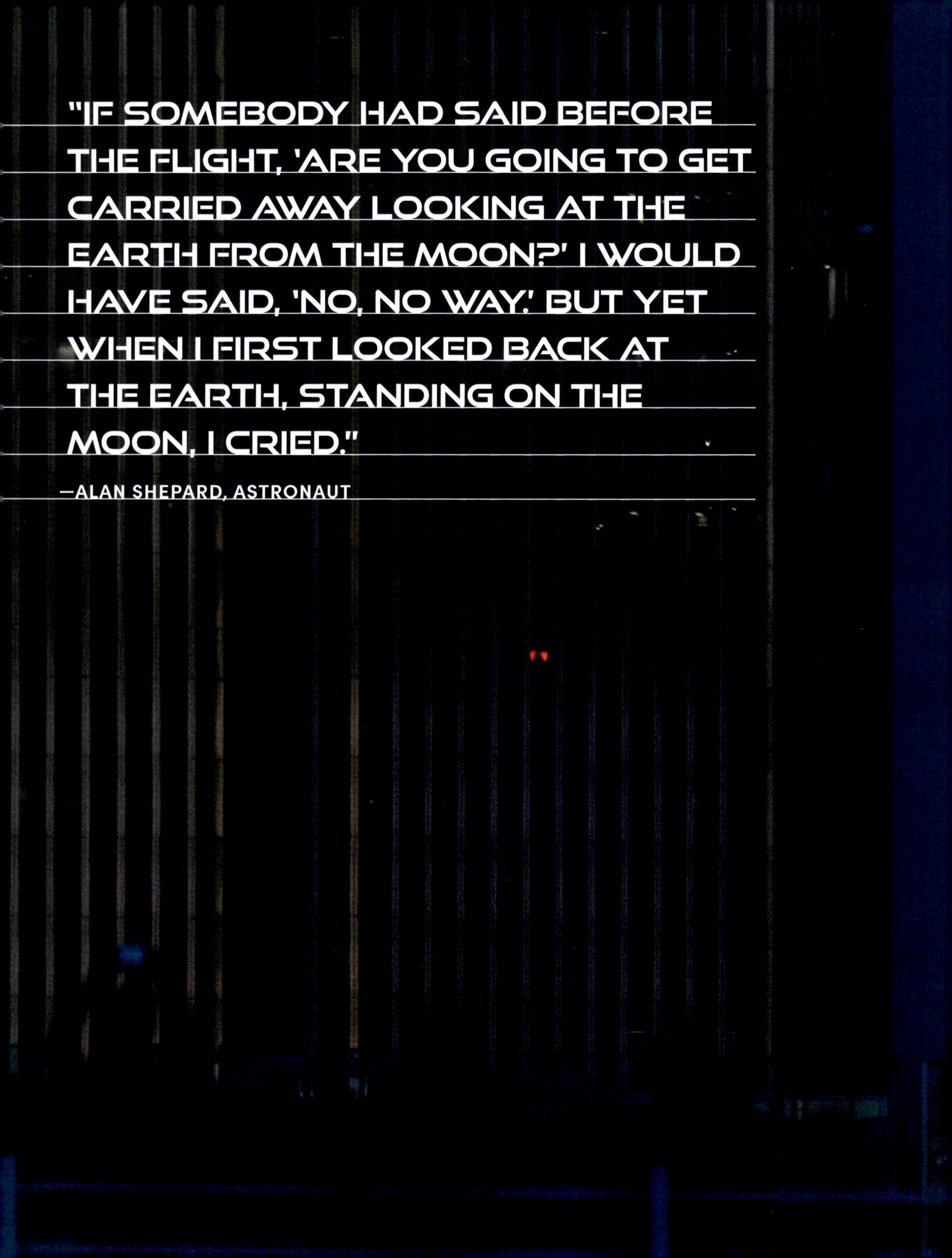

"IF SOMEBODY HAD SAID BEFORE THE FLIGHT, 'ARE YOU GOING TO GET CARRIED AWAY LOOKING AT THE EARTH FROM THE MOON?' I WOULD HAVE SAID, 'NO, NO WAY.' BUT YET WHEN I FIRST LOOKED BACK AT THE EARTH, STANDING ON THE MOON, I CRIED."

—ALAN SHEPARD, ASTRONAUT

Onlookers watch the super moon rise over London from Waterloo Bridge in 2024.

CURRENT FACT COUNT 0545

THE MOON AROUND THE WORLD

Humanity has attributed meaning—and human characteristics—to our celestial companion across time and geography.

- **1.** Moon Rabbit (East Asia):
  - Moon-gazers in Japan see a rabbit making rice cakes using a mortar and pestle.
  - In China and Korea, they see him, too—except he's mixing an immortality elixir in lieu of a starchy mixture.
  - Moon rabbits appear in Mesoamerican myths, one of which says that the gods tossed a rabbit at the Moon in jest, and there the creature stayed, visible on the lunar surface.
- **2.** Man in the Moon (Europe):
  - Many European cultures see an old man bearing a bundle of sticks when they look up at the Moon.
  - Christian lore holds that the man has been eternally condemned for violating the Sabbath.
- **3.** Handprints (India):
  - Astangi Mata, mother of all living things, sent her twins into the sky to be the Sun and Moon.
  - Her hands brushed Chanda's cheek in a poignant farewell.
- **4.** Man in the Moon (continental U.S.):
  - The face of a man is visible to many in North America.
  - Ancient lava flows form his eyes, nose, and mouth.
- **5.** Tree in the Moon (Hawaii):
  - A woman named Hina uses this banyan tree to make cloth for the gods.
  - *Mahina,* Hawaiian for "moon," is derived from her name.
- **6.** Woman in the Moon (New Zealand):
  - This is Rona, a Maori maiden who disrespected the Moon and must spend eternity on its surface as penance.

"THE SUN BECAME ANGRY AND THREW ASHES INTO THE TWO FACES OF THE MOON. THE ASHES WHICH THE SUN THREW INTO ONE OF THE FACES OF THE MOON CONTAINED FIRE. THE FACE OF THE MOON WAS BURNED OUT AND BECAME BLACK."

—MYTH OF THE NABALOI PEOPLE OF THE PHILIPPINES

VIEW FROM NORTHERN HEMISPHERE

VIEW FROM SOUTHERN HEMISPHERE

1

2

3

4

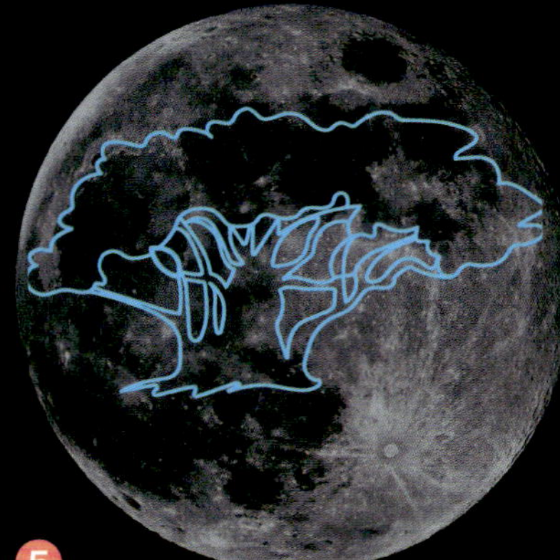

5

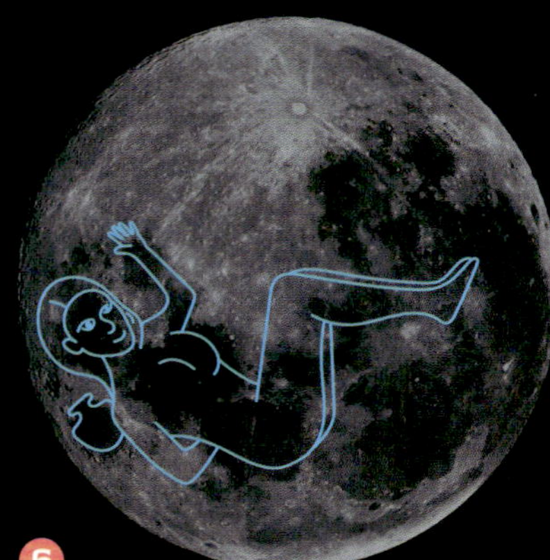

6

CURRENT FACT COUNT 0588

IN THE DARK: ECLIPSES

Solar eclipses occur when the Moon blocks out the sunlight illuminating Earth, and lunar eclipses occur when Earth blocks the sunlight illuminating the full moon.

- Solar eclipses happen because the Sun and Moon move into just the right distances in the sky when viewed from Earth that they appear as if they're the same size.

- **WHEN THREE CELESTIAL BODIES MOVE INTO A STRAIGHT LINE—THE ORIENTATION THAT PRODUCES AN ECLIPSE—IT'S CALLED A SYZYGY.**

- The word "eclipse" comes from the Greek *ekleipsis,* meaning "abandonment" or "forsaking," likely in reference to the disappearing sunlight.

- A solar eclipse occurs two to five times a year.

A photograph taken in 1854 of the first total eclipse of the Sun in North America since the invention of photography

- A total solar eclipse—when the Moon completely blocks the Sun from our view—is visible somewhere on Earth about every 18 months.

- A total solar eclipse is visible from a given place on Earth about every 400 years.

- **SO, IN OUR GLOBALIZED WORLD, SOLAR ECLIPSES ARE NO LONGER EXCEEDINGLY RARE TO WITNESS LIKE THEY WERE FOR OUR ANCESTORS.**

- Have you ever witnessed a total solar eclipse? If you're unsure, that likely means you haven't. When the Moon fully shrouds the Sun, a day steeped in sunlight abruptly turns to twilight. For those few unforgettable minutes, you may witness planets, comets, and other phenomena that are otherwise reserved for nocturnal hours.

- At under 99 percent totality, that meager one percent of sunlight is equivalent to the light pollution of about 10,000 full moons.

- We don't experience eclipses every month because the Moon does not orbit along a perfectly flat plane. Sometimes, it blocks Earth's view of the Sun a smidge more or less on each go-round.

- Although the Sun travels east to west in the sky, the path of a total eclipse is opposite, moving west to east along the path where it's visible.

- The environs darken during a solar eclipse because Earth is cast in the Moon's shadow.

- During a solar eclipse, the Moon's shadow races across Earth at more than 1,000 miles an hour.

- No other rocky planet in the solar system experiences total solar eclipses like Earth does.

- The gas giants—Jupiter, Saturn, Uranus, and Neptune—can experience Earthlike total solar eclipses with the help of their myriad moons, but watching them would be tough on a world without solid ground.

✦ SEEKING MEANING

- In ancient times, solar eclipses were seen as bad omens in some cultures, which sparked many rituals for supposed protection.

- Assyrians hid their king during solar eclipses, and a substitute would dress and eat like royalty, then be sacrificed.

- An Assyrian cuneiform tablet cites a solar eclipse, likely the one that occurred on June 15, 763 B.C.

- **IN ANCIENT MESOPOTAMIA (MODERN-DAY IRAQ), ECLIPSES PREDICTED THE DOWNFALL OF RULERS.**

- Ancient Chinese myth held that the Sun disappeared because a dragon consumed it.

- The Book of Amos in the Bible likely cites the 763 B.C. eclipse as well: "'And on that day,' says the Lord God, 'I will make the Sun go down at noon, and darken the earth in broad daylight'" (Amos 8:9).

- European superstitions around eclipses heightened after King Henry I died, as a total solar eclipse appeared two years prior, on August 2, 1133. It was retroactively declared an omen foretelling his passing.

- Not all cultures perceive total eclipses as portending doom: Various Native American tribes, such as the Southern Paiute people, view them as opportunities to reflect and welcome love.

✦ PAST, PRESENT, FUTURE

- Since the Moon is moving slowly away from Earth, total eclipses will become less frequent—and eventually stop entirely.
- **EARTH'S FINAL TOTAL SOLAR ECLIPSE WILL BE ABOUT 620 MILLION YEARS FROM NOW, AFTER WHICH THE MOON WILL BE TOO FAR AWAY TO FULLY BLOT OUT THE SUN.**
- 7 minutes 28 seconds: the duration of the longest total solar eclipse in the last 6,000 years, which took place in 743 B.C.
- 7 minutes 29 seconds: An even longer total solar eclipse will occur on July 16, 2186.
- The shortest total solar eclipse in the last 6,000 years occurred in February A.D. 919, lasting only nine seconds.
- The first photograph taken of a total solar eclipse was a daguerreotype made by Johann Julius Friedrich Berkowski in Königsberg, Prussia, on July 28, 1851.
- **MAY 29, 1919: PERHAPS THE MOST SCIENTIFICALLY MOMENTOUS ECLIPSE IN HISTORY. ARTHUR EDDINGTON USED THIS ECLIPSE TO DERIVE THE FIRST VERIFIABLE PROOF OF EINSTEIN'S GENERAL THEORY OF RELATIVITY.**
- Eddington premeasured the exact location of a specific star cluster that was predicted to appear at the moment and location of the eclipse. When the sky darkened and the cluster appeared, the location was slightly off—in precise accordance with Einstein's calculations of how gravity warps spacetime.
- April 8, 2024: the last solar eclipse visible from North America for the next 20 years

Eclipse glasses block harmful UV and infrared rays for safe eclipse viewing.

- In 585 B.C., the Battle of the Eclipse was fought in modern-day Turkey. On May 28, the Medes and Lydians ceased their five years of war after "day was all of a sudden changed into night," according to Greek historian Herodotus.
- In the beloved animated fantasy saga *Avatar: The Last Airbender*, a critical plot point hinges on an eclipse, which they call the Day of the Black Sun. On this day, "fire-benders," whose power is derived from the Sun, become powerless during the eight-minute eclipse.
- The Ugarit eclipse of 1223 B.C. is one of the earliest records of an eclipse in written history.
- The account was etched into a clay tablet and kept in a library that was unearthed in modern-day Syria during a decades-long excavation of the once lost city.

✦ LUNAR ECLIPSES

- A lunar eclipse occurs when Earth is positioned in line between the Moon and the Sun, throwing Earth's shadow onto the Moon's surface.
- Lunar eclipses can be seen by many more people than solar eclipses—they are visible to practically a whole hemisphere of Earth.
- **A LUNAR ECLIPSE OCCURS BETWEEN ONE AND FIVE TIMES A YEAR SOMEWHERE ON EARTH.**
- As sunlight bends around Earth's atmosphere, short frequency blue light scatters away, leaving only longer-wavelength red light to reach the lunar surface.
- About one-third of all lunar eclipses are total eclipses.
- **IF THE MOON'S ORBIT HAD ZERO TILT, THEN WE WOULD HAVE A TOTAL SOLAR ECLIPSE AND A TOTAL LUNAR ECLIPSE ON EVERY FULL MOON.**
- Stranded in Jamaica in 1504, Christopher Columbus knew of a coming lunar eclipse, thanks to a 1474 German astrological almanac, and he leveraged the knowledge against the Indigenous Arawak people, convincing them that his god would be wrathful and make the Moon disappear unless they gave him and his crew more food. A total lunar eclipse occurred, and the Arawak complied.

"THE SUN DOES NOT HAVE A SPECTATOR UNLESS IT IS IN ECLIPSE."

—SENECA THE YOUNGER, *NATURAL QUESTIONS*

A solar eclipse as seen in New Mexico in 2012

CURRENT FACT COUNT 0649

METEOROIDS, METEORS, AND METEORITES

When meandering space rocks encounter Earth's atmosphere, they can present as gentle gasps of streaking light, as violent impacts, or—at least in one instance—as global extinction.

- Meteoroids are space rocks still in space and range in size from a grain of sand to about three feet. (Once a meteoroid exceeds three feet, it's generally considered an asteroid.)
- Particles in space smaller than a meteoroid are referred to as space dust.
- **A METEOROID BECOMES A METEOR WHEN IT COLLIDES WITH EARTH'S MESOSPHERE AND BURNS ACROSS THE SKY.**
- As the meteor smashes into air molecules, the kinetic energy converts into heat and sets the particles ablaze as a "shooting star."
- A meteoroid becomes a meteorite when it survives the flaying descent through Earth's atmosphere and hits the ground.
- Most meteoroids vaporize long before they can become meteorites.
- Scientists have found more than 50,000 meteorites on Earth.
- **MANY METEORITES ARE OLDER THAN OUR PLANET. THEY COULD BE BROKEN BITS OF AN ASTEROID OR COMET, OR EVEN THE CRUST OF ANOTHER PLANET.**
- The oldest bits of a meteorite, known as chondrites, were the first solid materials to form in our solar system.
- As space rocks rush through Earth's atmosphere, whirls of hot gases and bits of molten meteorite can carve dimples and pits into the rock, called regmaglypts.

Meteorite ALH84001 is thought to contain fossil evidence for life on Mars.

✦ SKY ROCKS

- Most meteorites are magnetic.
- Rare nonmagnetic meteorites are called achondrites.
- On Earth, the amino acids that compose the building blocks of life are all left-handed. The amino acids found on meteorites are equal parts left- and right-handed.
- **SCIENTISTS WEREN'T SURE THAT ROCKS COULD FALL FROM SPACE UNTIL 1803, WHEN MORE THAN 3,000 STONES RAINED DOWN ON NORMANDY, FRANCE.**
- Earth plows through several hundred tons of meteors every single day—most no larger than a pea.
- Roughly 5,200 tons of space dust rain down on Earth each year.
- Meteors whiz through Earth's atmosphere at speeds from seven to 45 miles a second—that's about 25,000 to 160,000 miles an hour.
- The speed of meteors diminishes as they move through Earth's atmosphere.
- Less than 5 percent of the meteors that enter Earth's atmosphere survive the fall. Most are vaporized entirely.
- Bolides, also known as fireballs, are meteors closer to the size of a golf ball than a pea.
- They glow more brightly than other types of meteors and remain aglow for much longer as they streak across the sky.
- Bolides are sometimes mistaken for UFOs.
- **SOME BOLIDES END THEIR JOURNEY IN AN EXPLOSION.**
- By the time a meteorite hits the ground, it's neither hot nor glowing.
- Most meteorites are small—no bigger than a cherry.

Scientists collected 570 meteorite samples on the blue ice field of Antarctica's Miller Range in 2015.

- About 30 percent of known meteorites have been found in sandy deserts.
- A desert's barren topography helps distinguish space rocks from desert rocks, while the arid conditions preserve the meteorite.
- **ABOUT 65 PERCENT OF KNOWN METEORITES HAVE BEEN FOUND IN ANTARCTICA.**
- The East Antarctic ice sheet provides a white backdrop perfect for meteorite hunting.
- Ice flow concentrates meteorites in some regions of Antarctica.
- Estimates suggest 300,000 or more meteorites are still awaiting discovery in Antarctica.
- Some meteorites found in Antarctica came from the Moon and Mars.
- Climate change could cause thousands of meteorites to sink more quickly into the ice, hiding them from view.
- **AMONG THE MOST FAMOUS METEORITES EVER DISCOVERED IS A MARTIAN METEORITE ABOUT FOUR BILLION YEARS OLD.**
- It carries the uninspiring name ALH84001: "ALH" because it was discovered in Allan Hills, Antarctica, "84" because it was discovered in 1984, and "001" because it's the first discovery of that year and location.
- Analysis of the meteorite revealed potential markers consistent with a watery Martian environment and possibly even ancient life. Although experiments remain inconclusive, the rock sparked decades of ongoing debate about whether or not Mars ever harbored life and how we might find evidence to prove it.

A pallasite meteorite found in the Atacama Desert of Chile in 1822

✦ CHEMISTRY LESSON

- A meteorite's color is dictated by its chemical composition.
- Magnesium accounts for the most common color: blue-green.
- Calcium creates a more violet hue.
- Most meteorites contain iron and nickel.
- **WHILE EARTH HAS ROUGHLY 5,800 DIFFERENT MINERALS, ONLY ABOUT 480 HAVE BEEN FOUND IN METEORITES.**
- About 30 percent of the minerals found in meteorites don't form naturally on our planet.
- The prettiest meteorites are the sparkling pallasites: shiny metal balls studded with the green gem olivine.
- Some pallasites may form from asteroids smashing into a planetlike body, mixing molten bits of asteroid with crystalline rocks.
- The Tarda meteorite, which fell to Morocco in 2020, is one of a few space rocks found to contain amino acids: essential building blocks of life.
- **SOME THINK THAT METEORITES MAY HAVE SEEDED EARTH WITH THE COMPLEX ORGANICS NECESSARY FOR LIFE TO DEVELOP.**
- The Murchison meteorite, which fell to Australia in 1969, smelled strongly of denatured alcohol.
- More than 90 amino acids have been identified in the Murchison meteorite.
- Only 19 of the Murchison meteorite's amino acids can be found on Earth.
- The Murchison meteorite contains bits that likely date back billions of years.
- **SOME METEORITES ARE STUDDED WITH TINY DIAMONDS.**
- Some diamonds may have formed under the immense pressure of impacting Earth or even earlier in their lifetime upon collisions with other space rocks.
- 99.8: the percentage of Earth's meteorites that come from smashed asteroids. The rest come from Mars and the Moon.
- To date, the Hoba meteorite is the largest intact space rock found on Earth.
- It landed less than 80,000 years ago in today's Namibia.
- Hoba weighs about 60 tons and measures about nine feet across.
- It is considered the largest natural piece of iron on Earth.
- **HOBA STILL SITS IN THE SPOT WHERE IT CRASHED AND IS A TOURIST ATTRACTION.**
- In 1804, English geologist William Thomson was using a bit of acid to clean rust from an iron meteorite when a crosshatching—a pattern unique to iron meteorites—appeared.
- **THE CROSSHATCHING OF IRON METEORITES, KNOWN AS THE WIDMANSTÄTTEN PATTERN, CAN'T BE REPRODUCED BY SCIENTISTS IN A LAB. IT CAN ONLY FORM WHEN MOLTEN METAL COOLS EXTREMELY SLOWLY.**
- Though Thomson first found the pattern, it was named after the Austrian scientist who later independently found it: Count Alois von Beckh Widmanstätten.

CURRENT FACT COUNT 0777

ASTEROIDS

Asteroids formed with our solar system 4.6 billion years ago. Most are safely contained within the asteroid belt—but some stray and pass within striking distance of Earth.

✦ CLOSE ENCOUNTERS OF THE ASTEROID KIND

- One hundred tons of space dust and granules bombard Earth every day.
- NASA's Planetary Defense Coordination Office continually scans the skies for objects that could collide with Earth.
- About a million asteroids have been discovered in the solar system, two-thirds of them in the main asteroid belt between Mars and Jupiter.
- **MORE THAN 35,000 NEAR-EARTH OBJECTS (NEOs) HAVE BEEN DISCOVERED.**
- Some 99 percent of NEOs are near-Earth asteroids (NEAs).
- An asteroid is considered near-Earth when it approaches at least 28 million miles from Earth.
- There are four types of NEAs, defined by their orbits' intersections with Earth's orbit.
- Each type of NEA is named after an asteroid typical of that orbit.
- Amors: These NEAs orbit completely outside Earth's orbit but inside Mars's orbit.
- Apollos: A small portion of their orbits occur inside Earth's orbit.
- Atens: Most of their orbits occur inside Earth's orbit.
- Atiras: These NEAs orbit completely inside Earth's orbit.
- NEAs more than 460 feet across whose orbits bring them within 4.65 million miles of Earth are considered potentially hazardous asteroids (PHAs).
- More than 2,400 asteroids are considered PHAs, worthy of constant surveillance.
- **AS MUCH AS 60 PERCENT OF ALL NEAR-EARTH OBJECTS MAY BE DARK COMETS.**
- Unlike on standard bright comets, the ice of a dark comet is hidden beneath a thick, dark, rocky surface. Because their ice is not exposed to the surface, dark comets are far less reflective than standard high-luminosity comets and throw no tail.
- Scientists think these dark comets originated in the asteroid belt. If true, that means much more water lurks within the belt than ever before believed, which may help bolster the hypothesis that Earth's water was delivered via ancient comet impacts.
- **THERE ARE MORE THAN 1,000 ASTEROIDS LARGER THAN A HALF MILE ACROSS WHOSE ORBITS CROSS THAT OF EARTH.**
- While in orbit around the Sun, Earth plows through more than 100 tons of meteors per day, most harmlessly burning up in the atmosphere as shooting stars.
- Contrary to what is often portrayed in space dramas, the average distance between asteroids is about a half million miles.
- One of this book's authors has an asteroid named for him: 13123 Tyson.

✦ WHAT'S AN ASTEROID MADE OF?

- Some asteroids are mostly rocky. Others are mostly metallic.
- **THREE GENERAL CATEGORIES DESCRIBE ASTEROID COMPOSITION: C-TYPE, S-TYPE, OR M-TYPE.**
- C-type, or carbonaceous asteroids, contain large amounts of carbon.
- More than 75 percent of known asteroids are C-type.
- C-types are found mostly in the outer regions of the asteroid belt.
- Because of their coal-rich composition, they have very low albedos (they appear very dark with low reflectivity) and are thus much more difficult to find and track. These types of asteroids generally cannot be seen with the naked eye.
- S-type, or siliceous asteroids, are composed mainly of stony iron and magnesium silicates.
- **APPROXIMATELY 17 PERCENT OF KNOWN ASTEROIDS ARE S-TYPE.**
- They dominate the inner asteroid belt.
- They are much closer and more reflective than C-type asteroids, so they're much easier to view and track.
- **MOST ASTEROIDS THAT HAVE IMPACTED EARTH ARE SILICEOUS ASTEROIDS.**
- M-type, or metallic asteroids, contain high metallic content such as iron and nickel.

- Although most asteroids that have impacted Earth belonged to the S-type category, the most devastating impact of Earth's history, the Chicxulub impactor, which ended the reign of dinosaurs, was a C-type.

✦ TARGETING AN ASTEROID

- In 2021, NASA sent a spacecraft—Double Asteroid Redirection Test (DART)—toward an asteroid to test our ability to influence space rock trajectories.
- **DART TARGETED THE ASTEROID DIDYMOS AND ITS ORBITING MOONLET, DIMORPHOS.**
- It took DART more than 10 months to reach its goal, some seven million miles from Earth.
- On September 26, 2022, DART struck Dimorphos at a hypervelocity of about 14,000 miles an hour.
- The impact moved Dimorphos closer to Didymos, creating a smaller orbit and reducing orbit time by 32 minutes.
- **DART'S SUCCESS WAS THE FIRST TIME HUMANS HAVE EVER CHANGED A CELESTIAL OBJECT'S PATH.**
- The impact not only changed Dimorphos's orbit but also its shape.
- The film *Armageddon* explored the least practical method conceivable to avoid an oncoming impactor: nuking it. (The film may have also violated more laws of physics—per minute—than any other film.)
- A snarky young Ben Affleck once cheekily asked director Michael Bay during work on *Armageddon* why it was easier to train oil drillers to be astronauts than astronauts to be oil drillers. An exasperated Bay supposedly retorted, "Oh, just shut up."

✦ BENNU

- One of the most likely asteroids to hit Earth is Bennu, a rubbly world roughly a third of a mile wide.
- **BENNU PASSES BY EARTH ABOUT EVERY SIX YEARS.**
- In 2013, nine-year-old Michael Puzio won the Name That Asteroid contest by suggesting Bennu, an Egyptian deity linked to the Sun, rebirth, and Osiris.
- Before it was named Bennu, the asteroid was called 1999 RQ36, reflecting its year of discovery.
- Bennu is likely formed from the rubble of a demolished, primeval world dating back 4.5 billion years.
- Bennu didn't coalesce into a fully formed body until one to two billion years ago.
- **ASTEROID BENNU IS MORE THAN 1,600 FEET ACROSS. IF PLACED IN DOWNTOWN PARIS, IT WOULD STAND HALF AGAIN AS TALL AS THE EIFFEL TOWER.**
- September 24, 2182: the most likely day that asteroid Bennu will collide with Earth. But even then, the risk is only 1 in 2,700.
- Most people who play the lottery, with jackpot odds far worse than 1 in 2,700, do so expecting to win.
- On September 8, 2016, the Origins, Spectral Interpretation, Resource Identification, and Security-Regolith Explorer (OSIRIS-REx) launched toward Bennu.
- The successful mission utilized the most sophisticated technology and feats of orbital mechanics yet imaginable.
- OSIRIS-REx was launched from Earth, a moving platform, to intersect Bennu, a moving target. It performed a Touch-and-Go (TAG) sequence (landing and taking off again without coming to a complete stop) on the surface of Bennu to kick up loose material and secure it in a capsule.
- From there, OSIRIS-REx returned to Earth and deployed the capsule onto the rotating planet such that it would safely land in the Utah desert.
- But OSIRIS-REx wasn't done. It then continued its journey to encounter and sample asteroid Apophis.
- **OSIRIS-REX RETURNED 4.3 OUNCES OF MATERIAL FROM BENNU.**
- Analysis shows the samples from Bennu contain carbon, nitrogen, serpentine, and organic compounds including magnesium-sodium phosphate.

✦ CHICXULUB

- Some 65 million years ago, a massive asteroid struck Earth in the area now known as Mexico's Yucatan Peninsula.
- The asteroid's impact was so immense, it carved out miles of rock on the ocean floor.
- **IT LEFT A 125-MILE-DIAMETER CRATER, NOW CALLED CHICXULUB.**
- Calculations suggest the asteroid struck Earth at about 12.5 miles a second.
- Its speed was more than 58 times the speed of sound.
- The energy from the impact equaled that of 100 trillion tons of TNT.
- **CHICXULUB'S IMPACT WAS AT LEAST 4.5 BILLION TIMES MORE FORCEFUL THAN THAT OF THE ATOMIC BOMB IN HIROSHIMA.**
- The impact vaporized sulfur-rich rocks, which unleashed as much as 100 billion tons of sulfur into the atmosphere along with dust and soot.
- It drove superheated winds that ignited everything within roughly a 621-mile radius.
- Molten debris from the impact rained down, igniting wildfires across 70 percent of Earth's forests.
- **THE AFTERMATH OF CHICXULUB'S IMPACT CAUSED THE EXTINCTION OF MOST PLANT AND ANIMAL SPECIES ON EARTH AND ENDED THE REIGN OF THE DINOSAURS.**
- Although Chicxulub ended the dominion of the dinosaurs, it didn't eradicate them all.
- All birds we see today are the evolved ancestors of the feathered dinosaurs that survived the apocalypse.
- Scientists consider the asteroid impact a turning point between two geological eras.
- Called the K-T boundary, it expresses the vast differences in fossil remains between the Cretaceous and the Tertiary periods.
- **WHAT IF THE CHICXULUB ASTEROID HAD NEVER STRUCK?**
  - Dinosaurs might still dominate the planet.
  - Flowering plant species might not have developed.
  - Many mammal species might never have emerged.
  - Humans may never have evolved.

✦ DANGEROUS STRIKES

- Collision speeds between asteroids and Earth typically range from 25,000 to 50,000 miles an hour.
- **HISTORICAL ACCOUNTS SUGGEST A NUMBER OF INSTANCES WHERE ASTEROIDS KILLED OR INJURED PEOPLE.**
- The most ancient account, in the Chinese *Book of Sui,* cites 10 people killed and a siege tower toppled by a shooting star in January A.D. 616. (It might have been a military attack instead.)
- Several Chinese accounts record stones falling from the sky, some the size of goose eggs, and killing thousands of people in 1490.
- A meteor shower rained down on Pultusk, Poland, in 1868. Some 100,000 stones fell from the sky, including one weighing 20 pounds, the heaviest ever recorded. No deaths were reported.
- In 2013, a meteorite exploded over Chelyabinsk, Russia, injuring 1,200 people.
- **THE RUSSIAN METEOR WAS THE SIZE OF A HOUSE WHEN IT ENTERED THE ATMOSPHERE.**
- It was traveling at about 11 miles a second.

"ASTEROIDS ARE OUR OLDEST AND MOST NUMEROUS COSMIC NEIGHBORS."

—PHYSICIST CARRIE NUGENT

Arizona's Barringer Meteor Crater

- It exploded about 14 miles above Earth's surface.
- **THE SHOCK WAVE DAMAGED STRUCTURES IN SIX CITIES.**
- It shattered windows for 200 square miles.
- Damage occurred up to 55 miles on either side of the meteorite's trajectory.
- Many people were knocked off their feet, but nobody was killed. People experienced burns to their skin and eyes.
- **THE BEST PRESERVED METEORITE IMPACT CRATER IN THE WORLD IS BARRINGER METEOR CRATER IN CENTRAL ARIZONA.**
- Barringer is nearly a mile wide and 750 feet deep—enough to swallow a 60-story building. Its diameter is about 4,000 feet wide.
- The impact that formed the crater likely occurred some 50,000 years ago.
- The hole formed when a metal meteorite crashed to the surface, exploding with the force of 2.5 million tons of TNT.
- Iron-rich rocks found in the vicinity are called Canyon Diablo meteorites.
- In 1908, a falling space rock exploded above Earth with a boom so powerful it toppled some 80 million trees across hundreds of miles.
- We now call it the Tunguska asteroid impact event, after the central Russian location where it occurred.
- In 1954, an 8.5-pound meteorite smashed through the roof of Ann Hodges's Sylacauga, Alabama, home, whacking her right side.
- Hodges is the only human known to have been hit by a meteorite and to survive it.
- A piece of the same meteorite, found several miles away, is now displayed in the Smithsonian Museum of Natural History in Washington, D.C.
- In 2021, a 2.8-pound meteorite plunged through Ruth Hamilton's British Columbia roof and narrowly missed her as she slept.
- In 2019, stargazers watching a lunar eclipse noticed a flash as a meteor smashed into the Moon.

In 1897, the crew of the S.S. *Hope* gently maneuvered the 34-ton Ahnighito meteorite into the ship's hold.

✦ THE FANTASTIC FIRST FOUR

- The first four asteroids discovered—Ceres, Pallas, Vesta, and Juno—were presumed to be planets.

- **CERES IS THE ONLY ASTEROID LARGE ENOUGH FOR ITS GRAVITY TO SHAPE IT INTO A SPHERE. THIS GOT IT PROMOTED IN 2006 TO THE STATUS OF DWARF PLANET.**

✦ THE PERFECT BALANCE

- Leading and trailing Earth in its orbit around the Sun are pockets of space where centrifugal forces and gravity balance precisely, attracting space dust and what are called Trojan asteroids.

- **THE HEAVILY CRATERED MOON WAS NOT UNLUCKY. EARTH IS JUST BETTER AT COVERING UP THE EVIDENCE OF IMPACTS. WEATHERING AND EROSION ALL OVER EARTH'S SURFACE PREVENT EARTH FROM LOOKING LIKE THE MOON.**

- The four round, plate-shaped landing pads of the first Apollo lunar modules were much wider than the ones used later, after we'd learned that there isn't a deep bed of lunar dust on the Moon's surface, due to the accumulation of pulverized rocks from billions of years of asteroid and meteor collisions.

✦ THE FIVE BIGGEST SPACE HUNKS ON EARTH

- **1. HOBA**
  - 60 tons, found in Grootfontein, Namibia
  - Meteorite enthusiasts and scientists have chipped away some six tons of Hoba's mass.
  - Made of 82.3 percent iron, Hoba is considered the largest natural piece of iron on Earth.

- **2. EL CHACO**
  - 37 tons, found in Gran Chaco Gualamba, Argentina
  - El Chaco is the largest meteorite among a collection called Campo del Cielo.
  - The first recorded discovery of the Campo del Cielo meteorites dates to 1576.
  - The Campo del Cielo site includes at least 26 craters.

- **3. AHNIGHITO**
  - 34 tons, found in West Greenland
  - The Ahnighito meteorite is part of a much larger, 200-tonner called Cape York that broke apart in Earth's atmosphere.
  - Ahnighito is now on display at New York's American Museum of Natural History.
  - The Cape York meteorite dates back 4.5 billion years, as do most meteorites, to the origin of the solar system.

- **4. ARMANTY**
  - 22 to 28 tons, found in Xinjiang, China
  - Armanty is one fragment among many in a meteorite field called Aletai.
  - Aletai was first discovered in 1898.
  - Pieces of Aletai meteorites can be found for sale for as much as $1,125.
  - An even bigger meteorite was reported found in the same area in 2011.

- **5. BACUBIRITO**
  - Nearly 20 tons, found in Sinaloa, Mexico
  - At more than 13 feet, Bacubirito is the longest meteorite ever found.
  - It was discovered in El Camichin, Mexico, in 1863.
  - Bacubirito is now on display at Centro de Ciencias de Sinaloa (Science Center of Sinaloa) in Culiacán, Mexico.

CURRENT FACT COUNT 0836

COMETS

These dirty snowballs harbor ancient ice that might yield clues to how the solar system formed and how water may have arrived—and thus, how life may have formed—here on Earth.

- While asteroids are made of rocky material, comets are an amalgamation of rock, ice, and dust.
- Astronomer Fred Whipple was the first to describe them as "dirty snowballs" in 1950. Prior to his hypothesis, comets were believed to be globs of sand and pebbles.
- **COMETS VAPORIZE AS THEY PASS BY THE SUN, CREATING THEIR VISIBLE TAILS.**
- Blown back by the solar winds and radiation pressure, comets' tails always point away from the Sun.
- Comets can be divided into four parts: the nucleus, the coma (or surrounding atmosphere), and two tails.
- One tail of a comet is made of plasma, the other one of dust.
- The plasma tail glows blue as ultraviolet light ionizes its gases.
- The tail is made up of tiny particles, but if you compressed them to the average density of sea-level air, it would fill a half-mile cube—about the size of the average warehouse.

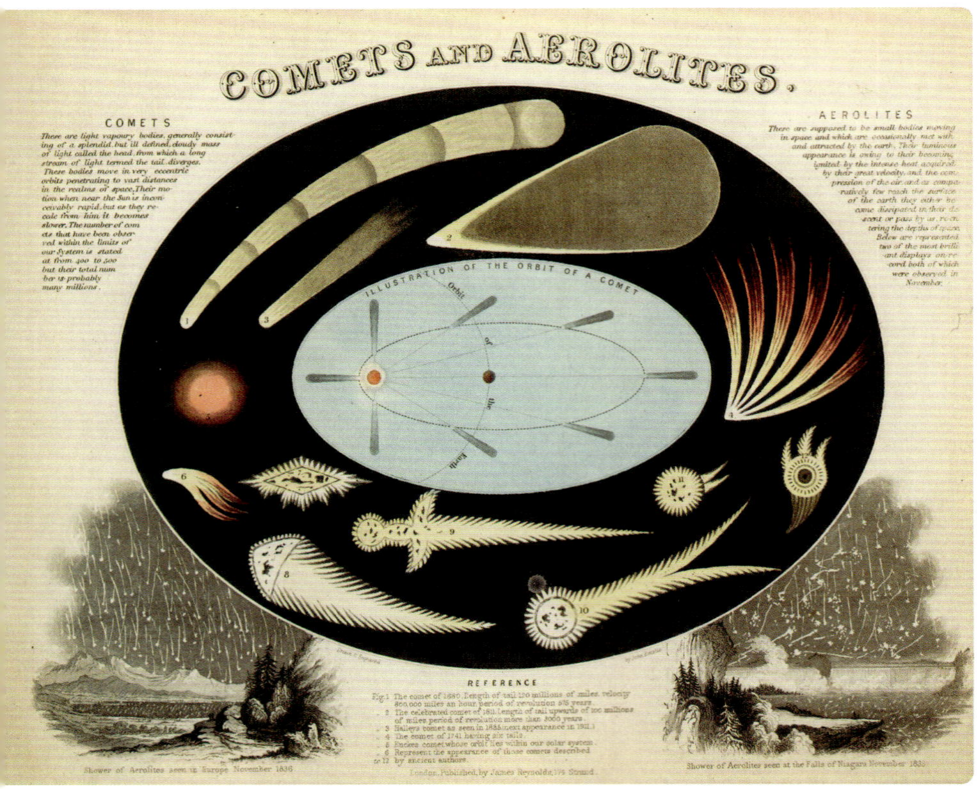

A circa 1851 illustration of the forms of comets by James Reynolds & Sons

- A comet's tail can extend up to 100 million miles in length.
- Most comets in our solar system are too small or far away to be seen from Earth.
- Large rocky bits freed from a comet continue orbiting as a dusty cloud along the comet's path.

✦ FAMOUS COMETS

- Swift-Tuttle is about twice the size of Chicxulub, the asteroid that ended the age of dinosaurs.
- Comet Swift-Tuttle is named for the two astronomers, Lewis Swift and Horace Tuttle, who discovered it independently of one another in 1862.
- The comet's full astronomical name is 109P/Swift-Tuttle.
- The *p* in its name stands for "periodic," signaling that its orbit around the Sun takes less than 200 years.
- **THE DEBRIS FROM SWIFT-TUTTLE FUELS THE ANNUAL PERSEID METEOR SHOWER.**
- Like a cosmic firecracker, comet Hyakutake passed by Earth, close and bright, earning the nickname the Great Comet of 1996.
- An amateur Japanese astronomer, Yuji Hyakutake, spotted it using powerful binoculars.
- The comet came within 9.3 million miles of Earth.
- Sky-gazers could see Hyakutake with the naked eye for three months.
- **DON'T WAIT UP: HYAKUTAKE WON'T REEMERGE FOR AT LEAST ANOTHER 70,000 YEARS.**

The comet ATLAS in the sky behind the Magellan Telescope at Carnegie's Las Campanas Observatory in Chile in 2025

- Comet Bernardinelli-Bernstein's nucleus is the largest ever seen—larger than Rhode Island.
- On September 30, 2016, a NASA/European Space Agency mission landed the Rosetta spacecraft on comet 67P/Churyumov-Gerasimenko.
- The comet's name combines the names of its two discoverers with the fact that it was the 67th periodic comet to be discovered.
- The Rosetta mission discovered exposed water ice on the comet's surface, as well as glycine and phosphorus in the comet, all key components of life.
- Rosetta also discovered molecular nitrogen and oxygen in the comet.
- In 2017, the first known interstellar visitor—the first recorded object from outside our solar system—zipped by: an object dubbed 'Oumuamua.
- **ITS NAME MEANS "A MESSENGER FROM AFAR ARRIVING FIRST" IN HAWAIIAN.**
- Scientists believe it's been roaming space for hundreds of millions of years, unattached to any star or system—a rogue asteroid.
- 'Oumuamua measured just 62 feet wide, about the length of a bowling lane, but around a quarter mile long.
- The cigar-shaped object, reddish in hue, moved as if propelled by its own engine.
- It has no dust or gas coma that normally helps a comet move through the solar system. That means it must be composed entirely of densely packed rock and/or metal.
- The comet's acceleration may therefore be attributed to the release of trapped molecular hydrogen (H_2).
- **IN THE HIT FANTASY BOOK AND TV SERIES *GAME OF THRONES,* A BRIGHT RED COMET ILLUMINATES THE DAYTIME SKIES ACROSS MULTIPLE CONTINENTS.**
- While exceedingly rare, large comets have been witnessed in daylight hours while their paths streak near the Sun—but they're never a bright red hue, as described in the books and shown in the TV show, unless viewed through sensitive infrared telescopes.
- Most comets appear a glowing bluish white to the naked eye.

Comet NEOWISE as seen from Nevada's Valley of Fire State Park in 2020

✦ COMET 2I/BORISOV

- The first confirmed interstellar comet and second interstellar visitor to our solar system was comet 2I/Borisov, observed in late 2019.

- **COMET 2I/BORISOV SPED THROUGH THE SOLAR SYSTEM AT 110,000 MILES AN HOUR.**

✦ THE APOCALYPSE COMET

- Hale-Bopp—one of the 20th century's brightest comets—is named for its discoverers: amateur astronomers Alan Hale and Thomas Bopp.

- **HALE-BOPP WAS VISIBLE FROM EARTH FOR 18 MONTHS.**

- In a cosmic coincidence, the band Hanson released "MMMBop" in April 1997, the same month that Hale-Bopp reached its peak brightness.

- The last time Hale-Bopp's orbit brought it this close to Earth was 4,200 years ago.

- Hale-Bopp was huge, therefore easily visible, because its nucleus was estimated to measure around 20 miles across.

- An apocalyptic cult called Heaven's Gate saw Hale-Bopp as an alien visitor and a sign of end times. The leader and 38 followers died by suicide, believing they would enter a higher realm.

✦ HALLEY'S COMET

- English astronomer Edmond Halley theorized that three recorded comets were, in fact, a recurrent appearance of the same comet.

- Halley introduced the idea that comets can orbit the Sun.

- The European Space Agency's Giotto space probe chased Halley's comet in 1986, coming within 370 miles of its surface.

- At its greatest distance, Halley's comet orbits beyond Neptune.

- Halley's comet bookended Mark Twain's life. Twain said, "I came in with Halley's comet in 1835; it's coming again next year [in 1910], and I expect to go out with it." And he did. Twain died one day after its closest approach to the Sun.

- The first record of Halley's comet was from Chinese astronomers in 239 B.C.

- The appearance of Halley's comet in A.D. 451 was taken as an omen of Attila the Hun's defeat.

- **IN 837, HALLEY'S COMET CAME SO CLOSE TO EARTH THAT ITS TAIL STRETCHED ACROSS MORE THAN A THIRD OF THE SKY.**

- In 1066, the comet was thought to portend the Norman conquest of the English.

- When it streaked westward in the skies of 1222, Genghis Khan took it as a sign he should turn west and invade southeastern Europe.

- **HALLEY'S COMET MAKES AN APPEARANCE ONCE EVERY 76 YEARS (GIVE OR TAKE).**

- The comet swung closest to Earth most recently on April 11, 1986.

- Halley's comet won't return until the summer of 2061.

- On each pass through the inner solar system, Halley leaves behind a trail of debris.

- Earth, in its orbit around the Sun, plows through that debris trail annually, manifesting the Eta Aquarid and Orionid meteor showers.

CURRENT FACT COUNT 0873

METEOR SHOWERS

When Earth encounters the dusty remnants of a large comet or asteroid, the atmosphere is alight with hundreds or thousands of shooting stars.

- We call them shooting stars, but they aren't stars at all. They're meteors.

- Most meteoric illuminations are made by debris no larger than grains of sand.

- **METEOR SHOWERS HAPPEN WHEN OUR ATMOSPHERE INTERCEPTS A COMET'S DEBRIS FIELD AS WE ORBIT THE SUN.**

- Why are most meteor showers best viewed after midnight? It's not only because the sky is dark. Earth's night side rotates into the stream of debris, increasing the appearance of the meteor's speed. Imagine driving through a rainstorm—more raindrops splatter against the front windshield than the side windows or rear windshield.

- Only the fastest meteors are visible before midnight. They form long, lazy streaks across the sky because they move with the direction of Earth's rotation.

- **THE POINT FROM WHICH METEORS APPEAR TO EMANATE DURING THEIR PEAK IS CALLED THE RADIANT POINT.**

- A viewer looking directly into the radiant would see debris approaching from a single distant point toward all directions: left, right, above, and below—like staring into the barrel of a confetti gun.

- But the radiant is an optical illusion crafted by the perspective of our vantage point from Earth. All the bits of meteor debris move in the same direction, parallel with one another.

✦ THE BEST AND BRIGHTEST

- There are more than 100 cataloged meteor showers.

- Some meteor showers are best viewed in the Southern Hemisphere, while others are best viewed in the Northern Hemisphere.

- **THE FOLLOWING ARE SOME OF THE BRIGHTEST METEOR SHOWERS THROUGHOUT THE YEAR:**
  - In the Northern Hemisphere, the Quadrantids occurs in early January, sometimes with as many as 20 meteors an hour in the wee hours of the night.
  - The Lyrids becomes visible in April.
  - The Eta Aquarids reaches maximum visibility in early May.
  - The debris field forming the Eta Aquarids derives from the most famous comet in history: Halley's comet.
  - The Delta Aquariids and the Alpha Capricornids peak in late July.
  - The Perseid meteor shower, a summertime favorite, is visible the second week of August.
  - Each of the Perseids meteors in the annual shower is a remnant of the comet Swift-Tuttle, which orbits the Sun every 133 years.
  - The Orionids may be seen for several nights in late October.
  - The Taurids is active through October and November, especially early November.
  - The Leonids becomes visible in mid-November, but every 33 years, it's timed with the peak debris trail left by comet 55P/Tempel-Tuttle.
  - The Leonids claims the fastest meteors of any annual shower, reaching speeds near 160,000 miles an hour.
  - The Geminids appears in mid-December with a strong, dependable show.
  - The Ursids shows up just before Christmas each year.
  - During the Geminid meteor shower, you can witness more than 100 meteors per hour at its peak, assuming perfect viewing conditions.
  - Unlike most other meteor showers, the Geminids originates from an asteroid rather than a comet.

- Most Southern Hemisphere meteor showers are visible in the winter, between July and September.

- Some of the major Southern Hemisphere meteor showers include the Alpha Centaurids, Gamma Normids, Pi Puppids, Piscis Austrinids, Delta Aquariids, Alpha Capricornids, Dec Phoenicids, and Puppid/Velids.

- **THE ALPHA CAPRICORNIDS SHOWER IS ON ITS WAY TO BEING CROWNED THE BRIGHTEST AND STRONGEST METEOR SHOWER YET—200 YEARS FROM NOW, WHEN EARTH FINALLY ENCOUNTERS THE BULK OF ITS DEBRIS.**

- Most meteor showers are more like meteor dribbles, peaking at about 12 an hour.

- In 1833, the most intense meteor shower ever recorded brought 50,000 to 150,000 meteors each hour—the equivalent of 2,500 per minute.

- **IT'S REFERRED TO AS "THE NIGHT THE STARS FELL."**

- A witness said it was "magnificent beyond conception," according to an 1833 article in the *New York Evening Post*.

- At the time, no one knew what caused a meteor shower, and its intensity made many fear it marked the end of the world.

An illustration, published in 1888, of the 1833 Leonid meteor storm

- In 1833, the Leonids was so intense, displaying thousands of meteor streaks per hour, that 24-year-old Abraham Lincoln was woken by a pastor who was sure the end of the world was imminent. He was calling for everyone to repent their sins before meeting their maker.

- **LINCOLN, A SELF-TAUGHT LEARNED MAN, LOOKED UP, SAW THAT ALL THE GRAND CONSTELLATIONS HE REMEMBERED WERE STILL THERE, AND CONCLUDED THAT THE ACTUAL STARS COULD NOT HAVE BEEN FALLING FROM THE SKY. HE PROMPTLY WENT BACK TO SLEEP.**

- The 1833 meteor storm became one of the first citizen science efforts when astronomer Denison Olmsted posted a call for information from anyone who witnessed it.

- His subsequent study of the storm finally overturned Aristotle's "bubble" hypothesis, which had persisted for more than 2,000 years.

"I WOULD RATHER BE A SUPERB METEOR, EVERY ATOM OF ME IN MAGNIFICENT GLOW, THAN A SLEEPY AND PERMANENT PLANET."

—JACK LONDON

The Geminid meteor shower streaks above Joshua Tree National Park in California.

CURRENT FACT COUNT 0941

SEEING STARS

When Galileo turned his telescope to the heavens, he discovered that the fuzzy, glowing band of nebulous matter they called the Milky Way was composed of countless individual stars.

- On a perfectly clear, dark night, in pristine conditions, you could see about 6,000 stars.
- Light pollution diminishes that number exponentially. Most people see less than 200 stars. Some may only see a dozen.
- Ancient Greek and Chinese cultures produced the first star catalogs we know of, both around 2,000 years ago.
- Ancient Chinese astronomers assigned numbers to stars within a constellation, such as Fang 1, Fang 2, Fang 3, and Fang 4, all within the constellation Fang.
- Most modern star names derive from translations of ancient Greek to Arabic, medieval Latin, and modern languages.
- The stars appear to rise in the east and set in the west, a function of Earth's rotation.
- **THE STARRY CEILING OF NEW YORK CITY'S GRAND CENTRAL STATION IS BASED ON JOHANN BAYER'S *URANOMETRIA*, A 17TH-CENTURY CELESTIAL ATLAS.**
- *Uranometria* was the 17th and 18th centuries' standard reference for the heavens.
- But because of the way Bayer's map was projected onto the ceiling, the Grand Central ceiling version is reversed.
- When a commuter pointed out the mistake, one official's (inane) response was that the sky was painted from God's perspective.
- Meanwhile, the constellation Orion faces forward among all the backward constellations. There's no official response to that one.
- Venus makes an appearance in Vincent van Gogh's masterpiece "The Starry Night," along with the Moon.

A *Uranometria* illustration of the Andromeda constellation (1624)

- Van Gogh painted at least 20 variations of "The Starry Night." One famous variation, "La Nuit étoilée," now resides at the Musée d'Orsay in Paris.

✦ STAR LIGHT, STAR BRIGHT

- The 11 brightest stars in the night sky and their constellations:
  - The Sun, of course, tops this list (no associated constellation)
  - Sirius A, aka the Dog Star (Canis Major)
  - Canopus (Carina)
  - Rigil Kentaurus, also known as Alpha Centauri A (Centaurus)
  - Arcturus (Boötes)
  - Vega (Lyra)
  - Capella (Auriga)
  - Rigel (Orion)
  - Procyon (Canis Minor)
  - Achernar (Eridanus)
  - Betelgeuse (Orion)
- The brightest stars, as they appear to us on Earth, aren't necessarily the most inherently luminous. Some of the brightest stars are small but nearby, while some of the dimmest pinpricks of light in the night sky are massive but distant.
- **HOW FAR AWAY ARE THE STARS? IF YOU COULD SHRINK THE SUN TO THE SIZE OF THE PERIOD AT THE END OF THIS SENTENCE, THE NEXT NEAREST STAR, ALPHA CENTAURI, WOULD BE EIGHT MILES AWAY.**
- Imagine four bumblebees randomly released across the entire continental United States. The chances of two of them accidentally colliding with one another are higher than the chances of another star crashing into our Sun.
- When the star Betelgeuse explodes—that is, when it goes supernova—it will shine as bright as the full moon and will appear fully visible in daylight skies. The dazzling cosmic event is forecasted for sometime

"The Starry Night" by Vincent van Gogh (1889)

between next week and a hundred thousand years from now.

✦ THE DIPPERS

- Ursa Major and Ursa Minor—both named after the Latin word for "bear"—are also known as the Big Dipper and Little Dipper.
- **MOST STARS IN THE BIG DIPPER ARE ADOLESCENTS, ROUGHLY HALF A BILLION YEARS OLD.**
- Ursa Major and Ursa Minor each have stars that represent their long, bushy tails. But bears don't have long, bushy tails—a sign of human creativity while connecting the dots.
- The Big Dipper and Little Dipper aren't constellations but asterisms: patterns of stars that are within constellations or part of several constellations.
- The stars of the Big Dipper are slowly drifting in space, so the pattern will look quite different 100,000 years from now.
- Next time you look at the Big Dipper, see if you can pass the eye acuity test once administered by our ancient ancestors. Notice the point of light that creates the bend of the handle. Do you see that it is really two points of light very close together? If so, congratulations! If you lived in the Roman Empire 2,000 years ago, your eyesight might get you into the imperial army.
- The brighter of these two stars is Mizar (Arabic for "veil" or "cover"), and the dimmer is Alcor (Arabic for "forgotten" or "neglected").
- With the help of telescopes and, later, spectroscopy, astronomers discerned that Mizar and Alcor are not two stars but six—a sextuple star system.
- Alcor itself is one binary star system in a lazy orbit around Mizor, which contains four stars as two separate binary star systems.

Star trails framed by the Alabama Hills outside Lone Pine, California

✦ THE SCIENCE BEHIND THE TWINKLE

- "Twinkle, Twinkle Little Star" says it right: Night sky lights that twinkle are stars; those that don't are usually planets.
- **STARS ARE SO DISTANT FROM EARTH THAT WE PERCEIVE THEM AS MERE PINPRICKS OF LIGHT, THOUGH SOME OF THEM ARE MANY TIMES THE SIZE OF THE SUN.**
- After journeying many trillions of miles across empty space, starlight must traverse the particle-dense sea of Earth's atmosphere before piercing a stargazer's eyeballs.
- The turbulent air molecules and particles within the layers of Earth's atmosphere refract the starlight hither and thither, causing the pinprick of light to disappear from and reappear in our sight line. That's why a star appears to twinkle.
- **THE SCIENTIFIC WORD FOR A STAR'S TWINKLE IS "SCINTILLATION."**
- Planets will appear to twinkle when they're viewed through the extra-thick atmosphere near the horizon.
- Planets don't twinkle because they're so much closer to Earth and therefore appear larger than stars.
- The light we see from planets undergoes atmospheric distortion, but far less perceptibly.
- Stars and planets viewed from space both shine steadily bright.

✦ THE BRIGHTEST OF THEM ALL

- You can't determine a star's distance by its brightness alone. A distant, massive star may shine more brightly than a nearby, dim star.
- The inherent luminosity of a star typically depends on its mass and age.
- A star's apparent brightness depends on its luminosity and its distance from you.
- Luminosity is defined by the total energy emitted by a star.
- A star's luminosity remains the same no matter its distance from the viewer.
- **SIRIUS A, THE BRIGHTEST STAR IN OUR SKY, IS 25 TIMES MORE LUMINOUS THAN THE SUN AND 8.6 LIGHT-YEARS AWAY.**
- The star RMC 136a1 in the Large Magellanic Cloud is 6.7 million times more luminous than the Sun. But, as it lies 163,000 light-years away, you can only view it through a telescope.
- Star brightness can change over time.
- **POLARIS, OR THE NORTH STAR, MAY HAVE BEEN MORE THAN FOUR TIMES AS BRIGHT IN ANCIENT TIMES.**
- Records suggest Polaris got about two and a half times brighter in the last two centuries.
- Polaris's brightness waned in the early 2000s but has increased again since 2014.

✦ THE NORTH STAR

- Unlike other stars that gradually travel in a circle around the night sky, Polaris—the North Star—stays in almost exactly the same place.
- It sits right above the North Pole.
- **ITS CONSTANT POSITION AND LOCATION DUE NORTH HAS MADE IT AN IMPORTANT NAVIGATIONAL AID FOR CENTURIES.**
- Polaris hasn't always been the North Star in human history. Earth's wobble means that Polaris wasn't always positioned right above the North Pole, but other stars once were.
- When the earliest Egyptian pyramids were being built, the star Alpha Draconis was considered the North Star.
- Vega was once the North Star, and it will be again some 12,000 years from now.
- After Vega becomes the North Star, it will take 26,000 more years for Polaris to become the North Star again.
- Right now, there's no southern equivalent to the North Star.
- At the Equator, the North Star just crests the horizon.
- **IN THE NORTHERN HEMISPHERE, THE ELEVATION OF POLARIS ABOVE THE HORIZON IS YOUR LATITUDE ON EARTH.**
- For Santa Claus, the North Star always appears directly above his head.

"AFTER MY DEATH, THE MOLECULES OF MY BEING WILL RETURN TO THE EARTH AND THE SKY. THEY CAME FROM THE STARS. I AM OF THE STARS."

—CHARLES LINDBERGH, *AUTOBIOGRAPHY OF VALUES*

CURRENT FACT COUNT 0967

CONSTELLATIONS: PATTERNS IN THE SKY

Inspired by their traditions and religions, ancient people identified patterns among their night skies. But beyond serving as a window into past cultures, constellations carry no scientific significance.

- The word "constellation" is derived from Latin words for "together" and "stars": *con + stella.*
- **THE STARS WE GROUP INTO CONSTELLATIONS ACTUALLY HAVE NO SPATIAL RELATIONSHIP TO ONE ANOTHER.**
- Constellations are pictures created by humans, connecting unrelated stars into recognizable patterns.
- Many different cultures have seen the stars as a giant "connect the dots" puzzle, naming their own groups of constellations.
- **AS THEY ARE HUMAN CONSTRUCTS, CONSTELLATION PATTERNS DIFFER ACROSS CULTURES AND CHANGE OVER TIME.**
- 88: the number of officially named constellations, adopted by the International Astronomical Union (IAU) in 1922
- The IAU recognizes regions of the sky, not drawn figures, for each constellation.

✦ THE PLEIADES

- The Pleiades, or Seven Sisters, is not a constellation at all but an asterism: a fancy nickname for a more interesting pattern of stars within a less interesting constellation.
- Astronomers call the Pleiades the star cluster M45.
- The cross-cultural fascination with the number seven has infused many legends, which went on to declare that the Pleiades asterism contains seven stars, though only six are visible to the naked eye.
- **IN FACT, THE PLEIADES CLUSTER CONTAINS THOUSANDS OF STARS.**
- You might recognize the Pleiades's configuration of stars as the logo for the popular car brand Suburu, whose name means "unite" in Japanese.
- Influenced more by objective truths than by legend, the logo for Subaru contains six, not seven, stars.
- Other asterisms include: the Big Dipper (Ursa Major), the Little Dipper (Ursa Minor), the Teapot (Sagittarius), and the Northern Cross (Cygnus).
- The largest constellation in the sky was Argo Navis, representing the ship sailed by Jason and the Argonauts, though it was informally broken up throughout the 18th century and formally broken up by the IAU in 1930.
- The smallest constellation in the sky is the Southern Cross. Your fist, held at arm's length, covers this constellation completely.
- The Southern Cross contains four principal stars, each forming the corner of a rhombus.
- With no central star to represent the transept of a cross, perhaps it should be renamed the Southern Rhombus.
- Some star groupings are recognized across multiple cultures: the Pleiades, the Southern Cross, and Orion's belt, to name a few.
- Petroglyphs carved into the stones of Peru's Machu Picchu represent the Southern Cross, what the Inca called Chakana.
- **THE ANCIENT GREEKS PROJECTED THEIR MYTHS ONTO THE STARS: ORION THE HUNTER CHASING THE SEVEN SISTERS OF THE PLEIADES, FOR EXAMPLE.**
- The constellation Lyra is named for the stringed instrument used by the musician Orpheus of ancient Greek legend. As the myth goes, the lyre was the first musical instrument ever built. Upon Orpheus's death, Zeus placed it in the sky for all to witness.

✦ ORION'S BELT

- The stars of Orion vary in their distance from Earth: The closest, Bellatrix, on Orion's western shoulder, is 243 light-years away; the farthest, Alnilam, the middle of his belt, is 1,360 light-years away.
- **BELOW ORION'S BELT IS THE ORION NEBULA, ONE OF THE BRIGHTEST NEBULAE IN THE SKY AND THE MIDDLE STAR IN HIS SWORD.**
- The ages of stars in familiar constellations can vary greatly. Stars in the Orion Nebula, a stellar nursery, are newborns, about five million years old.
- Stars in the Pleiades are toddlers, a mere 100 million years old.

Orion's famous belt and sword amid the swirling dust clouds of the Orion Nebula

CURRENT FACT COUNT 1045

THE ZODIAC

Throughout the year, the Sun appears against a background of a different constellation each month. The Babylonians bestowed meaning upon the phenomenon. The ancient Greeks adopted the concept and called it the zodiac.

- The zodiac represents an ancient division of the entire sky, each segment represented by a different constellation.
- As Earth orbits the Sun, each of the segments is visible at night. Thus, to the ancients, each constellation represented a different time of year or season.
- Since it was conceived of during a time that held a geocentric worldview, these seasonal changes were described in terms of the Sun moving through segments of the sky, but it was actually Earth's annual orbit.
- The zodiac we commonly think of today in Western culture was configured differently in different cultures.
- The ancient Babylonian zodiac had 18 segments, but by 500 B.C., it had decreased to the 12 we know today.
- The zodiac's 12 constellations roughly correspond to the months of the year.

- Some of the Babylonian demarcations are still considered zodiac signs today: the Twins (Gemini), the Scales (Libra), and the Crab (Cancer).
- **THE ANCIENT GREEK ASTRONOMER PTOLEMY HELPED POPULARIZE THE ZODIAC SIGNS.**
- The traditional dates of the zodiac no longer align with the corresponding constellation.
- This shift, called the precession of the equinoxes, is due to a cyclical wobbling in Earth's axis.
- **LIKE A SLOW-SPINNING TOP, EARTH'S WOBBLY AXIS TAKES NEARLY 26,000 YEARS TO GO AROUND ONCE.**
- The wobble causes the Sun's position relative to the starry background we see on Earth to drift over time.
- According to current astronomical principles, there should be a 13th constellation—but Ophiuchus (the Serpent Bearer) has yet to appear in the popular horoscope lineup.

✦ THE 12 CONSTELLATIONS

- The zodiac is one of the most ancient systems for describing the starry skies.
- Though no science supports the prophetic nature of the zodiac, the meaning ascribed to these constellations can offer a glimpse into the societies of our ancient ancestors.

✦ 1. ARIES

- Aries (the Ram) is traditionally the first sign of the zodiac because in ancient times, the Sun appeared in this constellation's segment when spring arrived in the Northern Hemisphere.
- Although the start of Aries once marked the day of the spring equinox, the precession of Earth's axis has shifted it out of alignment.

- The four main stars of the constellation Aries have inspired many stories. In Chinese astronomy, they represent part of a lasso and the emperor's hunting partner; in some Polynesian cultures, they form part of a porpoise.
- For the ancient Greeks, Aries was the ram whose golden fleece was stolen by Jason and the Argonauts.

✦ 2. TAURUS

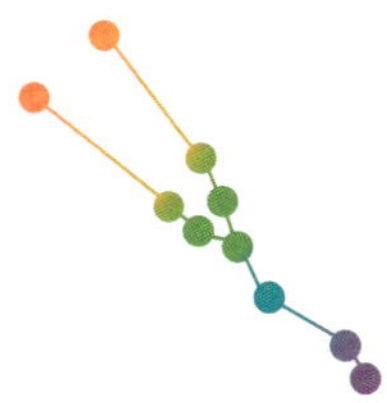

- The constellation Pleiades (Seven Sisters) marks the shoulder of Taurus (the Bull).
- The three stars of Orion's belt point to the red star Aldebaran, the eye of the bull.

✦ 3. GEMINI

- The constellation Gemini (the Twins) appears as two stick figures side by side.

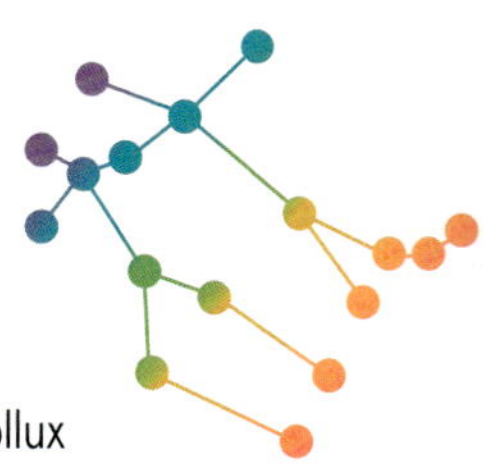

- In the ancient Greek myth, the twins were named Castor and Pollux, brothers of Helen of Troy.
- When Castor died, Pollux begged Zeus to bring him back to life. Zeus agreed, on the condition

that they spend half the year on Earth and half their time in the heavens among the stars.

- Indeed, Gemini is visible in the sky for much of the year.

✦ 4. CANCER

- Best viewed on moonless nights, Cancer (the Crab) is the dimmest of all constellations in the zodiac.

- According to Greek myth, when a giant crab sent to distract Hercules was killed, the goddess Hera rewarded it with a place in the heavens.
- The bright star cluster M44 (also called Praesepe or the Beehive cluster) is visible within the constellation Cancer.

✦ 5. LEO

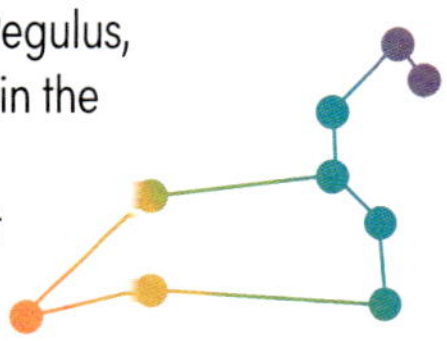

- Observations of Regulus, the brightest star in the constellation Leo, appear in ancient Babylonian tablets.
- The ancient Egyptians connected the annual rising of the Nile River with the Sun's appearance in Leo.

✦ 6. VIRGO

- The constellation Virgo is often depicted as a maiden holding a sheaf of wheat, representing its association with harvest time.
- Within the constellation Virgo's borders lies the Virgo cluster, a gigantic collection of galaxies spanning 15 million light-years.

✦ 7. LIBRA

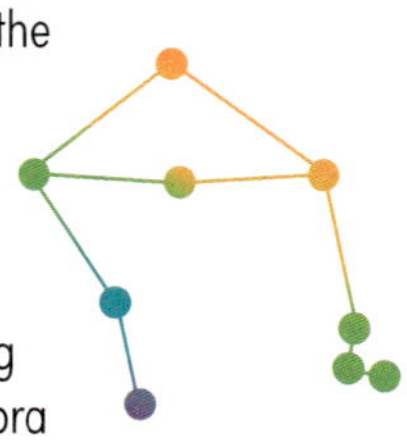

- The familiar sign for pound (lb) comes from an abbreviation of the constellation Libra (the Scales).
- While today the picture representing the constellation Libra is a balance scale, the ancient Greeks believed it to be the claws of the constellation Scorpius.
- Therefore, the names of Libra's two brightest stars, Zubeneschamali and Zubenelgenubi, come from the Arabic meaning "northern claw" and "southern claw."
- Zubenelgenubi and Zubeneschamali have the longest names of all stars in the sky.
- The ancient Greeks saw in this constellation the scorpion that killed Orion the Hunter. Afterward, Zeus placed the two on opposite sides of the sky.

✦ 8. SCORPIO

- Antares, 15th brightest star in our sky, sits at the heart of Scorpio.
- A red supergiant, Antares is 10,000 times more luminous than our Sun.

✦ 9. SAGITTARIUS

- The ancient Greeks and Romans looked skyward and saw the constellation Sagittarius as a bow-wielding centaur.

- The eight central stars of Sagittarius can also be seen as the shape of a teapot.

✦ 10. CAPRICORN

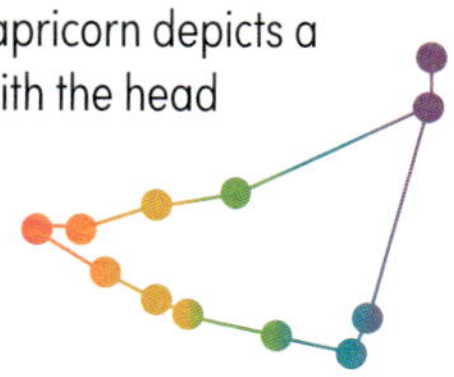

- The constellation Capricorn depicts a mythical creature with the head and front legs of a goat and the tail of a fish.
- This creature appears in Babylonian legends dating to 2100 B.C.

✦ 11. AQUARIUS

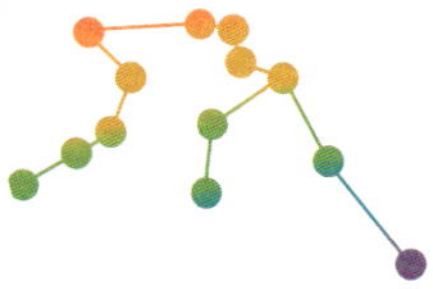

- It requires some serious imagination and creative liberties to see this smattering of stars as a man pouring water from a jar. And yet these stars have long been associated with rain in European and North African cultures.

✦ 12. PISCES

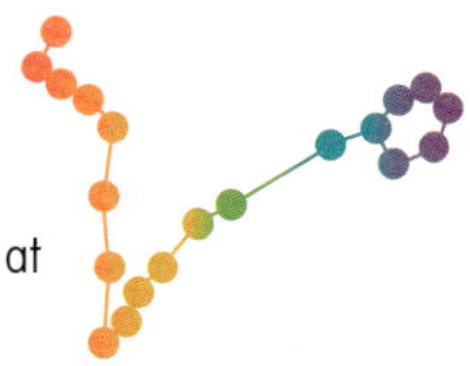

- The giant V-shaped constellation Pisces represents two fish bound by a long cord at their tails.
- This image of a celestial fish dates back to the Babylonians.
- The ancient Greeks saw the fish as the goddess Aphrodite and her son Eros, transformed as they fled a sea monster.
- **THE HISTORY AND PERSISTENCE OF THE ZODIAC SIGNS REVEAL HUMANITY'S QUEST TO FIND MEANING IN THE STARS ABOVE.**

✦ INCA CONSTELLATIONS

- The Inca saw constellations of a llama, fox, snake, and toad in the night sky.
- **EACH INCA CONSTELLATION MARKED THE SEASONS WHEN THAT DESIGNATED ANIMAL APPEARED OR GAVE BIRTH.**
- The Inca considered Chakana—their name for the Southern Cross—to be the center of the universe.
- The Inca of South America had both light and dark constellations.
- Yacana, a dark constellation representing a llama, figured prominently for the Inca, with eyes made of the stars Alpha and Beta Centauri.
- Other Inca constellations included a serpent and birds native to the Andes.

The Milky Way above El Castillo pyramid (the Temple of Kukulkán) in Mexico

✦ MEANING IN THE MILKY WAY

- A number of cultures saw shapes and animals in the dark areas of the Milky Way, where dust and gas block faraway stars from our view.
- Some Aboriginal traditions perceive these dark spaces as water holes or caves housing animals or evil spirits.
- The Maya of Mesoamerica interpreted the Milky Way's dark zones as an entrance to the underworld, the open jaws of a crocodile, and a crocodile-canoe.
- The Zulu people of South Africa and the Polynesians also incorporated the dark areas of the Milky Way into their cosmological picture.
- **MULTIPLE CULTURES VIEWED THE MILKY WAY AS A CELESTIAL WATERWAY HARBORING LOCAL ANIMALS.**
- Some islanders of the Torres Strait near Australia saw the Milky Way as a cloud of sea-floor debris stirred up by a scrounging shark.
- The Inca saw the Milky Way as earthly water drawn into the sky for a heavenly refreshing before returning to Earth.
- One important Aboriginal constellation is the Emu in the Sky, celebrating the large, flightless bird native to Australia.
- The emu's body stretches along the dark lanes of the Milky Way, the negative space forming the shape rather than the stars themselves.
- If the ancient Romans had named their thoroughfares "streets" rather than "ways," the Milky Way might have instead been named Milky Street.
- The word "galaxy" ultimately derives from the Greek word for "milk."
- Galactose is the name of a naturally occurring milk sugar in mammals.

"Emu Sky," a linocut by New South Wales Barkandji artist Badger Bates, 2008

✦ NATIVE AMERICAN SKIES

- Many Native American constellations are connected to animals, natural elements, and values of human life.

- **NAVAJO ANIMAL CONSTELLATIONS INCLUDE THE PORCUPINE, BUTTERFLY, BEAR, GILA MONSTER, AND TURKEY.**

- In the Lakota tradition, stars from Ursa Major, Aries, and Triangulum form constellations of a sacred pipe and the medicinal herbs smoked ceremonially.

- In the Ojibwe tradition, the constellation now known as Orion, connected to other nearby stars, represents Biboonkeonini, the Wintermaker—appropriate, since the constellation comes into view primarily in the winter.

- Navajo (Diné) people see Dilyéhé—"pinlike sparkles"—in the Pleiades constellation and a Gila monster in Andromeda.

- The Ojibwe interpret the Pleiades as a spiritual gateway to the heavens.

- In Navajo tradition, Náhookòs Bikò (the star Polaris) represents a home's central fire.

- Náhookòs Bikà' (Male Revolving One; also called the Big Dipper) is known as a leader, warrior, and father.

- Náhookòs Bi'áád (Female Revolving One; also called Cassiopeia) provides stability, food, and growth as a mother and partner.

- **TOGETHER, NÁHOOKÒS BIKÒ, NÁHOOKÒS BIKÀ', AND NÁHOOKÒS BI'ÁÁD DEPICT THE BONDS OF HOME AND FAMILY, WITH FAMILY REVOLVING AROUND THE FIRE.**

- The Anishinabe and Ojibwe cultures see a panther with a curly tail around the constellation now considered Leo.

- The constellation identified as Cygnus (the Swan) is a salamander in Dakota and Lakota culture and a combined crane and skeleton bird in Ojibwe culture.

"WHEN I TRACE . . . THE WINDINGS TO AND FRO OF THE HEAVENLY BODIES . . . I STAND IN THE PRESENCE OF ZEUS, HIMSELF, AND TAKE MY FILL OF AMBROSIA."

—CLAUDIUS PTOLEMY

CURRENT FACT COUNT 1057

THE NORTHERN SKY

"Summer" constellations and "winter" constellations exist in the sky throughout the entire year. But as Earth continues its orbit, some constellations shift into daylight hours, where sunlight obscures our view of them.

- **1.** Corona Borealis (the Northern Crown): A semicircle of stars so named for a laurel crown worn by the daughter of King Minos. The Cheyenne saw a campfire circle, Australian Aboriginals saw a boomerang, and in Arab mythology it is called the Bowl of the Poor People.
- **2.** Northern Cross:
  - The Northern Cross is really an asterism (see opposite page, top) within the larger constellation Cygnus the Swan.
  - The six-star formation appears as a cross or a bird in flight.
  - Chinese mythology interprets it as a bridge of magpies that unite a mortal with his divine lover after a goddess separated them with a raging white river (the Milky Way).
- **3.** Delphinus (the Dolphin): A curved, kitelike pattern of stars seen by the ancient Greeks as a dolphin, messenger of Poseidon, the god of the seas. This interpretation is shared by some Polynesian cultures.
- **4.** Hercules: A large constellation containing two globular clusters (M92 and M13), visible with binoculars in the northern sky. It was so named by the Greeks for their hero. Even earlier—as far back as 2500 B.C.—the Babylonians associated it with their hero, Gilgamesh.
- **5.** Ursa Major (the Great Bear) and Ursa Minor (the Little Bear): Perhaps the best known pair of star formations, named for two bears in classical Greek and Roman myths. Other cultures have seen in these stars a plow,

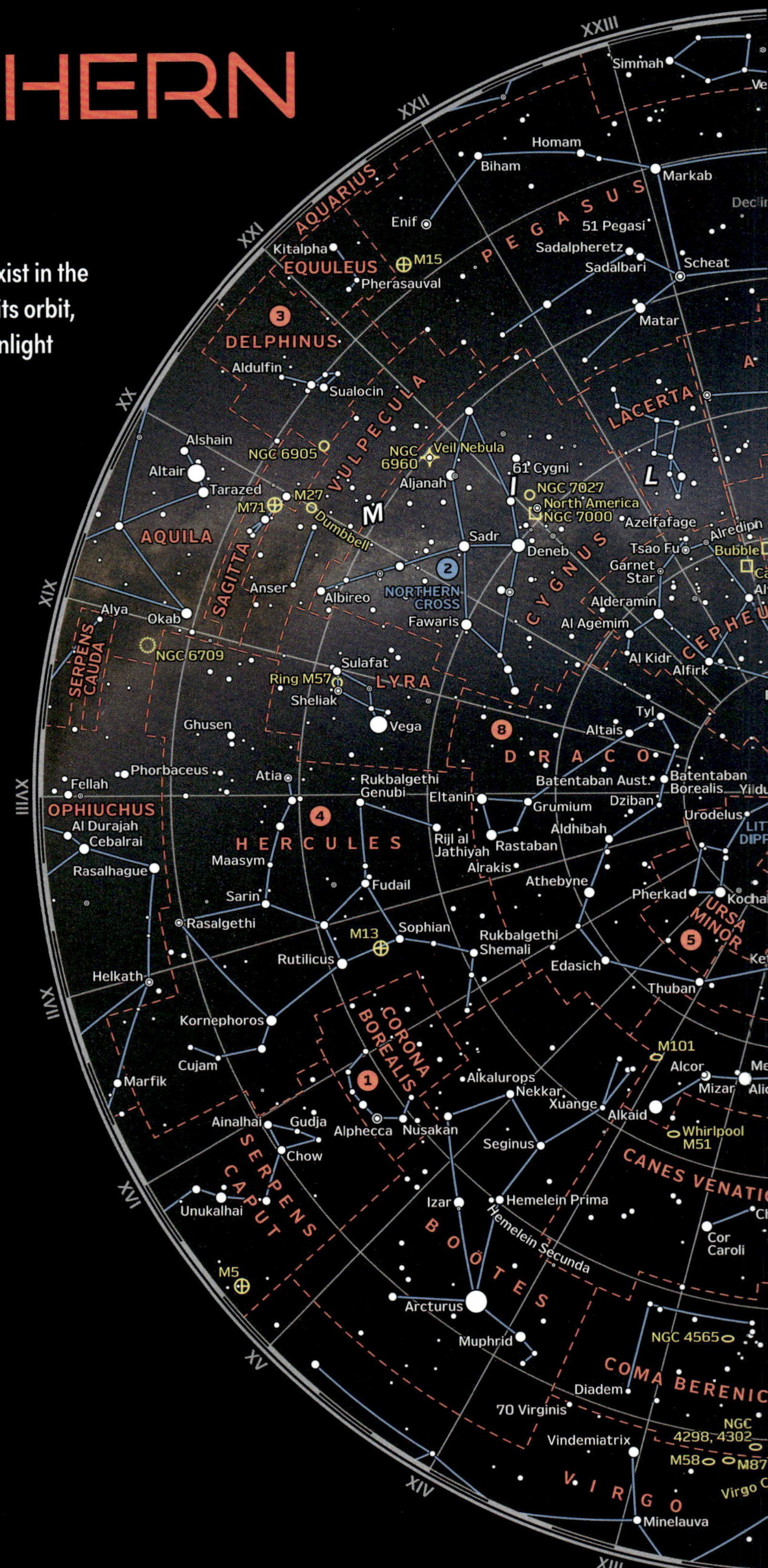

Constellation
An official division of the sky named in Latin and bounded by celestial coordinates, it serves to locate objects in space.

Asterism
Often differing across cultures, it is an imaginary pattern of stars—such as of a bear or dipping ladle.

wagon, coffin, skunk, camel, shark, canoe, bushel, sickle, and hog's jaw.

- **6.** Canis Major (the Great Dog) and Canis Minor (the Lesser Dog): Two small constellations said to run at the heels of the larger constellation Orion the Hunter (see number 9 below). Sirius, the brightest star in our sky, forms the eye of Canis Major (see map on p. 82).
- **7.** Cassiopeia (the Queen): A W-shaped constellation representing the vain queen of Aethiopia who claimed to be more beautiful than sea nymphs. Poseidon banished her to the stars for her arrogance, dooming her to circle the star Polaris for eternity.
- **8.** Draco (the Dragon): A chain of stars winding around the North Star. The ancient Sumerians and Greeks envisioned it as the dragon that guarded the golden apples of the Hesperides.
- **9.** Orion (the Hunter): A constellation known to both hemispheres, because it straddles the celestial equator. In ancient Hindu texts it is called Mriga (the Deer); in Hungary, the Reaper.
  - It was carved into a mammoth tusk more than 30,000 years ago; the artifact was discovered in a German cave in 1979.

DEEP-SKY OBJECT

- Open star cluster
- Globular star cluster
- Galaxy
- Diffuse nebula
- Planetary nebula
- Supernova remnant

STAR MAGNITUDE

-1 0 1 2 3 4 5

Variable star

CURRENT FACT COUNT 1080

THE SOUTHERN SKY

The South Pole points to the heart of the Milky Way. Southern stargazers can see more stars, brighter stars, and more constellations than what Northern Hemisphere viewers can behold.

- **1.** Carina (the Keel): One of the three constellations that once formed Argo Navis, the legendary ship of Jason and his Argonauts. Carina was the largest constellation in the southern sky before being divided up in the 18th century.
- **2.** Centaurus (the Centaur):
  - A constellation once seen in the ancient northern sky that now appears in the southern sky due to the shifts in Earth's orientation on its axis.
  - Centaurus includes Alpha Centauri, the nearest star to the Sun, four light-years away.
- **3.** Crux (the Southern Cross): Called "Te Punga" (the Anchor) by New Zealand's Indigenous Maori, this constellation appears on the flags of both Australia and New Zealand.
- **4.** Dorado (the Swordfish): This constellation harbors a rich trove of celestial objects, including the Large Magellanic Cloud.
- **5.** Eridanus (the River): The second longest constellation in the sky, after Hydra.
- **6.** Lepus (the Hare):
  - The brightest of its main four stars is Arneb, Arabic for "hare."
  - Other Arabic legends interpreted the stars as four camels crossing the desert.
  - Ancient Egyptians saw them as the boat of Osiris, god of the dead.
- **7.** Tucana (the Toucan):
  - A small, faint constellation containing the Small Magellanic Cloud.

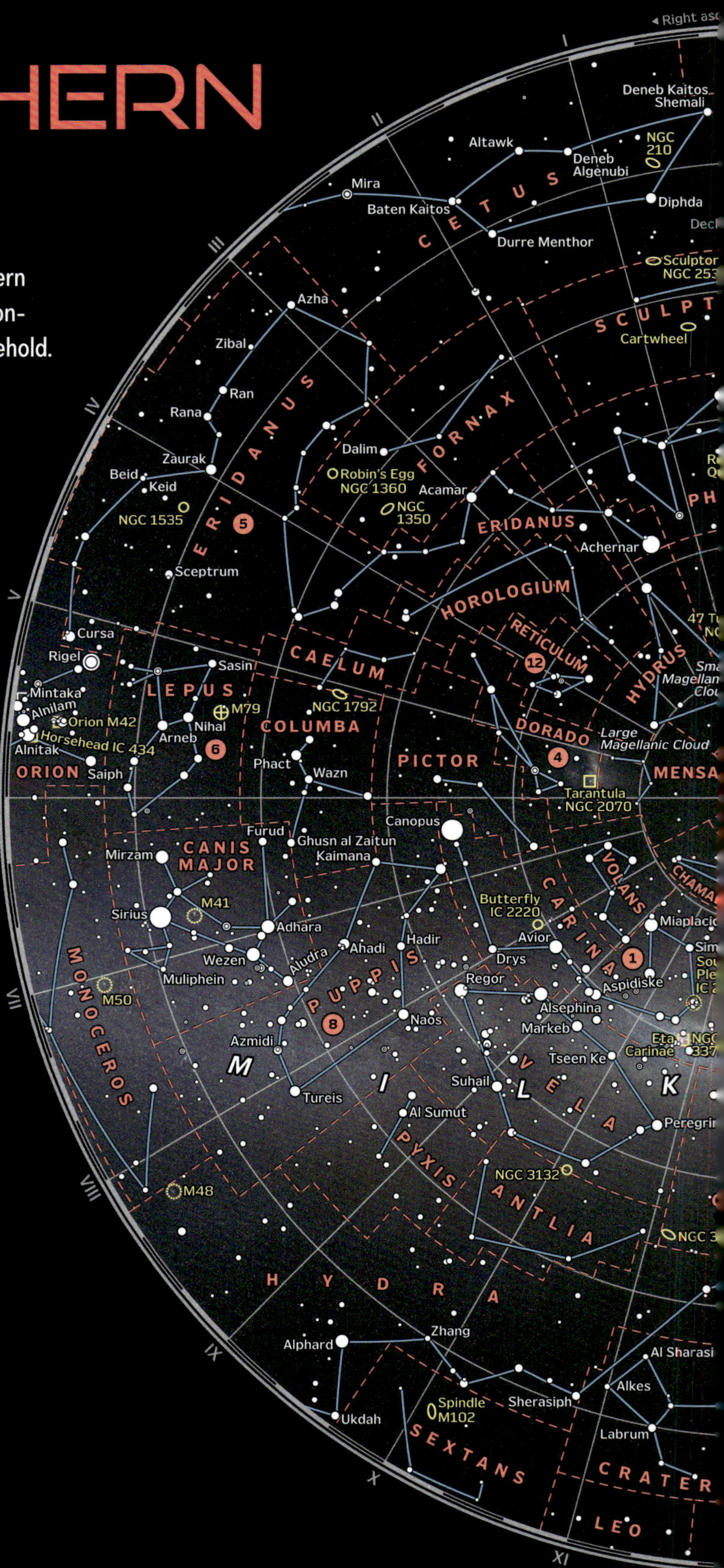

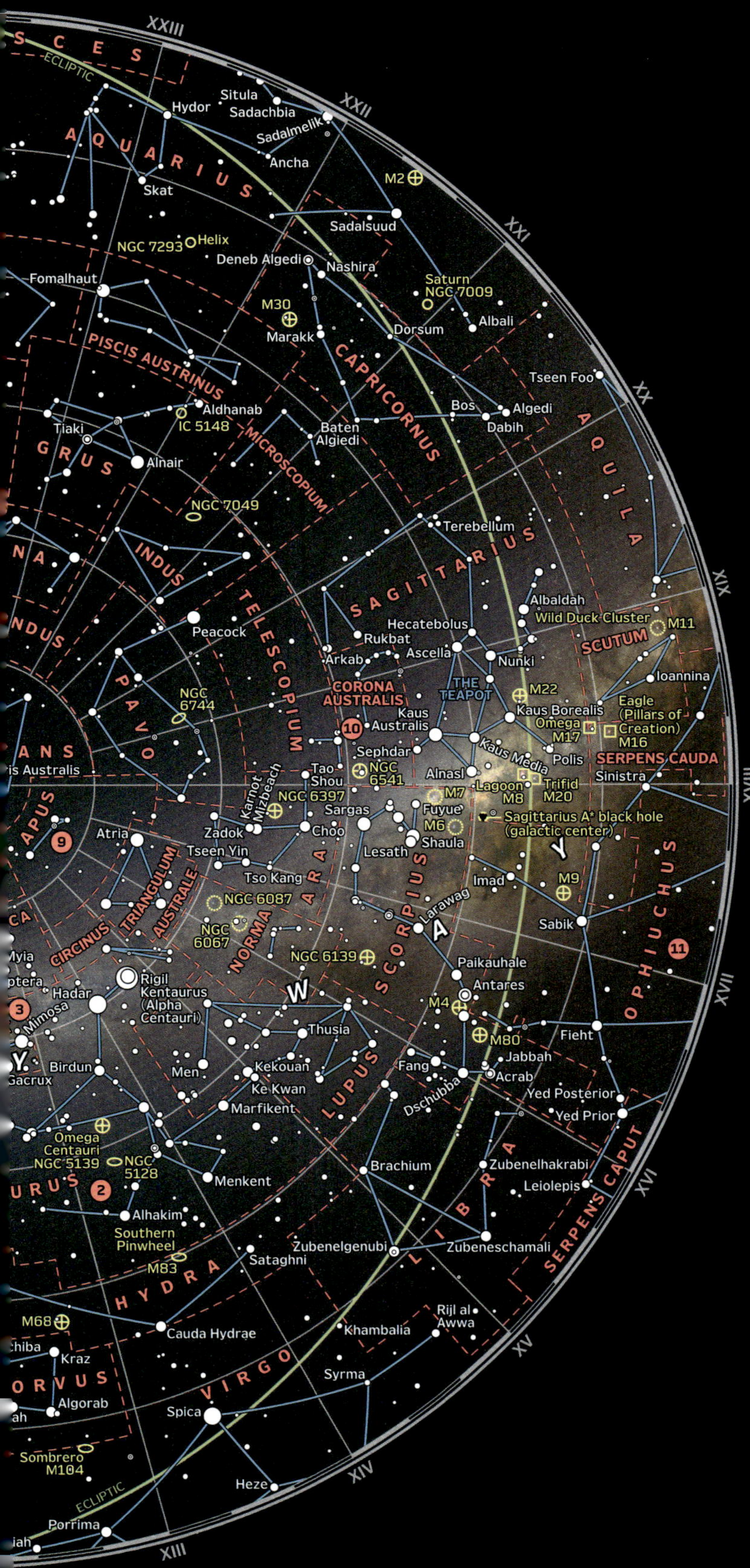

- It was the target of the Hubble Space Telescope when it produced one of its farthest views, the Hubble Deep Field South.

- **8.** Puppis (the Stern): The rear portion of the larger constellation Argo Navis, the ship of Jason and the Argonauts, which French astronomer Nicolas-Louis de Lacaille declared too large and divided into three parts in the 18th century.

- **9.** Apus (the Bird of Paradise):
  - It is located near the South Celestial Pole in an area of sky that harbors three star systems with confirmed exoplanets.
  - Apus perhaps requires more imagination than any other constellation in the sky to see the fully plumed animal the stars are purported to represent.

- **10.** Corona Australis (the Southern Crown):
  - Counterpart to Corona Borealis (the Northern Crown), the Southern Crown displays a nearly semicircular pattern.
  - The San people of the Kalahari Desert saw it as people sitting around a campfire.
  - The ancient Chinese saw it as a giant turtle.
  - Ancient Arab astronomers saw it as a tent or ostrich nest.

- **11.** Ophiuchus (the Serpent Bearer) and Serpens (the Serpent):
  - Once considered a single constellation but was officially split in the 20th century.
  - The ancient Greeks associated it with Asclepius, their god of medicine, and the serpent wrapped around his staff.
  - The Rod of Asclepius symbol is still used today to represent modern medicine.

- **12.** Reticulum (the Reticle):
  - A modern addition to the roster of constellations, Reticulum was recognized by the International Astronomical Union in 1922.
  - It is named after the grid of silk threads that were placed in telescope eyepieces to assist with mapping the stars.

2

After hundreds of thousands of years looking skyward, humanity finally progressed from stargazing to spacefaring. New discoveries and inventions shattered the illusions and delusions that had shackled us to Earth's surface—and unveiled the hidden languages spoken by the universe.

In 2013, NASA astronaut Mike Hopkins completed two spacewalks to repair a faulty water pump on the exterior of the International Space Station.

LOOKI

NG UP

CURRENT FACT COUNT 1123

LEARNING TO SEE THE HEAVENS

The first telescopes revolutionized seafaring, warfare, and trade as much as they revolutionized stargazing. No aspect of civilization was left untouched by the magnification of the universe.

✦ BRINGING THE COSMOS CLOSER

- Our atmosphere blocks most high-frequency wavelengths, but it is perfectly transparent to visible light.
- **THAT SIMPLE BUT PROFOUND FACT IS WHY WE ARE ABLE TO SEE STARS AND PLANETS AT ALL.**
- That fact is also why our eyeballs evolved to perceive visible light—an otherwise oddly specific, tiny fraction of the electromagnetic spectrum—and why our first telescopes were designed to see in only this frequency.
- The oldest known instrument that may have had an optical use is the ancient Assyrian Nimrud lens dating back to the eighth century B.C.
- It is a thin, polished piece of clear ground quartz an inch and a half across.
- The Nimrud lens may have been used for magnification, decoration, or, most likely, to start fires.
- Written around 300 B.C., Euclid's *Optics* laid out the basic rules of vision.
- **EUCLID WAS THE FIRST TO TREAT VISION AND THE INTERACTION OF LIGHT AND MATTER MATHEMATICALLY.**
- The Alexandrian-Roman scientist Claudius Ptolemy (ca A.D. 100) used Euclid's work when he wrote his own *Optics*.

An illustration of Galileo explaining the Moon's topography to skeptics

- The original version of Ptolemy's *Optics* is entirely lost; only heavily translated fragments remain.
- **WHAT'S LEFT OF *OPTICS* IS PART OF A LATIN TRANSLATION, DATED A.D. 1154 AND TRANSLATED FROM GREEK, WHICH WAS TRANSLATED FROM ARABIC, WHICH WAS TRANSLATED FROM THE ORIGINAL GREEK.**
- In the 10th century, Ibn Sahl was the first Islamic scientist to analyze Ptolemy's *Optics.*
- Ibn Sahl worked out the mathematics of curved mirrors and convex lenses.
- Later, Ibn al-Haytham (known as Alhazen in Europe, A.D. 985–1040) published the *Book of Optics,* an accomplishment for which he earned the nickname "the father of modern optics."
- **IBN AL-HAYTHAM DEVELOPED A VERSION OF THE MODERN SCIENTIFIC METHOD THAT DEMANDED HYPOTHESES BE SUPPORTED BY EXPERIMENTS.**
- Al-Haytham designed experiments to prove that light travels in a straight line.
- He also argued that vision occurs when light enters the eye from outside, contradicting the Greek theory of the time that light originates in the eye.
- With the above concepts, al-Haytham posited that light could not be instantaneous but traveled at a finite (but very fast) speed.
- **THAT SPEED, HE CORRECTLY ARGUED, VARIES WITH THE MEDIUM THROUGH WHICH THE LIGHT TRAVELS.**
- Liu An, a Chinese king in the second century B.C., pioneered the use of multiple mirrors to see around walls—a technique that would later become important in telescope design.

✦ WINDOWS TO THE UNIVERSE

- At the turn of the 17th century, two inventions blew open our access to the natural world and awakened us to the untold ways the universe could be revealed.

- **FIRST, THE INVENTION OF THE MICROSCOPE AROUND 1600, FOLLOWED BY THE TELESCOPE IN 1608.**

- Prior to these technologies, our worldview was entirely shaped by—and biased to—what we could see. After all, why would God create a world our bodies couldn't perceive?

- In 1608, Dutch spectacle maker Hans Lippershey built the first known telescope.

- Lippershey said his instrument was for "seeing things far away as if they were nearby."

- Galileo Galilei (1564–1642) didn't invent the telescope, but he improved it. His first telescope magnified things by three.

- **GALILEO'S ULTIMATE TELESCOPE MAGNIFIED THINGS 30 TIMES.**

- The telescope Galileo used to revolutionize astronomy was no better than a good set of opera glasses today.

- An urban legend claims Galileo conned an official of the Arsenal of Venice (the Pentagon of its day) into a contract to purchase his telescopes, knowing cheaper instruments from Holland were on their way.

- When Galileo showed his telescope to the Doge in Venice, he received a lifetime lectureship for double his salary.

- Galileo's most important achievement was turning his telescope to the heavens.

- What he saw contradicted the accepted geocentric cosmology put forward by Ptolemy and widely accepted in Europe, informed by the religious assumption that the Bible is literally correct and unerring.

An illustration of Claudius Ptolemy from the 15th century

- Galileo saw mountains and craters on the Moon, which was supposed to be a perfect crystal sphere.

- He observed sunspots on the Sun, also supposed to be perfect and unblemished.

- Galileo saw that Venus undergoes phases, just like the Moon, and concluded it must orbit the Sun and not Earth.

- Most egregiously, Galileo saw moons orbiting Jupiter, an impossibility with Earth at the center of the universe.

- Rebel astronomer of the 17th century, Galileo wrote his results in Italian, the language of the people, not Latin, the language of the aristocracy and Catholic Church.

- His book was titled *Sidereus Nuncius,* translated in English as *The Starry Messenger.*

- Another book got Galileo into trouble: He was tried as a heretic for promoting a heliocentric solar system in his 1632 *Dialogue Concerning the Two Chief World Systems.*

- The two "world systems" discussed in his book were Ptolemaic and Copernican: Earth-centered and Sun-centered.

- A cardinal representing the pope warned Galileo in 1616 not to promulgate the Copernican system.

- Summoned to Rome in 1633, Galileo was tried by the Inquisition.

- **UNDER PRESSURE, HE PUBLICLY RENOUNCED THE HELIOCENTRIC UNIVERSE BUT IS SAID TO HAVE MADE THIS STATEMENT AS HE DID: "*EPPURE SE MUOVE*—AND YET IT MOVES."**

CURRENT FACT COUNT 1 1 5 0

THE ELECTRO-MAGNETIC SPECTRUM

Before the discovery of the invisible infrared, humans did not and could not conceive of a universe made of invisible stuff emitting invisible information. We now know the universe communicates almost exclusively in wavelengths our eyeballs cannot detect.

✦ MORE THAN MEETS THE EYE

- When astronomer Sir William Herschel accidentally discovered the invisible infrared wavelength in 1800, he threw open a new window to the universe, akin to Galileo's perfection of the optical telescope.

- During Herschel's usual solar observations, he noticed that the amount of light and heat shining through colored filters varied with each color. This gave him an idea.

- **HERSCHEL PASSED SUNLIGHT THROUGH A PRISM TO FORM A RAINBOW AND TOOK THE TEMPERATURE OF EACH COLOR. HE FOUND NOT ONLY THAT THE TEMPERATURE INCREASED FROM BLUE TO RED, BUT THE PATTERN CONTINUED BEYOND THE VISIBLE PATCH OF RED.**

- The only conclusion was that a kind of light we cannot see heated the thermometer below the red band of the rainbow. He called it "light unfit for vision."

- **TO HERSCHEL'S CONTEMPORARIES, "INVISIBLE LIGHT" WAS AN OXYMORON. TODAY, WE RECOGNIZE SEVERAL FORMS OF INVISIBLE LIGHT IN THE INFINITE EM SPECTRUM (EM).**

An 1801 drawing of Sir William Herschel's telescope

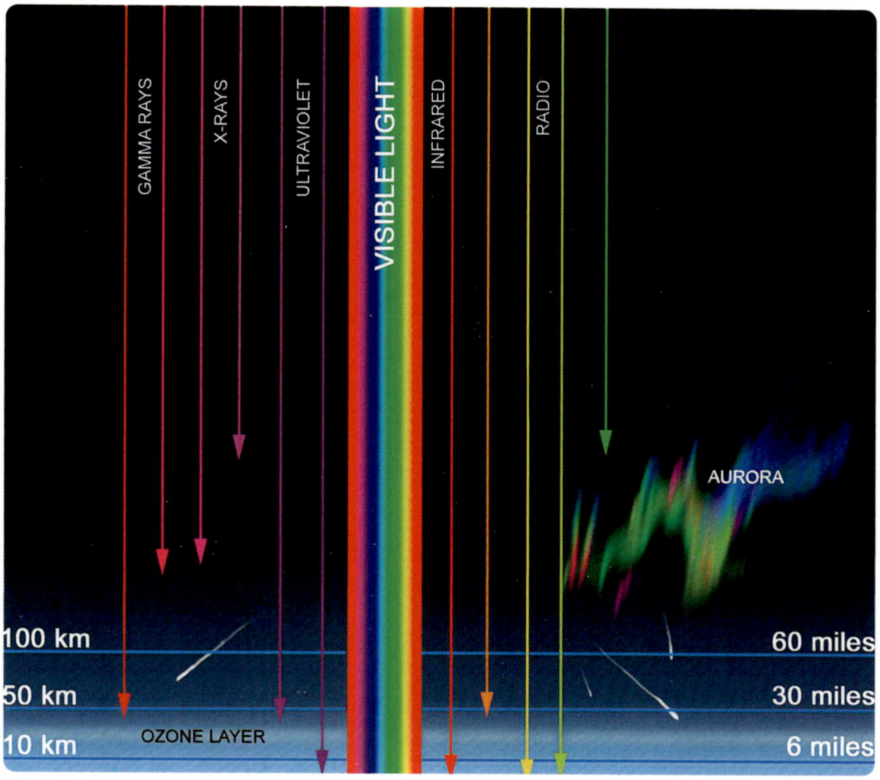

This illustration demonstrates the distances at which Earth's atmosphere blocks different electromagnetic wavelengths.

- From lowest frequency to highest frequency, the EM is divided into radio waves, microwaves, infrared, visible light, ultraviolet, x-rays, and gamma rays.

- To get a fuller picture of the universe, astronomers need to see across different frequencies of light. Modern telescopes can search for light in every part of the spectrum.

- Earth's atmosphere is so efficient at blocking wavelengths outside the visible spectrum that, despite tremendous advances in ground-based telescopes, we did not make great progress observing the universe until we put telescopes into space.

- Telescopes on Earth are often placed atop mountains. The air is thin, and fewer molecules cloud the view of distant phenomena. In space, no molecules interrupt the view.

- German scientist Hermann Oberth first proposed launching a telescope into space in a 1923 paper.

- NASA launched the first space telescopes in 1966 and 1968 as large satellites that sensed ultraviolet light: the Orbiting Astronomical Observatory, nicknamed "Stargazer."

- The mission was overseen by Nancy Grace Roman, for whom NASA's new infrared space telescope is named.

- The Nancy Grace Roman Space Telescope will study dark energy, exoplanets, and more.

- **AT THE TURN OF THE CENTURY, NASA LAUNCHED THE NEXT WAVE OF SPACE TELESCOPES, CALLED THE FOUR "GREAT OBSERVATORIES":**
  - The Hubble Space Telescope, launched on April 24, 1990
  - Hubble detects visible, ultraviolet, and infrared light.
  - The Compton Gamma Ray Observatory, launched on April 5, 1991
  - Compton was designed to sense gamma rays.
  - The Chandra X-ray Observatory, launched on July 23, 1999
  - Chandra sees x-ray emissions.
  - The Spitzer Space Telescope, launched on August 25, 2003
  - Spitzer senses infrared light.

- Together, these four space-based observatories sense a wide range of the electromagnetic spectrum.

- *Star Trek*'s Geordi La Forge, helmsman of the U.S.S. *Enterprise*, was born blind. With his cybernetic device, called a VISOR (Visual Instrument and Sensory Organ Replacement), La Forge could see nearly the entire band of the electromagnetic spectrum—far beyond the narrow band of visible light.

- With a VISOR, one might see things our most advanced telescopes are designed for: the ancient remnants of billions-of-years-old supernovae or the formation of black holes.

- Using a constellation of ground- and space-based telescopes, astrophysicists have crafted our version of La Forge's VISOR.

- We can combine and overlay images to see phenomena across the spectrum.

Star Trek's Geordi La Forge dons his VISOR.

CURRENT FACT COUNT 1194

RADIO WAVES

The longest wavelengths of light, containing the lowest energy of all bands

- Radio waves fall at the farthest end of the electromagnetic spectrum, while gamma rays lie at the opposite end.
- "Radio" is the generic term for waves below the more specifically defined "infrared" and "microwave" wavelengths of the spectrum.
- The radio spectrum has the longest wavelengths. Peak to peak, one wave can span hundreds of miles or more.
- Most radio wave frequencies tend to bypass Earth's atmosphere, unlike higher frequencies of the electromagnetic spectrum that never make it to ground observatories.
- **THEIR LONG WAVES MEAN THEY CAN TRAVERSE GREAT DISTANCES BEFORE LOSING ENERGY, SO THEY'RE IDEAL FREQUENCIES TO TUNE IN TO THE SOUNDS OF THE DEEP, ANCIENT UNIVERSE—INCLUDING POTENTIAL CALLS FROM ET.**

- Since the first radio transmission in the late 19th century, humans have leaked a continuous cascade of radio emissions into space, cocooning ourselves in an ever growing, invisible radio bubble.
- That bubble grows at the speed of light (at the rate of one light-year per year) in every direction all around Earth.
- The bubble has washed over dozens of star systems and hundreds of known exoplanets more than 100 light-years beyond Earth.
- "One, two, three, four. Is it snowing where you are, Mr. Thiessen?" Those were the first words ever transmitted over radio waves and the first human-uttered sentence an eavesdropping civilization might hear, if the message wasn't garbled on its way into space by our own ionosphere.
- The first official search for alien intelligence—Project Ozma in 1960—used radio waves.
- Radio astronomers study radio emissions from stars, galaxies, nebulae, and other celestial objects like quasars and pulsars.
- Radio telescopes were the earliest telescopes built to look for invisible light.
- **THE FIRST RADIO TELESCOPE WASN'T BUILT TO DETECT OBJECTS BEYOND EARTH BUT TO PINPOINT SOURCES OF INTERFERENCE FROM STANDARD RADIO TELECOMMUNICATION.**
- In 1932, American engineer Karl Guthe Jansky accidentally stumbled across an interstellar source of radio waves while investigating radio interference for Bell Labs in New Jersey.
- Jansky's radio antenna picked up a strange hiss every 23 hours 56 minutes (every rotation of Earth).
- He rightly supposed the signal must be coming from the constellation Sagittarius,

The Karl G. Jansky Very Large Array (VLA) in the New Mexico desert where the antenna pointed during each hissing fit.

- Jansky realized the signals weren't emitted from Sagittarius but the heart of our Milky Way, traversing the constellation on their way to Earth.

- He didn't know it yet, but Jansky was detecting the heartbeat of the supermassive black hole at the center of our galaxy.

- **HIS DISCOVERY OF RADIO WAVES COMING FROM THE MILKY WAY LAUNCHED THE FIELD OF RADIO ASTRONOMY.**

- Because the long wavelengths of radio waves penetrate gas and dust, radio telescopes have historically been some of the best probes of gas-rich galaxies.

- In the 1950s, radio surveys of the sky detected quasars—the highly energetic cores of newborn galaxies, powered by supermassive black holes as they dine upon clouds and stars that wander too close—for the first time.

- Astronomers were stumped by the source and true identity of these powerful, distant emissions and originally named them "quasi-stellar radio sources"—quasars—for short.

- The first quasar, called 3C 273, is so named because it was the 273rd object in the Third Cambridge Catalogue of Radio Sources.

✦ FAST

- FAST (Five-hundred-meter Aperture Spherical Telescope) is the world's largest single-dish radio telescope, completed in 2016.

- It sits in a natural geologic depression in Guizhou, China.

- FAST is a mile in circumference and spans an area bigger than 30 American football fields.

- Its diameter is wider than the Empire State Building is tall.

- If aliens send radio signals to Earth, the first ones to detect the call will likely be astrophysicists in Guizhou.

✦ ARECIBO OBSERVATORY

- Before FAST, the U.S. boasted the largest single-dish observatory in the world: the Arecibo Observatory in Puerto Rico.

- Arecibo's dish measured 1,000 feet across.

- In 1974, a high-powered radio message was sent from Arecibo toward star cluster M13.

- The three-minute burst of radio waves was designed to travel interstellar distances.

- The Arecibo message is coded mathematically with 1,679 binary digits.

- This number was chosen because it is the product of two prime numbers: 73 and 23.

- The message reveals a human's shape and dimensions, information about our DNA, and the position of planet Earth.

- **THE FIRST EXOPLANET EVER DISCOVERED WAS FOUND BY ARECIBO IN 1992.**

- Arecibo starred in the James Bond film *GoldenEye* and in Robert Zemeckis's *Contact*.

- Arecibo collapsed in 2020 due to negligence when cable failures caused its 900-ton platform to plunge into the dish, shattering it.

✦ JANSKY VLA

- In 1980, a very large array of 27 dish antennae was constructed just outside Socorro, New Mexico, across 22 miles of desert.

China's FAST radio telescope

- Astronomers, ever the pragmatists, called it the Very Large Array (VLA). It was later renamed to honor Karl Guthe Jansky.

- The 1997 film *Contact,* based on Carl Sagan's novel in which Ellie Arroway makes contact with aliens, was filmed in part at the VLA.

- The VLA represents a class of radio telescope called interferometers: multiple smaller telescopes that work together to form a single, precise pattern.

- Interferometers combine several small telescopes to deliver information about distant phenomena with the same or greater precision as giant single dishes.

- **ANOTHER INTERFEROMETER IS THE EVENT HORIZON TELESCOPE. IT COMBINED DATA FROM TELESCOPES ACROSS CONTINENTS FOR THE FIRST EVER IMAGE OF A BLACK HOLE IN 2019.**

CURRENT FACT COUNT 1227

MICROWAVES

The universe is suffused with microwave energy: the afterglow left behind by the Big Bang itself.

- Between radio and infrared in the spectrum, we find microwaves.
- **"MICROWAVES" WERE DUBBED SO BECAUSE THEIR WAVELENGTHS WERE MICRO—RANGING FROM A MILLIMETER TO A FEW FEET ACROSS—COMPARED WITH INFINITELY LONG RADIO WAVES.**
- Microwaves interact heavily with water molecules, as evidenced by microwave ovens.
- Liquid water absorbs and converts microwave frequencies into heat.
- Water vapor molecules, such as those prevalent in Earth's atmosphere, absorb 10 times more microwaves than in liquid form.
- That's why microwave observatories are placed in the highest, driest locations.

We interact with microwave radiation every day via our cell phones.

- The prime location for such observation is atop a mountain in Chile's Atacama Desert: the second driest place on Earth.
- More modern applications of microwave radiation: cell phones, walkie-talkies, radar, garage door openers, and remote sensing

✦ CMB RADIATION

- In 1964, scientists Arno Penzias and Robert Wilson were using a radio telescope to study the Milky Way and picked up a persistent hiss they did not recognize.
- The hiss did not have a single origin; it seemed to come from everywhere, all the time, day and night, all year round.
- The sources they considered to explain the hiss included nuclear radiation, noise from New York City, and even pigeon poop.
- In actuality, they had accidentally discovered cosmic microwave background (CMB) radiation.
- **CMB IS THE PRIMORDIAL RADIATION LEFT OVER FROM THE FORMATION OF THE UNIVERSE 380,000 YEARS AFTER THE BIG BANG—THE DETECTABLE ECHOES OF THE UNIVERSE'S VERY BIRTH.**
- 380,000 years is to the universe what a single day is to a person with a 100-year life span.
- Their discovery of CMB settled an astronomical debate about the origin of the universe.
- It supported the Big Bang theory, which scientists predicted would leave a cosmic imprint in the form of radiation spread uniformly across the universe.
- Today's studies of the CMB are primarily conducted using data from microwave-sensitive ground- and space-based telescopes.
- Planck, a European Space Agency space-based observatory, was dedicated to the study of CMB from 2009 to 2013.
- Planck observed far-infrared, microwave, and high-frequency radio waves of light.
- From 2003 to 2020, the Spitzer Space Telescope studied the infrared spectrum, providing views of space not visible to optical telescopes.

✦ ALMA

- Single-dish telescopes have now given way to arrays of smaller telescopes working together.
- Set in the high, dry desert of northern Chile, the Atacama Large Millimeter Array (ALMA) consists of 66 radio telescopes that act as one.
- ALMA perches 16,404 feet above sea level.
- It also captures microwave and some infrared wavelengths.
- ALMA discovered one of the faintest objects ever observed: the A2744_YD4 galaxy.
- **THE LIGHT ALMA OBSERVES FROM A2744_YD4 HAS TAKEN 13 BILLION YEARS TO REACH EARTH.**
- That means what we see of A2744_YD4 represents a time when the universe was only 600 million years old—just 4 percent of its age today.

The South Pole Telescope, perched amid the frigid landscape of Antarctica

✦ SOUTH POLE TELESCOPE

- Antarctica's South Pole, the driest place on Earth, is home to the South Pole Telescope (SPT).
- It's located on Amundsen-Scott South Pole Station at an elevation of nearly 10,000 feet.
- The high altitude and dry location, combined with lack of light pollution and six continuous months of pure darkness, offer this telescope the clearest, most consistent images of the night sky anywhere on Earth.
- It was designed to measure CMB.
- **IT CONTRIBUTED TO THE EVENT HORIZON TELESCOPE DATA ARRAY THAT IMAGED THE M87 BLACK HOLE IN 2019.**
- Why don't we put more telescopes on the South Pole? Observatories require operation and repairs made by human astrophysicists. Antarctica's hostile environment poses logistical, operational, budgetary, and purely psychological hurdles for scientists to overcome.

"WE HAVE SEEN WHAT WE THOUGHT WAS 'UNSEEABLE.' WE HAVE SEEN . . . A BLACK HOLE."

—ASTRONOMER SHEP DOELEMAN

CURRENT FACT COUNT 1274

LIGHT IN THE INFRARED

Infrared light carries lower energy than visible light. We can't see it but can detect it as heat on our skin.

- **MOST OBJECTS ON EARTH'S SURFACE, INCLUDING YOU AND EVERYONE YOU LOVE, RADIATE IN THE INFRARED, ALSO KNOWN AS THERMAL RADIATION.**
- Every adult human radiates between 80 and 100 watts, entirely in the infrared.
- Our atmosphere traps most incoming or outgoing infrared light. We call it the greenhouse effect.
- When visible light penetrates Earth's atmosphere and strikes molecules on Earth's surface, those molecules absorb the visible light and convert the energy to infrared.
- When we feel the warmth of the Sun on our faces, we feel the infrared energy converted and emitted by our skin molecules.
- That energy radiates back into space, where greenhouse gases absorb the radiation.
- Water vapor and carbon dioxide are especially efficient at trapping infrared radiation.
- We like a little greenhouse effect to keep Earth just warm enough to sustain life, but humanity's current rate of greenhouse gas emissions is driving Earth toward irreversible, catastrophic warming.
- Earth will keep spinning, of course. But life as we know it will suffer dire consequences.
- **HUMANS HAVE DERIVED ALL KINDS OF CREATIVE USES FOR INFRARED RADIATION.**
- Infrared lamps keep your McDonald's french fries warm for hours before you buy them. The lamps glow a visible red, but it's the invisible infrared doing the work.
- Traditional TV remote controls use infrared light to change channels at a distance.
- Military night vision goggles allow soldiers to see infrared light. If humans were cold-blooded creatures, those goggles would be useless for spotting other humans.
- Thermal imaging reveals infrared rays emitted from the subject.
- An object's heat directly corresponds to its wavelength on the electromagnetic spectrum.
- Foliage is green because of chlorophyll, a photosynthetic pigment used by most plants, algae, and some bacteria that reflects green visible light as well as near-infrared.
- The more chlorophyll a plant contains, the healthier it is. And the more chlorophyll, the more infrared it emits.
- Space satellites can help measure the health of forests by how much total near-infrared they emit—their chlorophyll fluorescence.
- **ASTROBIOLOGISTS, SCIENTISTS WHO STUDY LIFE IN THE UNIVERSE, ARE SEARCHING FOR SIGNS OF CHLOROPHYLL FLUORESCENCE ON OTHER PLANETS, A KIND OF BIOSIGNATURE.**
- Telescopes sensitive to infrared light help see stellar nurseries inside thick galactic gas clouds.
- Infrared telescopes can also reveal the atmospheric composition of exoplanets.

✦ SPITZER SPACE TELESCOPE

- After feeble attempts to measure infrared light emitted by the universe, humanity finally put an infrared telescope into space: the Spitzer Space Telescope.

- The Spitzer Space Telescope blew open the windows to the infrared universe.
- It launched on August 25, 2003.
- **EVERY OBJECT IN THE UNIVERSE, IT TURNS OUT, EMITS INFRARED LIGHT.**
- Even objects we think of as freezing, or below freezing, emit infrared energy. That includes telescopes.
- So, an infrared telescope must maintain a temperature cooler than the object it is measuring, lest it detect itself.
- Spitzer remedied this problem with a cryogenic telescope assembly that kept it at a chilly five degrees above absolute zero.
- **THAT'S MINUS 450 ON THE FAHRENHEIT SCALE.**
- It carried with it 42 liters of superfluid liquid helium to cool the telescope when needed.
- The telescope is coated in a shell of materials that either reflect or reradiate away oncoming photons.
- Spitzer trailed Earth's orbit at a far enough distance to minimize oncoming infrared from Earth itself.
- It explored star-forming regions, nascent planetary systems, and exoplanet atmospheres.
- The telescope had another name when it was launched: the Space Infrared Telescope Facility (SIRTF).
- In December 2003, the telescope was renamed for Lyman Spitzer, Jr., an American astrophysicist.
- Lyman Spitzer championed the idea of positioning telescopes in space, free of interference from Earth's atmosphere.
- Spitzer's 1946 paper "Astronomical Advantages of an Extra-terrestrial Observatory" advocated placing large telescopes in space.
- Spitzer likened the impact of capturing starlight through Earth's atmosphere to a "time exposure of a baby engaged with its rattle."
- Spitzer was the first telescope to observe light from planets outside the solar system, a breakthrough in the search for exoplanets.
- In 2009, Spitzer discovered a new ring of Saturn.
- In 2010, Spitzer found WASP-12b, the first known carbon-rich planet orbiting a star.
- In 2017, Spitzer found the TRAPPIST-1 system, containing seven Earth-size planets orbiting a star.
- **IN MAY 2009, SPITZER RAN OUT OF CRYOGENIC HELIUM—ESSENTIAL TO KEEPING ITS TEMPERATURE SO LOW.**
- NASA reimagined its mission, renaming it the Spitzer Warm Mission.
- The Spitzer Space Telescope continued to make significant observations for almost 17 years until it was decommissioned in January 2020.
- **KEY DATA FROM SPITZER—AND ALL NASA'S GREAT OBSERVATORIES, WHETHER STILL OPERATIONAL OR NOT—ARE ESSENTIAL TO CONTINUING EXPLORATIONS OF THE UNIVERSE.**
- In 2021, Spitzer was succeeded by the James Webb Space Telescope (JWST), not only the greatest infrared telescope in history, but also the most sophisticated, complex, and powerful telescope of all time.

CURRENT FACT COUNT 1339

THE JAMES WEBB SPACE TELESCOPE

The largest and most complex telescope ever launched, JWST can see deeper and farther into the reaches of the ancient universe than ever before.

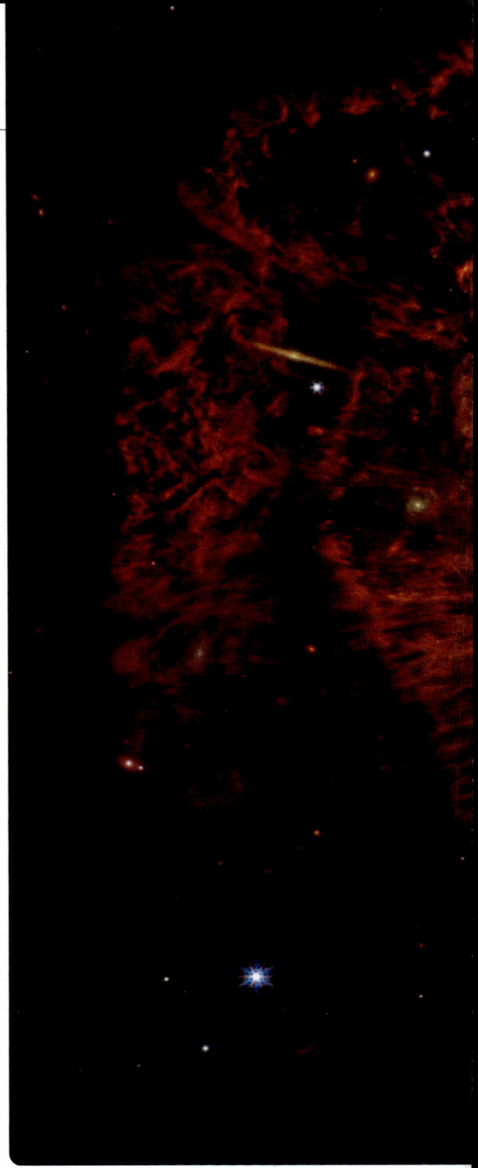

- NASA launched the James Webb Space Telescope (JWST) on Christmas Day in 2021.
- JWST can detect objects 100 times fainter than its predecessor, the Spitzer Space Telescope.
- It took 30 days for JWST to reach its destination, a point about four times the distance from Earth to the Moon.
- JWST was named for James E. Webb, administrator of NASA from 1961 to 1968.
- JWST is especially tuned to perceive infrared light.
- The most ancient, distant objects in the universe emit in the infrared.
- That's because their wavelengths stretch as the universe expands. A galaxy or nebula that was once suffuse with gamma radiation is now only faintly visible in infrared.
- **JWST IS LITERALLY GAZING BACK IN TIME, VIEWING THESE OBJECTS NOT AS THEY ARE NOW, BUT AS THEY ONCE WERE.**
- The $10 billion JWST project is an international collaboration among NASA, the European Space Agency, and the Canadian Space Agency.
- The JWST mission duration is five to 10 years, though it may operate longer.
- JWST can see much fainter objects, much farther away, at much sharper resolutions than any previous telescope.
- By seeing more distant objects, JWST is gazing farther back in time.
- **JWST'S SENSITIVITY ALSO ENABLES IT TO PROBE THE CHEMICAL COMPOSITION OF EXOPLANET ATMOSPHERES.**
- What space telescopes can see depends on where they are situated in orbit.
- While Hubble sits in low Earth orbit, JWST orbits the Sun.
- JWST sits one million miles from Earth and remains aligned with our home planet as it orbits the Sun—but it's complicated.
- JWST is parked at a point in space called the Earth-Sun L2, the second Lagrange point.
- Lagrange points are balance points that exist around any two orbiting objects where all forces are equalized.
- Around Earth and the Sun there are five such Lagrange points. Scientists take advantage of their special properties to position space telescopes and satellites there.
- At L2, for instance, JWST's centrifugal force is balanced by the gravitational pull of the Sun and Earth.
- JWST actually orbits L2, moving in sync with Earth as it orbits the Sun.
- At this position, JWST points its focus outward to the cosmos without the interference of heat or light from Earth or the Sun.
- **SOME JWST IMAGES REVEAL THE UNIVERSE IN THE FIRST FEW HUNDRED MILLION YEARS AFTER THE BIG BANG.**
- These images give us a glimpse of the earliest galaxies in the universe, some 13.5 billion years old.

One of the James Webb Space Telescope's first images was of NGC 3132, the Southern Ring Nebula, approximately 2,500 light-years away.

- JWST's cameras can penetrate dense veils of gas and dust that have hidden young stars inside star nurseries from the view of other telescopes.
- **THE TELESCOPE'S DATA AND IMAGERY COULD RESHAPE OUR THINKING ABOUT THE EARLY UNIVERSE AND THE FORMATION OF GALAXIES.**

✦ THE FIRST FIVE

- The first five JWST images were released in a July 2022 event hosted by President Joe Biden and Vice President Kamala Harris.
- SMACS 0723, a galaxy cluster, was one of the first images released.
- This image of SMACS 0723 is often called Webb's First Deep Field.
- A second image represented the atmosphere around WASP-96b, an exoplanet about 1,150 light-years away.
- A third image provided a new view of the Southern Ring Nebula, including a star in its center that was previously undetected.
- A fourth image revealed previously unknown details of the five galaxies called Stephan's Quintet, offering new data on how galaxies interact.
- The final image released during the event gave a look at the landscape-like edge of a stellar nursery in the Carina Nebula.
- **IMAGES RELEASED IN 2023 SHOWED FAR MORE GALAXIES IN THE EARLY UNIVERSE THAN EXPECTED.**
- They also revealed galaxies up to 100 times more massive than predicted by current cosmological models.
- In 2025, an image of the Cat's Paw Nebula showed newborn stars with masses 10 times greater than our Sun.

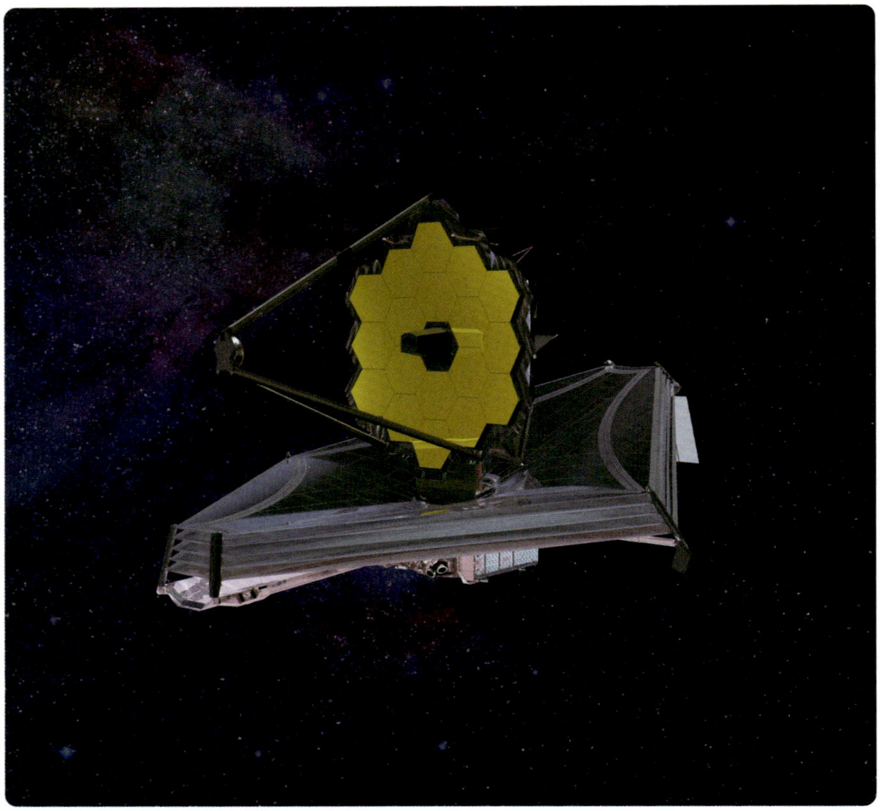

An illustration of the sunshield and mirror in action

✦ THE SUNSHIELD AND MIRROR

- JWST's mirror and sunshield were folded up for launch and deployment, then unfolded once in space.
- The sunshield keeps the telescope protected from the intense light and heat of the Sun.
- It's about 69.5 by 46.5 feet, or roughly the size of a tennis court.
- **JWST REACHES RADICALLY DIFFERENT TEMPERATURES ON EACH SIDE.**
- On the Sun-facing side, it reaches 185°F.
- On the side facing away from the Sun and protected by the sunshield, the telescope stays at 40 kelvins (around minus 388°F)—40 degrees above absolute zero.
- The sunshield protects the telescope's mirrors and delicate instruments, which require the cold to operate correctly.
- JWST engineers had to design everything, even adhesives and metals, to survive that same cold environment.
- **THE TECHNOLOGY USED TO MEASURE AND REFINE THE CURVATURE OF JWST'S MIRRORS IS ALSO USED TO MEASURE EYES DURING LASIK SURGERY.**
- The measurements are accurate to the nanometer, or one-billionth of a meter.

✦ JWST'S MISSIONS

- **JWST'S OBSERVATIONS CONTRIBUTE TO FOUR KEY SCIENTIFIC MISSIONS:**
  - Search out the earliest stars and galaxies.
  - Study the evolution of galaxies.
  - Observe the formation of stars and planetary systems.
  - Explore the potential for life in other planetary systems.
- JWST excels at seeing in infrared light, which is difficult to see from Earth because our atmosphere absorbs it.
- With its superb infrared vision, JWST penetrates dense veils of dust and gas that often block our view of new stars forming in stellar nurseries.
- Astronomers can now see previously hidden baby stars in the Cosmic Cliffs.
- The Cosmic Cliffs are part of the Carina Nebula, a star-making region roughly 7,600 light-years away.
- JWST's capabilities also open new frontiers in the exploration of exoplanets.
- JWST observations help study the chemical composition of faraway planetary atmospheres.
- **WITH THAT KNOWLEDGE, ASTROBIOLOGISTS CAN CONSIDER THOSE PLANETS' POTENTIAL TO HARBOR LIFE.**
- In addition to exploring exoplanets, JWST looks at our own solar system.
- JWST discovered a plume of water vapor shooting from Saturn's moon Enceladus.
- The plume extends more than 6,000 miles, about the distance from Los Angeles, California, to Buenos Aires, Argentina.
- Estimates suggest the water vapor is gushing at a rate of 79 gallons a second—fast enough to fill an Olympic-size swimming pool in two hours.
- **SCIENTISTS BELIEVE THIS PLUME SUPPLIES WATER TO SATURN'S SYSTEM, INCLUDING ITS RINGS.**
- The smallest object found by JWST, in 2023, is an asteroid orbiting the Sun between Mars and Jupiter.
- It's about the length of the Washington Monument in Washington, D.C. (554 feet).

An engineer examines JWST's gold-coated primary mirror, EDU.

CURRENT FACT COUNT 1411

THE VIEW FROM THE GROUND

Optical telescopes gather light from the spectrum visible to human eyes.

- Galileo's refractor telescope eventually magnified the world by 30-fold.
- Early models of refracting telescopes had a lens at each end of a tube. Light was refracted (bent) as it traveled through the first lens.
- **THE FIRST KNOWN DESCRIPTION OF A REFRACTING LENS APPEARS IN A 1589 LETTER WRITTEN BY GIAMBATTISTA DELLA PORTA, AN ITALIAN NATURAL PHILOSOPHER.**
- Since a lens refracts different wavelengths (or colors) of light slightly differently, it creates geometric distortions—a rainbow effect—and a fuzzy image.

- Efforts to solve these problems resulted in longer and longer telescopes.
- Dutch brothers Christiaan and Constantijn Huygens created a telescope 23 feet long that magnified objects by 100.
- German astronomer Johannes Hevelius created a 150-foot-long telescope—but it bent in the wind and proved entirely impractical.
- **TO FURTHER OUR VIEW OF THE COSMOS, ASTRONOMERS NEEDED A NEW DESIGN: THE REFLECTING TELESCOPE.**
- Reflecting telescopes focus light with mirrors instead of lenses and allow for shorter tubes.

✦ GREAT MINDS

- Sir Isaac Newton is credited with the invention of the reflecting telescope, but others built them as well.
- **NEWTON'S DESIGN PERFECTED THE CONCEPT WITH A SLIGHTLY CURVED MIRROR AND USED METAL TO REFLECT AND FOCUS LIGHT.**
- Newton revealed his new telescope to the Royal Society in 1672.
- All wavelengths of light reflect at the same angle, so the new design eliminated chromatic aberration (the rainbow effect).

- Parabolic mirrors eliminated spherical aberrations (geometric distortions).
- Newton's mirrors were made of speculum metal, a copper-tin alloy with arsenic added to allow for a shiny polish.
- In 1781 English astronomer William Herschel discovered Uranus with a reflecting telescope he built himself.
- **URANUS WAS THE FIRST PLANET TO BE "DISCOVERED." EVERY OTHER PLANET WAS VISIBLE TO THE NAKED EYE SINCE THE BEGINNING OF HUMANITY.**
- With no street lighting in his hometown of Bath, England, Herschel's backyard was dark enough to observe even Uranus.
- Herschel wanted to name the planet Georgium Sidus ("George's Star") after King George III.
- Herschel's name for Uranus never caught on, but he received a lifelong pension from George III.
- After the discovery of Uranus, Herschel built a new, 40-foot reflecting telescope.
- Herschel's new telescope saw first light in 1789.
- It was the largest telescope in the world for the next half century.
- In 1837, German astronomer Friedrich Bessel used a state-of-the-art refracting telescope and knowledge of the astronomical unit to measure the distance to 61 Cygni, a star about 10 light-years away.
- **WITH BESSEL'S DISCOVERY, OUR CONCEPTION OF THE SIZE OF THE UNIVERSE EXPANDED ENORMOUSLY.**

✦ LEVIATHAN OF PARSONSTOWN

- In 1845, the so-called Leviathan of Parsonstown saw first light.
- Leviathan was built by William Parsons, third Earl of Rosse, in County Offaly, Ireland.
- With a lens six feet in diameter and a focal length of 52 feet 6 inches, the Leviathan was the world's largest telescope until 1917.
- Problems with the speculum metal in mirrors, which tarnished quickly, hindered the advancement of reflecting telescopes in the 19th century.
- In the late 19th and early 20th centuries, astronomers perfected the technique of making and coating large parabolic glass mirrors.

✦ HOOKER TELESCOPE

- Mount Wilson's 100-inch Hooker telescope, completed in 1917, was the first instrument on Earth capable of seeing distant, ancient objects from the universe's beginnings.
- **AS TELESCOPES INCREASED IN SIZE, THEY COULD SEE FARTHER AWAY AND DIMMER OBJECTS.**
- Using the Hooker telescope, American astronomer Edwin Hubble made two of the most significant discoveries in the history of science: that the Milky Way is but one of many other galaxies and that the universe itself is expanding.
- The 100-inch Hooker telescope was the largest in the world for 30 years before the completion of the Hale telescope.

✦ HALE TELESCOPE

- The 200-inch Hale telescope saw first light on Mount Palomar in San Diego, California, in 1948.
- The Hale telescope's mirror alone weighed 14.5 tons.
- As the inverse square law prescribes, doubling the surface area will not double its collecting power. It will *quadruple* it. So, Hale was fourfold more powerful than its predecessor.
- **THE OBSERVATIONS MADE WITH HALE HELPED US BETTER UNDERSTAND STELLAR EVOLUTION—THE BIRTH, LIFE, AND DEATH OF STARS.**
- Hale was the last instrument to use a mirror built as a single solid piece of glass.

✦ ACTIVE AND ADAPTIVE OPTICS

- The bigger the mirror, the more light a telescope captures.
- But bigger mirrors are more vulnerable to the forces of gravity and environmental conditions.
- Enter active optics, the solution to the problem of distortion due to gravity.
- Active optical telescopes use thin mirrors with small motors attached.
- Computers in an active optical system assess the mirrors' curvature; the motors correct their deflection.
- All large modern telescopes use active optics.
- As telescopes became more sensitive, the turbulence in Earth's atmosphere caused more distortions.
- **TURBULENCE—THE MOTION OF MATERIAL IN THE ATMOSPHERE—MAKES STARS APPEAR TO TWINKLE.**
- Enter adaptive optics, the solution to the problem of atmospheric turbulence.
- An adaptive optical system monitors a light beam and adjusts the telescope mirrors to compensate for atmospheric motion.

- The incoming light beam can originate from a bright nearby star or from a laser beam shot into the atmosphere.
- **IN ESSENCE, ADAPTIVE OPTICAL SYSTEMS UNTWINKLE THE STARS.**
- Most of today's Earthbound telescopes rely on active and adaptive optical systems.
- In 2004, Keck II became the first telescope to use a laser-guided adaptive optics system.

Twinning: Lasers from the Keck telescopes beam skyward at Mauna Kea, Hawaii.

✦ KECK I AND II

- The twin Keck telescopes are regarded as the most scientifically productive telescopes on Earth.
- They detect both optical and infrared light.
- When they were built, the two telescopes were the largest in the world, surpassing the Hale telescope.
- With twice the collective surface area of Hale, the Keck telescopes each had four times the collecting power.
- The Keck telescopes are located in Hawaii, side by side atop the dormant volcano Mauna Kea.
- **THEY CAN SEE EVEN DEEPER INTO SPACE THAN THE SPACE-BASED HUBBLE TELESCOPE.**
- The Keck telescopes sit at an altitude of 13,600 feet.
- The nearby Pacific Ocean stabilizes the atmosphere at Mauna Kea, making for little light pollution—perfect for telescopes.
- Keck I saw first light in 1990; Keck II saw first light in 1996.
- The domes housing the Keck telescopes are kept at or below freezing, so the Hawaii heat doesn't deform the metal or mirrors.
- Both telescopes now have these systems, as do most large telescopes.
- Keck II can measure spectra from 1,200 objects simultaneously, which was at least five times more than any other optical telescope at the time it was built.

✦ VERA C. RUBIN OBSERVATORY

- Joining several other telescopes in Chile's Atacama Desert, the new Vera C. Rubin Observatory includes a 26-foot optical telescope.
- The Rubin Observatory's research focus is dark energy and dark matter.
- The observatory is named for American astronomer Vera Rubin, who pioneered the study of dark matter.
- Connected to the telescope is the world's largest and highest-resolution digital camera designed for astronomy.
- Rubin records lapsed images, like a movie of the night sky, a technique that helps us detect all the objects that go bump in the night.
- **THE SYSTEM GENERATES ABOUT 30 TERABYTES OF DATA EVERY NIGHT.**
- It will discover countless asteroids and other transitory phenomena that would easily go unnoticed without the time-lapsed images the telescope provides.

> "IS ANYTHING ORBITING IN THE SOLAR SYSTEM AN ALIEN SPACECRAFT? THAT'S ACTUALLY SCIENCE WE CAN DO [NOW]."
>
> — PLANETARY SCIENTIST MEG SCHWAMB, REGARDING THE RUBIN OBSERVATORY

CURRENT FACT COUNT 1447

VIEWS FROM ABOVE: THE HUBBLE

The most iconic visible light telescope is not on Earth but in space.

- The Hubble Space Telescope launched on April 24, 1990, aboard the space shuttle *Discovery*.
- Hubble, unfettered by the blurring of atmospheric disturbance, obtained higher resolution images than ever before.
- **IT FINALLY SETTLED THE QUESTION OF HOW OLD THE UNIVERSE IS BY LOOKING AT GALAXIES NEAR AND FAR AND GETTING AN ACCURATE MEASURE OF THE EXPANDING UNIVERSE.**
- Before Hubble, we could only place the age of the universe at between 10 billion and 20 billion years.
- Hubble helped narrow our uncertainty in the age of the universe to between 13.75 and 13.85 billion years.
- It was named for astronomer Edwin Hubble, who theorized an expanding universe with galaxies far beyond the Milky Way.
- Before Hubble's launch, we could only observe the universe about halfway back to the Big Bang, with the light of galaxies around seven billion light-years away.
- Hubble sits in low Earth orbit, about 320 miles above us.
- It is 43.5 feet long, about the length of a large school bus—or a *Tyrannosaurus rex*.
- Hubble's primary mirror weighs more than a full-size moose: a whopping 1,825 pounds.
- In addition to visible light, the telescope also senses ultraviolet and infrared light.
- Hubble's original deep field image regaled Earthlings with a detailed view of nearly 3,000 individual galaxies.
- Its observations allowed astronomers to redefine measurements of Cepheid variable stars.
- Thus, scientists began to understand the "cosmic distance ladder," a tool used to measure vast distances in the universe.

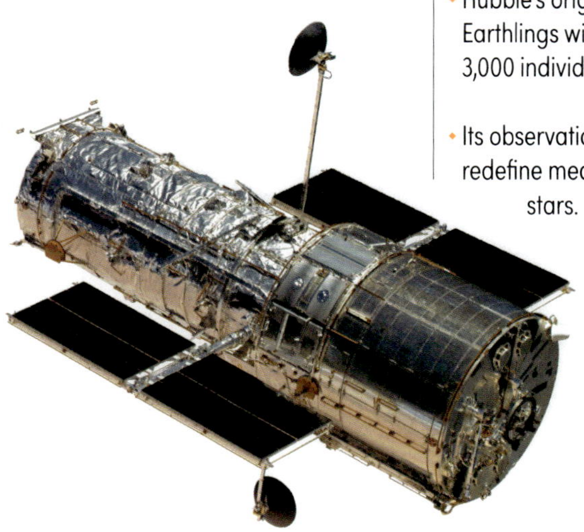

The Hubble Space Telescope

- Each rung, or method of measurement, on the cosmic distance ladder relies on the step before for calibration, so there's a cumulative effect, through to the origins of the universe.
- An error of about two-millionths of a meter caused seriously distorted images from Hubble after its 1990 launch.
- **NASA HAD TO SEND ASTRONAUTS INTO SPACE TO REPAIR THE ISSUE IN 1993.**
- Astronauts have serviced the telescope five times to replace, upgrade, and add instruments and parts that have extended Hubble's working life span.
- Without an operational Space Shuttle, no further hands-on service to the Hubble Space Telescope is possible.
- Early results in 2023 from the James Webb Space Telescope confirm Hubble's calculations of the crucial first rungs on the cosmic distance ladder.
- When Hubble launched, it had a 15-year predicted life span, which it far surpassed.
- **THE HUBBLE AND JAMES WEBB SPACE TELESCOPES COLLABORATE TO CAPTURE LIGHT FROM GALAXIES ROUGHLY 450 MILLION TO 750 MILLION YEARS AFTER THE BIG BANG.**
- The two telescopes excel at deep field, or "deep sky," surveys used to study very distant celestial objects like galaxies.
- Together, the two can see light from mid-ultraviolet to mid-infrared, a broader spectrum than ever before.

✦ CLAIMS TO FAME

- Hubble's Ultra Deep Field image of 2004 revealed about 10,000 galaxies.

For its 28th anniversary, Hubble captured a four-light-year portion of the Lagoon Nebula, which is 55 light-years wide and 20 light-years tall.

- The most distant galaxies in Hubble's Ultra Deep Field image lie some 13 billion light-years away.
- Hubble can pinpoint a distant location so accurately it's like shining a laser on a dime from 200 miles away.
- NASA likens Hubble's sharp-sightedness to seeing two fireflies in Tokyo—from Maryland.
- Hubble observations have confirmed the existence of dark energy.
- The telescope has photographed distant galaxies and star-forming nebulae.
- Hubble probed the atmospheres of planets that orbit other stars.
- **IT HAS ALLOWED US TO WITNESS A COMET CRASH INTO JUPITER, THE FIRST EVER OBSERVATION OF TWO ASTRONOMICAL OBJECTS COLLIDING.**
- Hubble has found previously unknown moons around the dwarf planet Pluto.
- Its eXtreme Deep Field (XDF) image of 2012 was the deepest image of the sky ever taken—until JWST surpassed it.
- The XDF image includes stunning spirals, remnants of galactic collisions, and some of the earliest objects that gave rise to later galaxies.
- The earliest galaxy found in Hubble's XDF image appears as it was when the universe was in its infancy, just 450 million years after the Big Bang.

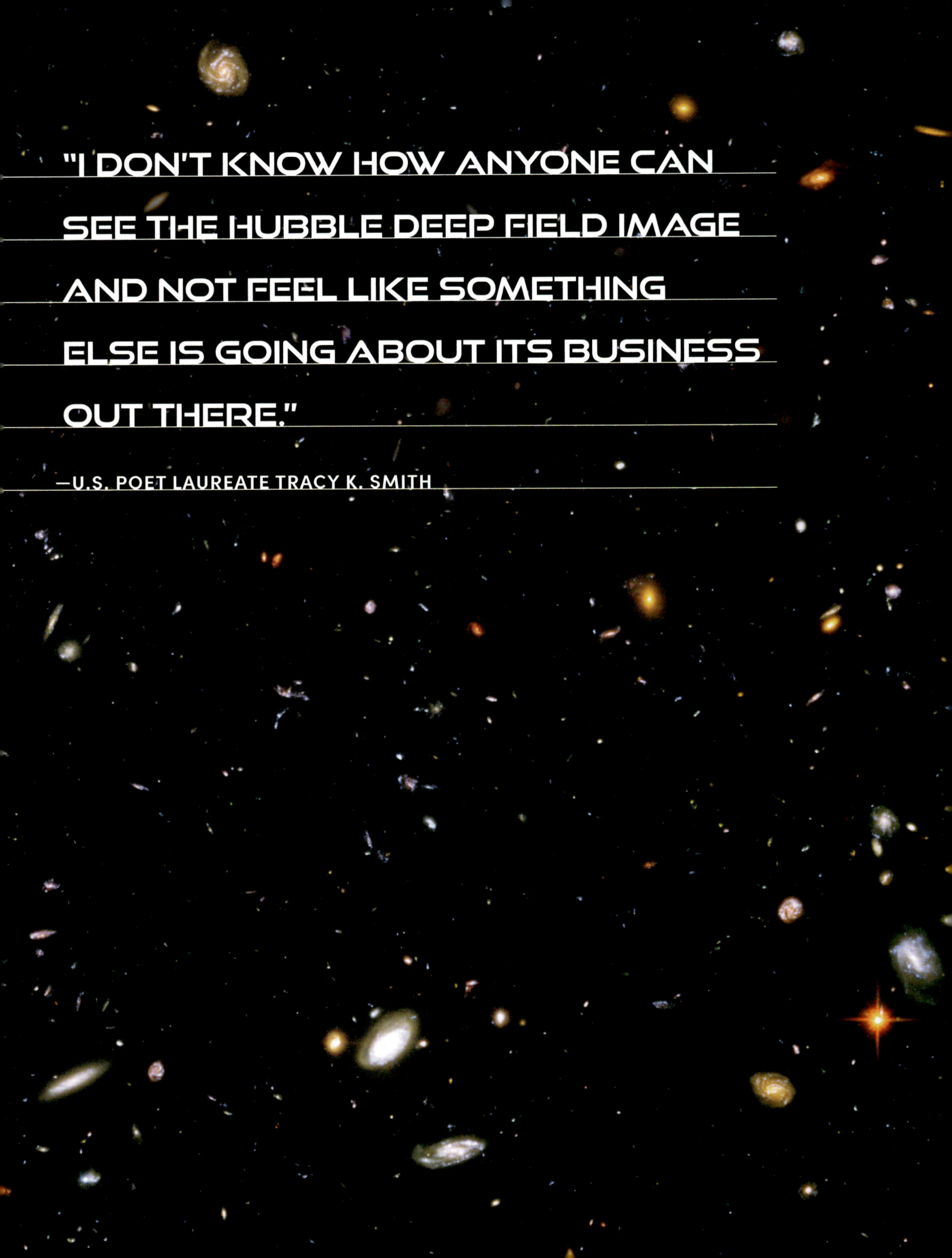

"I DON'T KNOW HOW ANYONE CAN SEE THE HUBBLE DEEP FIELD IMAGE AND NOT FEEL LIKE SOMETHING ELSE IS GOING ABOUT ITS BUSINESS OUT THERE."

—U.S. POET LAUREATE TRACY K. SMITH

Hubble's Ultra Deep Field image reveals nearly 10,000 galaxies.

CURRENT FACT COUNT 1471

ULTRAVIOLET

With its higher energy and shorter wavelengths than visible light spectra, UV energy is incompatible with life.

- Ultraviolet (UV) radiation was discovered in 1801 by Johann Wilhelm Ritter, one year after William Herschel's discovery of the infrared.
- Beyond the visible light spectrum, high-energy frequencies like ultraviolet radiation can break apart the molecules of life.
- **THANKS TO THE OZONE LAYER, OUR ATMOSPHERE EFFICIENTLY BLOCKS MOST HARMFUL UV RAYS.**
- That means our best UV-sensing telescopes are in space, above Earth's atmosphere.

- Hubble saw beyond the visible to the ultraviolet, which is the signature of highly energetic phenomena in the universe, such as exploding stars.
- UV is highly damaging to DNA. Science labs use UV lamps to sterilize their equipment and kill off any lurking microbes or bacteria.
- **OTHER TERRESTRIAL ENCOUNTERS WITH UV INCLUDE:**
  - Black light posters glow with "black light," a colloquial term for UV.
  - Many bugs, especially moths, are especially sensitive to ultraviolet light.
  - Evidence that humans are smarter than bugs: bug zappers. We lure bugs toward a high-voltage death with a simple ultraviolet light.
- **ABOUT 10 PERCENT OF THE TOTAL ENERGY OUTPUT OF THE SUN COMES TO US AS UV.**
- Ultraviolet light travels slowest through glass and other transparent media compared with visible and infrared light.
- The light sensitivity of a range of flying insects is shifted relative to humans.
- These insects can see deep into the ultraviolet light range, which humans cannot detect, but at the expense of red and orange.

Portable bug zappers attract mosquitoes and other insects by utilizing UV light.

- Old-style red-amber "bug bulbs" were effective, not because they repelled UV-sensitive insects, but because the insects simply couldn't see the light.
- Twilight is rich in the Sun's ultraviolet light, offering flying insects excellent illumination to begin their evening travels.
- UV, by convention, is divided into three ranges in order of decreasing wavelength and increasing energy: UV-A, UV-B, UV-C.
- **UV-A PENETRATES TO THE DERMIS. UV-B GETS ABSORBED AT THE EPIDERMIS. UV-C IS ABSORBED ALMOST ENTIRELY BY ATMOSPHERIC OZONE.**
- Both UV-A and UV-B can alter your skin's DNA, causing cancer.
- The skin pigment molecule melanin, common in darker-skinned people, absorbs solar UV, so DNA damage is minimized.
- The absence of melanin is why white people are 25 times more likely to contract skin cancer from the Sun than Black people.
- More than 90 percent of UV sunlight that reaches Earth's surface is UV-A. Up to 10 percent is UV-B.
- The ozone molecule (O_3) in Earth's stratosphere makes an effective absorber of high-energy ultraviolet light (UV-C).
- **UV PHOTONS FIRST BREAK APART THE ABUNDANT OXYGEN MOLECULE O_2, CREATING TWO FREE OXYGEN ATOMS. THEY EACH COMBINE WITH TWO OTHER OXYGEN MOLECULES, MAKING TWO O_3 OZONE MOLECULES, RIPE FOR ABSORBING UV-C.**
- UV telescopes are ideal for detecting interstellar phenomena such as the death throes of gas clouds that have wandered too close to black holes.

A peacock flower *(Caesalpinia pulcherrima)* is photographed using an ultraviolet-induced visible fluorescence (UVIVF) photography technique.

CURRENT FACT COUNT 1549

X-RAYS AND GAMMA RAYS

Only the most highly energetic phenomena in the universe emit x-rays and gamma rays. Humans have learned to harness these wavelengths to evolve and transform modern medicine.

- The first ever Nobel Prize went to Wilhelm Conrad Röntgen for his discovery of x-rays.
- Röntgen discovered that x-rays penetrate the soft tissue of our flesh and are absorbed by bones and other solid objects, making it ideal to probe the state of our skeleton.
- Röntgen took the first x-ray of a human—his wife's left hand—showcasing her wedding ring and bones underneath.
- Upon seeing the ghastly image of her own hand, Anna Bertha Ludwig proclaimed, "I have seen my own death."

Marie Skłodowska-Curie takes stock of her laboratory.

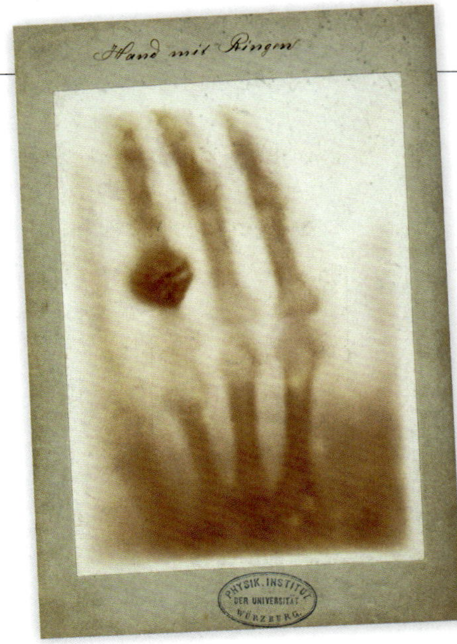

The first ever x-ray, taken by German physicist Wilhelm Röntgen of his wife Anna's hand

- Marie Skłodowska-Curie, who coined the term "radiation," paused her research to develop and deploy the first mobile x-ray units during WWI. She also trained about 150 women to operate them on the battlefield.
- **HER DEVICES, CALLED "LITTLE CURIES," ALLOWED DOCTORS TO NOT ONLY SEE BROKEN BONES BUT ALSO PINPOINT EMBEDDED BULLETS. THEY REVOLUTIONIZED MEDICINE.**
- Skłodowska-Curie's work with medical x-rays at the front lines, often overlooked among her multiple Nobel Prizes and other achievements, significantly increased her exposure to deadly radiation.
- She eventually died of radiation-induced aplastic anemia at the age of 66.
- In the century since their discovery, x-ray films have grown more sophisticated and machines more advanced, limiting necessary exposure time. X-rays remain the workhorses of the medical and dental community.
- While our atmosphere keeps us safe from most dangerous high-frequency wavelengths, it also makes it so that if we wish to see the universe properly, we have to put telescopes into space.

- **SOURCES OF X-RAYS BEYOND THE SOLAR SYSTEM:**
  - Supernova explosions
  - Very energetic stars
  - Black holes
  - Quasars
  - Rotating neutron stars
  - Intergalactic clouds of supremely hot gas

- The cores of some active galaxies also shoot out high-speed jets of material that give off x-rays and gamma rays.

- **WE DETECTED A BLACK HOLE FOR THE FIRST TIME VIA X-RAYS.**

- In 1964, long before the first x-ray telescope, a rocket that had launched to the edge of space detected x-ray radiation emitted from the constellation Cygnus.

✦ CHANDRA X-RAY OBSERVATORY

- The Chandra X-ray Observatory was launched by the space shuttle *Columbia* on July 23, 1999.

- **CHANDRA REPRESENTS ONE OF THE MOST REMARKABLE FEATS OF ENGINEERING IN MODERN ASTRONOMY.**

- It detects an x-ray after it ricochets off the observatory's mirrors at such a precise angle that the x-ray hits a tiny focal point about the width of a hair 30 feet away.

- To enable such exact reflection angles, the mirrors were polished to the smoothness of atoms.

- If the entire Earth were as smooth as Chandra's mirrors, then the tallest mountain would be about six feet tall.

- Because high-frequency telescopes require long focal lengths, they're heavier than any other telescope.

Scientists integrate high- and low-energy gratings into Chandra's optical bench.

- Chandra was the largest payload ever launched from *Columbia.*

- It was named for Indian astrophysicist Subrahmanyan Chandrasekhar, who won a Nobel Prize for pioneering physics in stellar evolution.

- Chandrasekhar's name means "crown of the moon" in Sanskrit.

- Chandra focuses on superhot areas of the universe, such as galaxy clusters, exploded stars, matter around black holes, and infant solar systems.

- It orbits more than one-third of the way to the Moon—86,500 miles above Earth.

- **CHANDRA OPERATES ON TWO KILOWATTS OF ELECTRICITY, ABOUT THE SAME AS A TOASTER.**

- It has helped us to better understand the nature of dark matter.

- When Chandra measured how stars settle after massive collisions, researchers realized that dark matter interacts with itself even less than previously believed.

✦ GAMMA RAYS

- High-frequency, short-wavelength gamma rays carry the highest energy per photon of the entire electromagnetic spectrum.
- Both x-rays and gamma rays bypass most solid objects, including standard telescope lenses that readily focus every other light of the electromagnetic spectrum.
- **DETECTING THESE HIGH-FREQUENCY EMISSIONS REQUIRED NEW FEATS OF ENGINEERING.**
- The first gamma-ray detection arose from the Cold War.
- In the early 1960s, the United States signed a test-ban treaty to stop nuclear testing as a tenuous peace deal with the Soviets. In the spirit of "trust but verify," the U.S. launched high-energy detection satellites.
- One of the satellites registered a high-frequency emission event but no ground-based observations confirmed nuclear activity. It couldn't be ascribed to a solar flare or supernova either. We scratched our heads.
- Eventually, astrophysicists out of Los Alamos National Laboratory (home of the Manhattan Project) solved the conundrum: The detected emissions were gamma-ray bursts emitted from deep space.
- Thus was born gamma-ray astronomy.
- Some of the most energetic explosions in the cosmos are gamma-ray bursts (GRBs).
- They are detected about once per day, emanating from the centers of galaxies across the universe.
- Gamma rays burst forth from giant star explosions and supermassive black holes that blast out high-energy jets that travel into space.
- These gamma-ray bursts are the hottest events in the universe.
- **GRBs ARE DETECTABLE BILLIONS OF LIGHT-YEARS AWAY.**
- A 10-second GRB sizzles with more energy than the Sun will release in its expected 10-billion-year lifetime.
- Though still not fully understood, gamma rays have been spotted around thunderstorms on Earth.
- Gamma rays are strong enough to penetrate a human body, ripping electrons from molecules, destroying body tissue, and damaging DNA on their way through.
- In the Marvel Universe, gamma radiation overwhelmed nuclear physicist Bruce Banner, transforming him into the massive, green, superstrong Incredible Hulk whenever things get intense.
- Apparently, high-dose gamma radiation also makes highly stretchy fabric. With only some ripping, somehow Dr. Banner's pants still fit him as the Incredible Hulk.
- In 2008, NASA's Fermi satellite recorded the most powerful gamma-ray burst ever observed at the time (GRB 080916C).
- More powerful than 9,000 supernova explosions, the 2008 GRB lasted more than 20 minutes before fading.
- **THE 2008 GRB TOOK PLACE MORE THAN 12.2 BILLION LIGHT-YEARS AWAY.**
- In other words, the explosion observed in 2008 occurred more than 12 billion years ago.
- In October 2022, the James Webb Space Telescope detected GRB 221009A, an even *brighter* GRB nicknamed the BOAT: Brightest of All Time.
- The 2022 GRB took place about two billion light-years from Earth.
- In other words, the explosion observed in 2022 occurred some two billion years ago.
- **BOAT WAS LIKELY THE BRIGHTEST GRB IN THE LAST 10,000 YEARS.**
- Even after traversing two billion light-years, the burst affected Earth's atmosphere with energy equivalent to a solar flare.
- Astronomers think an explosion this massive is likely the final death note of a collapsing star before it becomes a black hole.

✦ COMPTON GAMMA RAY OBSERVATORY

- The Compton Gamma Ray Observatory (CGRO) was named after Arthur Holly Compton, whose work in high-energy physics won him the Nobel Prize in 1927.
- The CGRO launched on April 5, 1991.
- **ITS INSTRUMENTS DETECTED THE HIGHEST ENERGY RADIATION IN THE UNIVERSE.**
- Among the CGRO's discoveries, it imaged two clouds of antimatter in the Milky Way galaxy.
- Gamma-ray detection could help determine the inhabitability of exoplanets.
- A planet receiving high-energy radiation from its host star is an unlikely candidate to harbor alien life (as we know it, anyway).

"LUMINOUS BEINGS ARE WE, NOT THIS CRUDE MATTER."

—YODA, *STAR WARS:* EPISODE V *THE EMPIRE STRIKES BACK*

Darth Vader's red lightsaber utilizes a (fictional) kyber crystal, a power source that's also used to power the Death Star's superlaser.

- Data from the CGRO is still essential to research in astrobiology.
- When one of the CGRO's gyroscopes failed, NASA decided to deorbit it.
- The CGRO was deliberately brought back to Earth on June 4, 2000.

✦ LIGHTSABER SCIENCE

- In the *Star Wars* galaxy, Jedi wield lightsabers: high-energy laser swords. While they could surely cut through steel, they're not likely to interact as swords do. Instead, they would simply pass through each other.
- At extremely high gamma-level energies, though, photons *can* bounce off one another.
- **CALLED GAMMA-GAMMA SCATTERING, THE PHENOMENON IS STUDIED AT PARTICLE ACCELERATORS.**
- Strategically placed magnetic fields could contain the energy to the length of the "blade" in the same way fusion reactors do.
- If lightsabers adhere to the laws of physics, then saber colors indicate different energy levels, as prescribed by the electromagnetic spectrum.
- The red sabers wielded by Darth Vader and others on the dark side contain the lowest level of energy.
- The blue sabers wielded by the Jedi, like Obi-Wan Kenobi and Luke Skywalker, would rank among the highest energy levels, followed closely by Yoda's green saber.
- **THE DEADLIEST AND MOST ENERGETIC SABER OF ALL WOULD BE THE VIOLET SABER WIELDED BY JEDI MASTER MACE WINDU.**

CURRENT FACT COUNT 1564

THE SATELLITES THAT STARTED THE SPACE RACE

In trading technological breakthroughs and defeats alike, the Soviet Union and United States launched the world's first human-made creations into space, taking human ingenuity where no one had traveled before.

- The Soviet Union officially kicked off the space race in 1957 with the successful launch of Sputnik, the first artificial object in orbit.
- **SPUTNIK, RUSSIAN FOR "FELLOW TRAVELER," MARKED MORE THAN A MERE SCIENTIFIC BENCHMARK. IT WAS INTERPRETED AS A SHOW OF FORCE FROM ONE WORLD POWER TO ANOTHER.**
- Russia surprised the world with its achievement, even though the U.S. had promised an Earth-orbiting satellite for the same year, deemed the International Geophysical Year by scientists around the world.
- The 24-inch satellite rode into orbit aboard the world's first intercontinental ballistic missile—a missile designed with one purpose: be powerful enough to deliver a nuclear warhead to the United States and other adversaries.
- Two months later, the United States rushed Project Vanguard, a smaller Sputnik-like satellite, to the launchpad. The mission failed spectacularly.
- Vanguard's rocket rose four inches before sputtering back down and exploding live on camera. News reporters added insult to injury when they jeeringly nicknamed the mission "kaputnik," "flopnik," and "dudnik."
- The United States responded with the creation of a new government agency dedicated solely to the development of space technology: the National Aeronautics and Space Administration. NASA was officially born.

A Soviet scientist works on Sputnik 1 in 1957.

- When the Soviet Union launched Sputnik 1 in 1957, many media outlets reported that Earth had an extra moon. Use of the word "satellite" for an artificial orbiting object was not yet in common use.
- America's first satellite, launched in 1958, was Explorer 1. It contained a scientific payload and was designed and built by California Institute of Technology's Jet Propulsion Laboratory.
- Beginning in 1959, the U.S. began the Corona spy satellite program and accelerated it after a Lockheed U-2 spy plane was shot down over the Soviet Union in 1960.
- **SPY SATELLITES ARE TYPICALLY GIVEN ORBITS THAT TRAVERSE POLE TO POLE, ALLOWING EARTH TO TURN WITHIN THEIR ORBITS, PROVIDING A VISTA ON THE ENTIRE GLOBE.**
- In 1962 and 1963, the U.S. launched Telstar 1 and 2, two similar communication satellites conceived to send television and radio signals between the U.S. and Europe while the satellite was high above Earth between the continents.

✦ FIRST WORDS

- The first utterance from Earth orbit was "bleep," as broadcast from Sputnik 1, launched on October 4, 1957.
- If you don't believe that hardware should hold the first-word-from-orbit record, then the first word would be "woof," as barked by Laika the dog aboard the second Soviet satellite, Sputnik 2, launched on November 3, 1957.
- If for some reason you feel this record should only be held by *Homo sapiens,* then the first words are "Поехали!—Let's go!" proclaimed by Russian cosmonaut Yuri Gagarin as the engines of Vostok 1 thrust him into orbit on April 12, 1961.

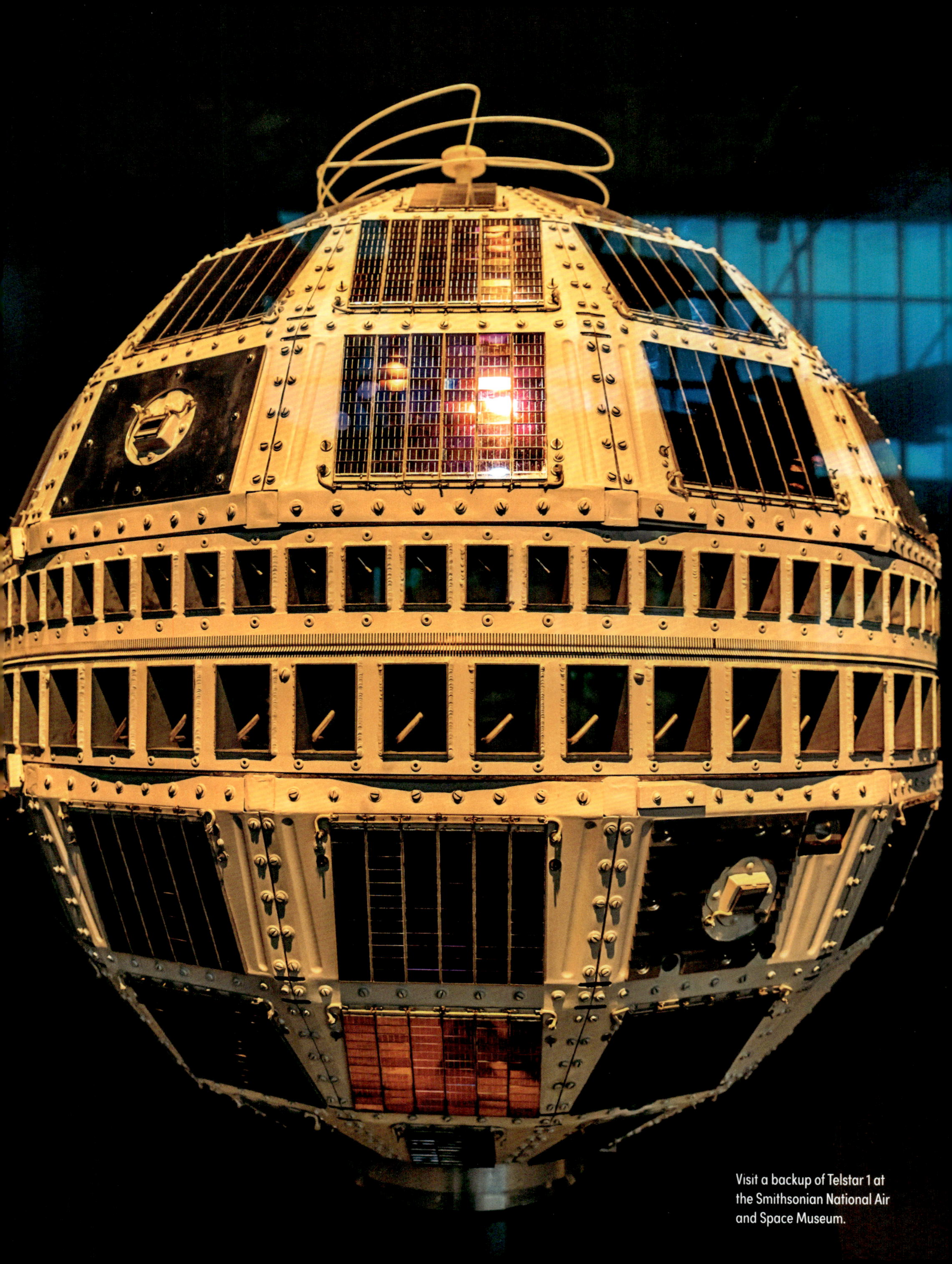

Visit a backup of Telstar 1 at the Smithsonian National Air and Space Museum.

CURRENT FACT COUNT 1586

BY THE NUMBERS: SATELLITES

80

inches: the length of rocket-shaped Explorer 1

22

inches: the diameter of Sputnik 1

3

the number of weeks Sputnik 1's transmitters operated

4

the number of months Explorer 1 sent data to Earth

8 TO 10

feet: the length of the four antennae that trailed from Sputnik 1 and sent data to Earth

92

days: the amount of time Sputnik 1 remained in orbit before it broke apart

1970

the year Explorer 1 entered Earth's atmosphere and burned up

200

pounds: the rough weight of Sputnik 1

1,400

the number of orbits Sputnik 1 completed around Earth

98

minutes: the amount of time it took Sputnik 1 to orbit Earth

1958

the year four more Explorer satellites launched

JANUARY 31, 1958

the date of the United States' first successful satellite launch, Explorer 1, three months after Sputnik 1

2,570

the number of Earth orbits made by Soviet dog Laika

1,500

miles: the distance to the horizon when viewed from low Earth orbit

4

the number of Earth orbits Laika made before she died of hyperthermia in orbit

150

the width, in degrees of longitude, of the Pacific Ocean at its widest, spanning 10,000 miles, or 40 percent of all Earth's longitude. For most satellites that are deorbited, what remains of the satellite after reentry falls into the Pacific.

18,000

miles an hour: the speed of a satellite in low Earth orbit

6,000

miles an hour: the speed of a satellite in geostationary orbit

15

the number of years (to within a day) between the first flight of the German V-2 rocket and the launch of Sputnik 1, the first satellite

16

the number of sunsets and sunrises an astronaut in low Earth orbit sees per orbit

914

miles an hour: the speed boost a satellite gets due to Earth's rotation when launched eastward from Cape Canaveral, Florida

321

the area code of Cape Canaveral

CURRENT FACT COUNT 1615

FIRST ONES UP

The journey to space, where no human had gone before, was rife with fits and starts.

- Russian cosmonaut Yuri Gagarin was the first human in space.
- On April 12, 1961, Gagarin circled Earth once aboard Vostok 1.
- Gagarin's flight lasted 108 minutes.
- He parachuted from the capsule about four miles above Earth.
- In 1968, Gagarin and his instructor died in a crash during a routine fighter jet practice flight.
- **THE GAGARIN CRATER ON THE FAR SIDE OF THE MOON IS NAMED IN HIS HONOR.**
- With Sputnik and Gagarin's unprecedented success, the space race was on—and Russia was ahead.
- In 1961, President John F. Kennedy declared the U.S. would put a man on the Moon and return him safely to Earth within the decade.
- **PROJECT MERCURY, WHICH RAN FROM 1958 TO 1963, WAS THE FIRST U.S. PROGRAM TO PUT HUMANS INTO SPACE.**
- Project Mercury consisted of six crewed missions.
- The initial mission, Freedom 7, was the first U.S. human spaceflight, piloted by Alan Shepard.
- On May 5, 1961, a Mercury-Redstone rocket lifted off, carrying the Freedom 7 capsule with Shepard in it.
- Strapped into the capsule through hours of launch delays and without any other options, Shepard was forced to pee in his space suit.

John Glenn enters the Mercury spacecraft Friendship 7 before launch in 1962.

- Since that day, every astronaut, including those aboard the ISS today, is outfitted with high-tech, high-absorption diapers.
- Shepard's flight lasted 15 minutes and 22 seconds.
- Launched from Cape Canaveral, Florida, Shepard reached an altitude of 116 miles.
- During the sixth Mercury mission, on February 20, 1962, John Glenn became the first American to orbit Earth.
- Glenn's spacecraft was named Friendship 7.
- Due to a glitch in his radio, Glenn didn't hear the famous phrase "Godspeed, John Glenn" uttered as he launched for what would become his historic orbit of Earth.
- Glenn went on to become a U.S. senator representing Ohio, serving for 25 years.
- **IN 1998, ABOARD THE SPACE SHUTTLE *DISCOVERY,* GLENN EARNED ANOTHER SUPERLATIVE WHEN HE BECAME THE OLDEST HUMAN EVER TO ORBIT EARTH.**
- Upon returning to Earth for the second time, Glenn again received a ticker tape parade up New York City's Canyon of Heroes route.

Russian coins (rubles) were released in 2001 with Yuri Gagarin's portrait.

Valentina Tereshkova, the first woman in space

- On June 16, 1963, Valentina Tereshkova became the first woman in space, and she remains the only woman to ever make a solo space journey.
- On March 18, 1965, Russian cosmonaut Aleksey Leonov completed the world's first spacewalk.
- Leonov stayed in space, tethered to the orbiting Voskhod, for 10 minutes.
- In March 1965, Virgil "Gus" Grissom and John Young piloted Gemini 3, orbiting Earth three times in the first U.S. Earth-orbital mission.
- Gemini 4, piloted by James McDivitt and Edward White in June 1965, included the first American spacewalk.
- **DURING THE FINAL GEMINI MISSION IN 1966, ASTRONAUTS JIM LOVELL AND EDWIN "BUZZ" ALDRIN COMPLETED THREE EXTRAVEHICULAR ACTIVITIES (EVAs).**
- The success of the Mercury and Gemini missions prepared NASA for their next undertaking: the Apollo program.

CURRENT FACT COUNT 1652

NASA

Founded in 1958, the National Aeronautics and Space Administration (NASA) has helped to expand the reaches of science and technology—and imagination.

✦ THE COST OF AN AGENCY

- The first "A" in NASA stands for aeronautics.
- Because NASA is a pronounceable abbreviation, it's officially an acronym. Because of this, elsewhere in the world they write it as "Nasa."
- NASA's budget outlay as a percentage of federal spending is now the lowest ever, falling below three-tenths of one percent of the federal budget.
- Funding peaked in 1966, during maximum ramp-up of the Apollo program, when it received 4.4 percent of the federal budget.
- **BETWEEN 1960 AND 1970, WHILE WE WERE GOING TO THE MOON, THE U.S. GDP PER CAPITA INCREASED BY 34 PERCENT. NO DECADE SINCE HAS RIVALED THIS.**
- For most of the 21st century, NASA's budget has languished below half of one percent of the federal budget.

Earth rises over the Moon's horizon in a photograph taken from Apollo 11.

- The year 2025 saw a constriction of NASA's capacity to do science, as well as its workforce shrinking to pre-Apollo levels.

✦ MISSION SUPPORT

- NASA has 10 centers in eight U.S. states, spreading its base wide.
- The eight states with NASA centers have tended to vote in a combination—four red–four blue; five red–three blue; three red–five blue—that ensures bipartisan support from one administration to the next.
- NASA was founded as a civilian space agency. Yet all but one of the Apollo astronauts have been officers in the U.S. military.
- **WHILE AMERICAN SPACE TRAVELERS ARE CALLED ASTRONAUTS,**
  - Russian space travelers are called cosmonauts.
  - European space travelers are called spationauts.
  - Chinese space travelers are called taikonauts.
  - Indian space travelers are called vyomanauts.
- **BETWEEN 1968 AND 1972, NINE APOLLO MISSIONS TRAVELED TO THE MOON:**
  - Apollo 8
  - Apollo 10
  - Apollo 11
  - Apollo 12
  - Apollo 13
  - Apollo 14
  - Apollo 15
  - Apollo 16
  - Apollo 17
- The famous Apollo 8 image of Earthrise over the Moon, taken in December 1968, had a profound effect on society's awareness and urge to protect Earth.
- The Environmental Protection Agency (EPA) and National Oceanic and Atmospheric Administration (NOAA) were both created during this time.

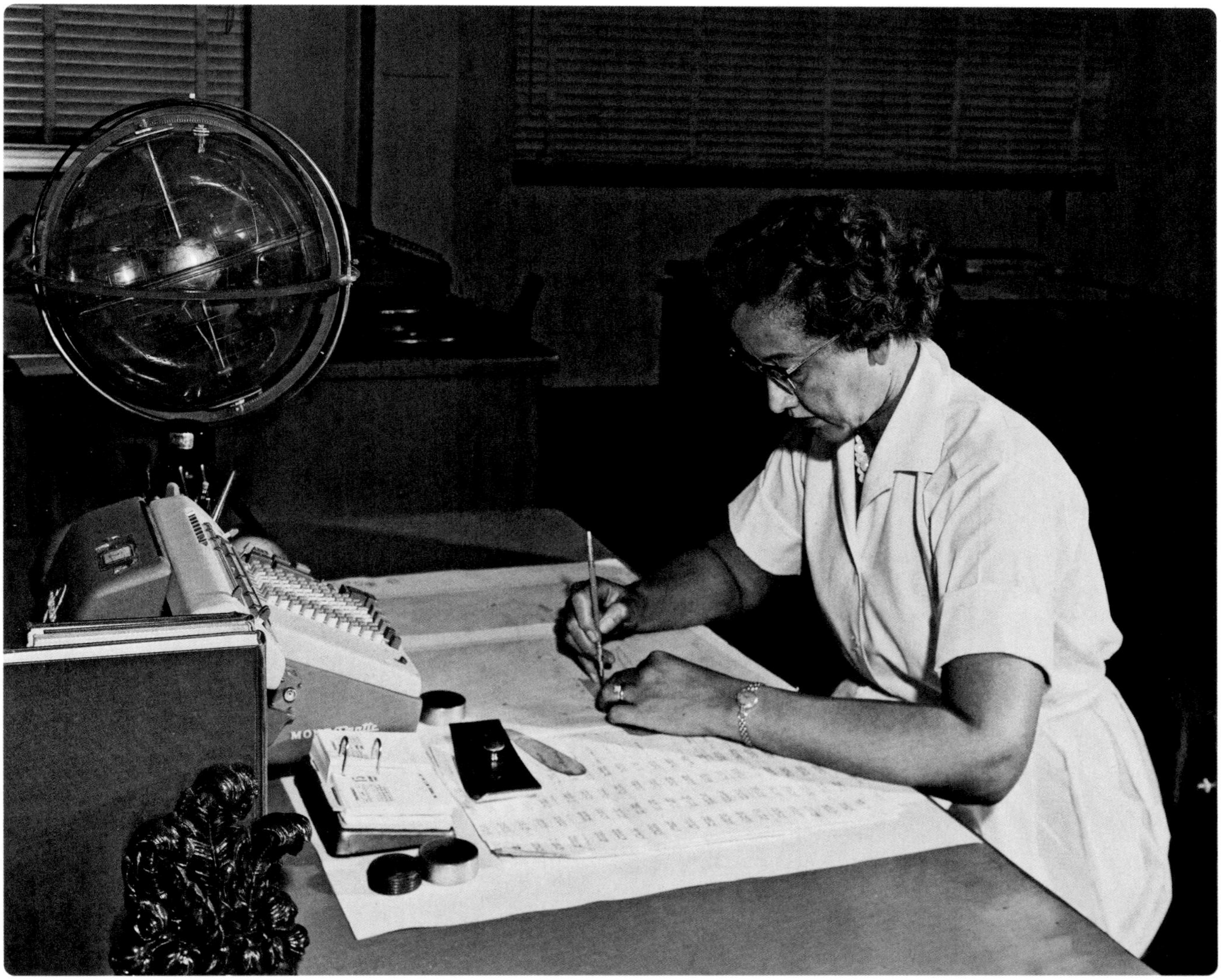

NASA research mathematician Katherine Johnson at her desk in 1962

- *"This is Ground Control to Major Tom, you've really made the grade."* Pop singer David Bowie's first big hit was "Space Oddity," released in 1969.
- Earth Day was founded in 1970.
- **THE APOLLO ASTRONAUTS CONDUCTED VARIOUS SCIENCE EXPERIMENTS ON THE MOON. BUT THE FIRST MISSION TO CARRY A SCIENTIST WAS APOLLO 17, THE LAST MISSION TO THE MOON.**
- The only Apollo astronaut without military rank—and the only scientist to fly on an Apollo mission—was geologist Harrison Schmitt.
- Famed astronomer Eugene Shoemaker was scheduled to fly to the Moon as an astronaut but a heart condition prevented it.
- A geologist by training, Shoemaker trained Apollo astronauts in Meteor Crater, Arizona, for what geological features to look for and how to identify them on the Moon.
- The Vehicle Assembly Building in Cape Canaveral, Florida, is one of the largest in the world by volume.
- The upward flip at the end of modern aircraft wings is for added aerodynamic efficiency and was a feature discovered by NASA engineers.
- **IN THE EARLY DAYS OF THE U.S. SPACE PROGRAM, THE COMPUTERS AND CALCULATORS WE RELIED UPON WERE NOT DIGITAL INSTRUMENTS BUT SMART WOMEN WHO EXCELLED AT SOLVING THE COMPLEX MATHEMATICAL EQUATIONS REQUIRED FOR SPACE TRAVEL.**
- The blockbuster 2016 film *Hidden Figures,* based on the book of the same name by Margot Lee Shetterly, chronicled three of these trailblazing mathematicians: Katherine Johnson, Mary Jackson, and Dorothy Vaughan.

CURRENT FACT COUNT 1676

THE PIONEER PROGRAM

NASA's Pioneer program was the first attempt by any country to send a spacecraft beyond Earth orbit.

- Pioneer was NASA's first space exploration program. It began in 1958.
- **THE FIRST FIVE MISSIONS ALL FAILED TO REACH THE MOON BUT GLEANED VALUABLE DATA.**
- In March 1959, Pioneer 4 became the first American spacecraft to fly beyond the pull of Earth's gravity.
- The U.S.S.R.'s Luna 1 spacecraft had beaten the United States to the feat two months earlier.
- In 1966, Pioneer 7 approached within 7.5 million miles of Halley's comet—then the closest of any spacecraft.
- Launched in March 1972, Pioneer 10 was NASA's first mission to the outer planets, designed to study Jupiter.
- Although Pioneer 10 was intended as a 21-month mission, it continued sending signals to Earth for three decades, finally falling silent in 2003.
- At the time of its last transmission, Pioneer 10 was 7.6 billion miles from Earth.
- **PIONEER 10 NABBED SIX AEROSPACE RECORDS:**
  - First spacecraft sent on a trajectory into interstellar space
  - First spacecraft to fly beyond Mars
  - First spacecraft to fly through the main asteroid belt
  - First spacecraft to fly past Jupiter
  - First spacecraft to fly beyond Neptune
  - First spacecraft to use all-nuclear electrical power
- During Pioneer 10's closest approach to Jupiter in 1973, it relayed detailed images of the planet's largest moons and of the storm known as the Great Red Spot.
- In April 1973, NASA launched Pioneer 11 on a similar path, designed to study Saturn.
- **IN 1979, WHEN PIONEER 11 MADE ITS CLOSEST APPROACH TO SATURN, IT REVEALED A NEW RING AND MOON.**
- Pioneer 11 provided the first data about the temperature of Saturn and its moon Titan.
- Its data also suggested an abundance of liquid hydrogen on Saturn.
- The Pioneer 10 and 11 spacecraft both carried golden plaques intended to relay Earth's address to any intercepting aliens, an idea proposed by Carl Sagan.
- Sagan worked on the concept with physicist Frank Drake and artists Linda Salzman (Sagan's then wife) and Jon Lomberg.
- The plaques include drawings of a man and a woman and of the solar system etched into 6-by-9-inch gold-anodized aluminum plates.
- The plaques also bear a map of distinctive pulsars that point in the direction of Earth.

A painting of Pioneer 10 as it crosses the asteroid belt

NASA engineers inspect the Pioneer 4 spacecraft in 1959.

CURRENT FACT COUNT 1796

FLY ME TO THE MOON

Humanity's mission to unmoor ourselves from Earth

- Russia's Luna program represented the first successful uncrewed space missions to the Moon.
- In January 1959, Luna 1 completed the first lunar flyby.
- **LUNA 2 WAS THE FIRST SPACECRAFT TO REACH THE MOON'S SURFACE IN SEPTEMBER 1959—WITH A CRASH LANDING.**
- Luna 3, in October 1959, was the first to take pictures of the far side of the Moon.
- Luna 9 completed the world's first soft landing on the Moon in January 1966.
- **THE FIRST PHOTOS OF THE LUNAR SURFACE CAME TO US FROM LUNA 9.**
- Surveyor 1, launched in May 1966, completed the first U.S. Moon landing.
- In 1968 the Soviet spacecraft Zond 5 carried the first Earthlings around the Moon—along with a couple of Russian tortoises and some insects, worms, and plants.
- Zond 5 was also the first spacecraft to circle the Moon and return to Earth. (Looks like the tortoise beat the hare this time, too.)
- In 2024, China's Chang'e 6 mission returned samples from the far side of the Moon—the first ever retrieved.

✦ THE U.S. JOINS THE SPACE RACE

- Gauntlet thrown: On May 25, 1961, President John F. Kennedy's address to Congress about "urgent national needs" included this statement: "I believe that this nation should commit itself to achieving the goal, before this decade is out, of landing a man on the Moon and returning him safely to the Earth."
- With five months to spare, the United States landed astronauts on the Moon before the decade was out.
- **GOING TO THE MOON WAS THE PRIMARY THING JFK WAS REFERRING TO WHEN HE FAMOUSLY SAID WE CHOOSE TO DO THINGS "NOT BECAUSE THEY ARE EASY, BUT BECAUSE THEY ARE HARD."**

✦ THE APOLLO MISSIONS

- NASA's Apollo program (1962–1972) answered JFK's call for an American man on the Moon by the end of the decade.
- The three original crew members selected for the Moon mission—Virgil "Gus" Grissom, Ed White, and Roger Chaffee—were tragically killed when a fire broke out in their capsule during a planned test launch in January 1967.
- **APOLLO 1 WAS THE NAME LATER GIVEN TO THE MISSION IN THEIR HONOR.**
- After the Apollo 1 tragedy, crewed spaceflight operations were halted for 20 months during the investigation and incorporation of revised safety protocols.
- There was no Apollo 2 or Apollo 3.
- Apollo 4, 5, and 6 were all uncrewed test flights preparing for travel to the Moon.
- Apollo 7, October 1968, was the first crewed mission after the Apollo 1 fire, completing 163 Earth orbits.
- **APOLLO 8, DECEMBER 1968, WAS THE FIRST U.S. CREWED MISSION TO ORBIT THE MOON AND RETURN.**
- Apollo 9 was an Earth-orbit test mission in March 1969.
- Apollo 10 in May 1969 was the dress rehearsal: A crew flew all the way to lunar orbit and back outfitted with the equipment to complete a real Moon landing mission.
- Apollo 10 astronauts even rehearsed the landing itself. Thomas Stafford and Eugene Cernan descended to about 10 miles above the planned Apollo 11 landing site before ascending and docking with the command module operated by John Young.
- They spent 2.5 days orbiting the Moon, snapping pictures and rigorously testing every facet of a real landing mission, minus the actual landing.
- Might Stafford and Cernan have been the first Moonwalkers if they'd only defied their orders and landed anyway? Likely not. Their ascent vehicle was slightly too heavy to ascend from the lunar surface.
- The Apollo 10 lunar lander was nicknamed "Snoopy" and the command module "Charlie Brown."
- **APOLLO 11, JULY 1969, MARKED THE FIRST TIME HUMANS LANDED AND WALKED ON THE MOON.**

- Apollo 13, April 1970, reached the Moon but had to cancel plans to land when an oxygen tank exploded.
- Apollo 15 was the first time humans drove a rover on the Moon, in July through August 1971.
- Apollo 17, in December 1972, was the last Moon mission of the 20th century.
- Apollo 18, 19, and 20 were due to launch in 1973 and 1974 but were canceled for budgetary reasons.
- Across the six Apollo missions that landed on the Moon, astronauts left behind 96 bags of human waste.
- Apollo astronauts trained for the Moon's low gravity by walking on a tilted wall while suspended sideways.
- **THE U.S. GEOLOGICAL SURVEY (USGS) SET OFF A SERIES OF EXPLOSIVES IN ARIZONA'S CINDER LAKES TO CRAFT 143 MOONLIKE CRATERS FOR ASTRONAUT TRAINING AND EQUIPMENT TESTS.**
- Apollo voyages to the Moon from Earth lasted three days, averaging about one mile per second en route.
- Up until the moment he exited the Eagle module on July 20, 1969, Neil Armstrong was still deciding what humanity's first words on the Moon would be.
- He intended to say, "That's one small step for *a* man, one giant leap for mankind"—but the "a" wasn't audible. He later explained that he didn't forget it but that people couldn't hear it.
- The bottom rung of the Apollo 11 lunar ladder was 3.5 feet from the Moon's surface, so Armstrong's first "small step" was actually quite a leap.
- Neil Armstrong left behind a gold replica of an olive branch as humanity's symbol of peace on the Moon's surface.
- Some claim the Apollo 11 Moon landing was faked because the flag they planted didn't ripple. But there's no wind on the Moon to make it wave. The flag hung stiffly and jutted straight out because a crossbar supported it.
- **THE LACK OF WIND IS ALSO WHY THE ASTRONAUTS' FOOTPRINTS ARE STILL PRESERVED IN MOONDUST.**
- Their footprints could last millions of years if left untouched by future missions. The lack of atmosphere, and thus wind or rain erosion, means no natural events short of an asteroid impact could wipe them away.
- When Apollo 11's lunar lander lifted off, it knocked over the American flag just planted.
- 21 hours 36 minutes: The time Neil Armstrong and Buzz Aldrin spent on the Moon
- The second statement Armstrong made after exiting the lunar lander informed scientists about the Moon's terrain: "The surface is fine and powdery. I can pick it up loosely with my toe. It does adhere in fine layers like powdered charcoal to the sole and sides of my boots."
- **APOLLO ASTRONAUTS DISCOVERED MOONDUST STICKS TO EVERYTHING DUE TO STATIC ELECTRICITY.**
- So scientists developed a Moon vacuum that uses a beam of electrons to defuse the charged lunar dust.
- Three new minerals were discovered in the Moon rocks brought back by Apollo astronauts: armalcolite, tranquillityite, and pyroxferroite. Since then, these minerals have also been found on Earth.
- Most iconic photos of the Moon landing are of Buzz Aldrin. Neil Armstrong was on camera duty.
- On Apollo 11, the first mission to land on the Moon, Neil Armstrong and Buzz Aldrin placed an array of mirrors on the surface. These allowed a precisely timed laser beam to be reflected between Earth and the Moon.
- The Apollo 11 mirrors on the lunar surface allow us to measure the Earth-Moon distance, accurate to centimeters.
- **WHEN APOLLO ASTRONAUTS RETURNED TO EARTH, THEY WERE QUARANTINED FOR 21 DAYS TO MONITOR FOR EXPOSURE TO LUNAR MICROBES OR HARMFUL ELEMENTS.**
- Neil Armstrong celebrated his 39th birthday in quarantine after returning from the Moon.
- In the months before Apollo 11 launched, the astronauts signed hundreds of autographs. The financial support was intended for their families in case disaster struck.
- Some of these autographs were sold in the 1990s for tens of thousands of dollars.

"TRUST YOUR INSTRUMENTS, NOT YOUR BODY, THE MODERN PILOT IS ALWAYS TOLD, BUT THIS BEAST IS BEST FELT. SHAKE, RATTLE AND ROLL!"

—NASA ASTRONAUT MICHAEL COLLINS, SPEAKING OF THE MOMENT OF LIFTOFF

Charles Duke collects samples during the first Apollo 16 extravehicular activity (1972).

✦ FURTHER APOLLO ACCOMPLISHMENTS

- On November 14, 1969, Apollo 12's Saturn V rocket was hit by lightning, 37 seconds into the mission.

- **APOLLO 12 CARRIED ON AND BECAME THE SECOND CREWED MOON LANDING ON NOVEMBER 19, 1969.**

- On April 13, 1970, on the third day of the mission, Apollo 13 astronauts heard a bang, then experienced a loss of power to the command module. One of the oxygen tanks had exploded.

- They had to reroute power from their lunar lander and cancel the planned landing.

- Its larger landing mission scrubbed, Apollo 13 continued around the far side of the Moon and then returned to Earth safely.

- During Apollo 14, Stuart Roosa carried hundreds of tree seeds as he orbited the Moon.

- **NUMEROUS "MOON TREES" NOW GROW FROM SEEDS CARRIED ABOARD APOLLO 14, INCLUDING AT:**
  - Goddard Space Flight Center in Greenbelt, Maryland, U.S.A.
  - Lockheed Martin in King of Prussia, Pennsylvania, U.S.A.
  - Lowell Elementary School in Boise, Idaho, U.S.A.
  - Camp Koch Girl Scout Camp in Cannelton, Indiana, U.S.A.
  - University of Arizona in Tucson, Arizona, U.S.A.
  - Fernwood Botanical Gardens in Niles, Michigan, U.S.A.
  - Château Lafayette in Chavaniac, France

- Apollo 15 landed on the Moon on July 30, 1971.

- Apollo 15 astronauts operated the first lunar rover on the surface of the Moon.

- Later, Apollo 16 astronauts took the lunar rover through a "grand prix" exercise of tight turns and sudden stops.

- Apollo 17 landed in an area of the Moon specially selected for its geological features.

- One of the Apollo 17 astronauts, Harrison Schmitt, was a geologist and NASA's first scientist-astronaut.

- While most of the Moon is shades of gray, Apollo 17 astronauts discovered a patch of orange dust: the ashes of an ancient volcanic eruption.

- **EACH OF THE SIX MOON LANDING APOLLO MISSIONS LEFT BEHIND THEIR LUNAR LANDER DESCENT VEHICLES.**

- After rendezvousing with their orbiting command module, the lunar ascent vehicles crash-landed back to the lunar surface.

- A seismic activity monitor left behind by the Apollo 17 mission picked up regular, repeating tremors emanating from the nearby abandoned lunar lander.

- **WILD TEMPERATURE SWINGS, FROM 250°F TO MINUS 200°F, MAKE THE APOLLO-ERA METALLIC INSTRUMENTS AND LANDERS EXPAND AND CONTRACT, CAUSING SMALL MOONQUAKES.**

- With Apollo 17, the Apollo program ended, having met JFK's challenge and beyond.

- **ALL TOLD, 12 APOLLO ASTRONAUTS (ALL AMERICAN MEN) WALKED ON THE MOON:**
  - Neil Armstrong (Apollo 11, 1969)
  - Buzz Aldrin (Apollo 11, 1969)
  - Charles "Pete" Conrad (Apollo 12, 1969)
  - Alan Bean (Apollo 12, 1969)
  - Alan Shepard (Apollo 14, 1971)
  - Edgar Mitchell (Apollo 14, 1971)
  - David Scott (Apollo 15, 1971)
  - James Irwin (Apollo 15, 1971)
  - John Young (Apollo 16, 1972)
  - Charles Duke (Apollo 16, 1972)
  - Eugene Cernan (Apollo 17, 1972)
  - Harrison Schmitt (Apollo 17, 1972)

The sycamore "Moon tree" at Goddard Space Flight Center in Greenbelt, Maryland

A statuette of Artemis, goddess of the hunt

✦ LUNA AQUA: THE ARTEMIS MISSION

- NASA's Artemis mission is the first program since the Apollo era designed to land humans on our Moon—for longer stopovers.
- **IN GREEK MYTHOLOGY, ARTEMIS IS THE TWIN SISTER OF APOLLO.**
- Apollo 11 aimed for the safest, smoothest, flattest region of the lunar surface that would best ensure a successful mission: a basaltic plain called the Sea of Tranquility.
- The Artemis program selected its target landing sites specifically for their proximity to the lunar south pole—and its vast expanse of permanently shadowed regions.
- As the name implies, a permanently shadowed region, also called a cold trap, never sees the light of day.
- At these regions, sunlight can't reach beyond the rims of deep craters. That means craters formed billions of years ago could hold troves of knowledge, frozen (literally) in time.
- These dark and freezing pockets are likely to contain ancient, undisturbed, and readily accessible volatile molecules—molecules that evaporate in even moderate temperatures. Molecules like frozen water.
- **MOON WATER ISN'T JUST AN INTERESTING SCIENTIFIC RESEARCH OBJECT, IT'S ALSO A CRITICAL IN SITU RESOURCE FOR SUSTAINED HUMAN VISITATIONS ... AND EVENTUALLY, HUMAN HABITATION.**
- A lunar water supply means a dramatic reduction of payload weight and, therefore, reduced mission costs.
- H_2O is critical for human survival, but it can also be separated into hydrogen and oxygen molecules—or rocket fuel.
- The Moon could serve as a useful filling station en route to the rest of the solar system.
- With only one-sixth Earth gravity and no atmosphere, launching from the Moon toward Mars would require significantly less fuel than launching directly from Earth's surface to Mars.
- **MOST OF THE FUEL REQUIRED TO GET TO THE MOON (OR MARS, OR ANYWHERE ELSE) IS REQUIRED TO ESCAPE EARTH'S GRAVITY. WITHOUT THAT BURDEN, OUR ENTIRE SOLAR SYSTEM OPENS UP BEFORE US LIKE A COSMIC PLAYGROUND.**
- In 1892, after careful observations, William Pickering concluded the Moon had no atmosphere and, thus, could sustain no water.
- After the Apollo program yielded no evidence of surface water (through physical observations and initial assessments of Moon rock samples), the Moon was considered waterless for three decades.
- NASA's Lunar Prospector orbital mission discovered evidence of water ice hiding deep within permanently shadowed regions of both the Moon's poles in 1998.
- Astronomer Eugene Shoemaker was the first person to have a lunar burial. Some of his ashes were aboard the Lunar Prospector spacecraft when it intentionally crash-landed into the Shoemaker crater after its mission.
- In 2008, careful reanalysis of Apollo samples revealed hydrogen hiding in bits of volcanic glass, indicating water existed on the Moon during its early existence and perhaps since its creation, when it was but a tumultuous volcanic landscape.
- **WITHOUT AN ATMOSPHERE, ANY LUNAR WATER EXPOSED TO DIRECT SUNLIGHT IMMEDIATELY VAPORIZES INTO THE VACUUM OF SPACE. SO, OUR BEST SHOT AT FINDING THE MOST REMAINING WATER MOLECULES? WE GOTTA GO WHERE THE SUN DON'T SHINE.**

✦ PREPPING FOR A MOONWALK

- NASA Artemis astronauts practice Moonwalking scenarios in a rocky volcanic field of Arizona's northern desert.
- The terrain was chosen for its lunar likeness.
- Apollo astronauts also conducted operational and technical aspects of scientific missions in the Arizona desert.
- Nobody has yet stepped foot on the Moon's rugged south pole, Artemis's mission target.
- The poles experience harsher lighting than the lit basins explored by Apollo astronauts.
- At the lunar poles, the Sun remains low on the horizon, throwing long, stark shadows across the cratered surface, creating a high-contrast, low-light landscape littered with hazards like boulders and pits.
- To simulate these otherworldly conditions, the Artemis team conducted experiments after sunset, illuminated by an enormous distant spotlight.

NASA astronauts Kate Rubins and Andre Douglas in a simulated Moonwalk in Arizona

CURRENT FACT COUNT 1833

THE VOYAGER MISSIONS

The Pioneer spacecraft made the first journeys to the reaches of the outer solar system. The Voyager probes followed, gifting humanity with the first ever detailed images of the outer planets.

- The Voyager interstellar mission was designed to explore the outer solar system, specifically Jupiter, Saturn, Saturn's rings, and the moons of both planets.
- Twin crafts, Voyager 1 and 2, launched from Cape Canaveral, Florida, in the summer of 1977.
- Built to last five years, the spacecraft have continued on ever since.
- **THEIR LONG LIVES HAVE EXTENDED THE MISSION TO INCLUDE URANUS, NEPTUNE, MORE PLANETARY MOONS, AND THE REALM WHERE THE SUN'S INFLUENCE TRANSITIONS INTO INTERSTELLAR SPACE.**
- The Voyager spacecraft are powered by radioisotope thermoelectric generators, using the heat from the radioactive decay of plutonium to power their instruments.
- The craft have at least three million times less memory than a modern cell phone.
- The Voyagers still transmit data back to Earth, but at a rate 38,000 times slower than a 5G connection.
- Each Voyager carries a golden record of human greetings, built to last a billion years.
- The record was inscribed with symbolic instructions on how to play it.
- Carl Sagan, the chair of the golden record committee, said that it would be played "only if there are advanced spacefaring civilizations in interstellar space."

Voyager's golden record (at right) and its cover

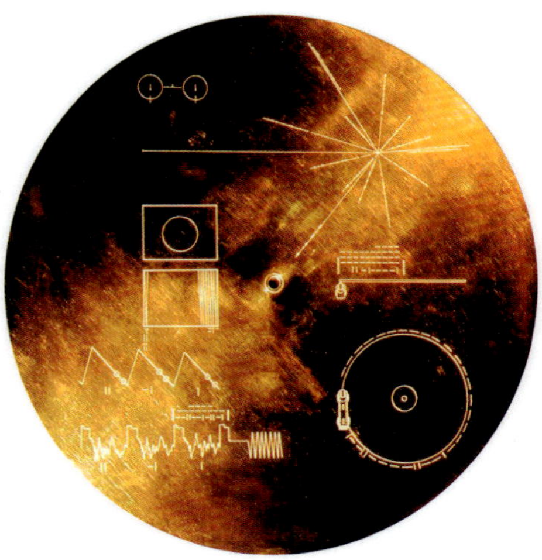

- **THE GOLDEN RECORDS INCLUDE:**
  - Greetings in 55 languages, from ancient Akkadian to modern-day Wu, a Chinese dialect
  - 115 images, including traffic in Thailand, children looking at a globe, houses in Africa and New Mexico, and the structure of DNA
  - 90 minutes of music, including Australian Aboriginal songs, Navajo chants, one of Bach's Brandenburg Concertos, a Peruvian wedding song, and Chuck Berry's "Johnny B. Goode"
  - "Per aspera, ad astra" in Morse code
  - Sounds from Earth, including thunder, a kiss, the Saturn V rocket launch, and the heartbeat of a human in love (actually a recording of Ann Druyan's heartbeat as she thought of her then fiancé Carl Sagan)
- Humpback whale songs, provided by biologist Roger Payne, who first discovered and recorded their singing, played behind the montage of human greetings in case aliens could better understand their language.
- The discovery that whales could sing launched the "save the whales" movement, which brought humpback whales back from the brink of extinction. Perhaps if altruistic aliens discover our record, they might launch a "save the humans" movement.
- In the 1986 film *Star Trek IV: The Voyage Home,* a mysterious spacecraft emits strange signals that wipe out all comms. When Lieutenant Spock realizes the signals match that of the extinct Earth species called humpback whales, he and Captain Kirk time-travel to bring a pair of humpbacks to the future, where they decipher the code and save the universe.
- **VOYAGER 2 WAS LAUNCHED FIRST, IN AUGUST 1977, FOLLOWED BY VOYAGER 1 IN SEPTEMBER.**
- Although it was launched second, Voyager 1 reached Jupiter first because it followed a shorter, faster trajectory.

Voyager 1 gave us this image—taken through color filters and recombined later—of Jupiter's Great Red Spot.

- In August 2012, Voyager 1 became the first human-made object to reach interstellar space, 35 years after it launched.
- **AT THAT POINT, VOYAGER 1 WAS 11 BILLION MILES FROM THE SUN.**
- Voyager 2, going more slowly and in a different direction, reached interstellar space in December 2018.
- Voyager 1 gave us the first close-up looks at Jupiter's Great Red Spot and Saturn's rings.
- Voyager 2 gave us the first close encounters with Uranus (1986) and Neptune (1989).
- Length of the scientific mission the Voyagers were designed for: four to five years
- They've lasted 10 times longer than their expected life spans.
- On November 14, 2023, Voyager 1 stopped sending readable messages back to Earth.
- Engineering adaptations fixed the problem, and on April 20, 2024, Voyager 1 resumed communications.
- **22.5 HOURS: THE TIME IT TOOK FOR A RADIO SIGNAL TO TRAVEL BETWEEN VOYAGER 1 AND EARTH IN 2024**
- As of the publication of this book, Voyager 1 is about 16 billion miles from Earth.
- And Voyager 2 is more than 13 billion miles from Earth.

✦ THE ULTIMATE SELFIE

- After Voyager 1 passed Neptune in 1990, its cameras turned around for one last look toward home.
- The photographs show Earth from a distance of 3.7 billion miles.
- **CARL SAGAN'S FAMOUS DESCRIPTION OF THE PHOTO CALLED EARTH A "PALE BLUE DOT."**
- Afterward, Voyager 1 switched off its cameras for good.
- It took nearly three months for the pictures to reach Earth.

CURRENT FACT COUNT 1866

COSMIC JOURNEYS

Since the first Moon landing, a veritable armada of spacecraft have left Earth to explore the solar system. Each line in these drawings represents one of those missions.

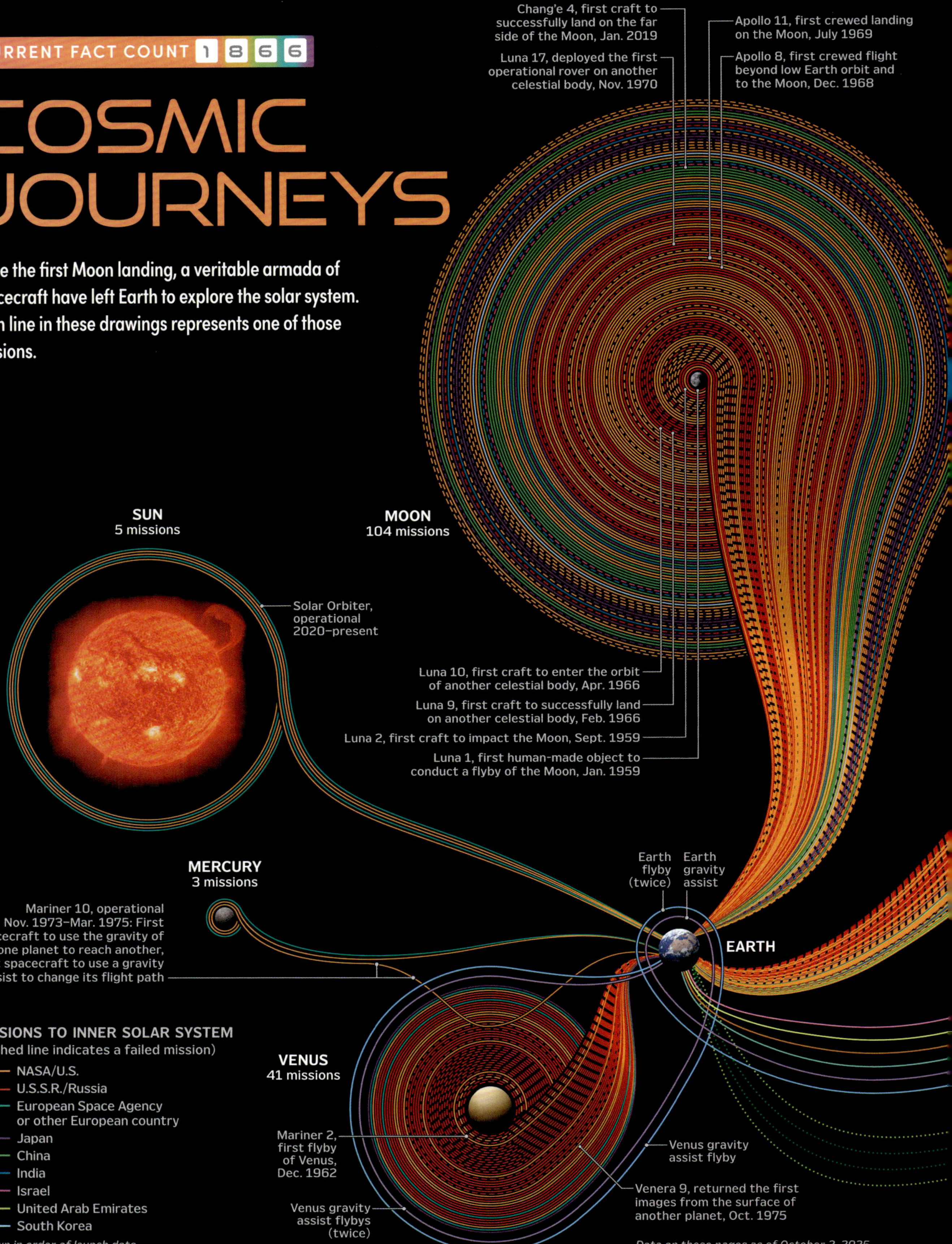

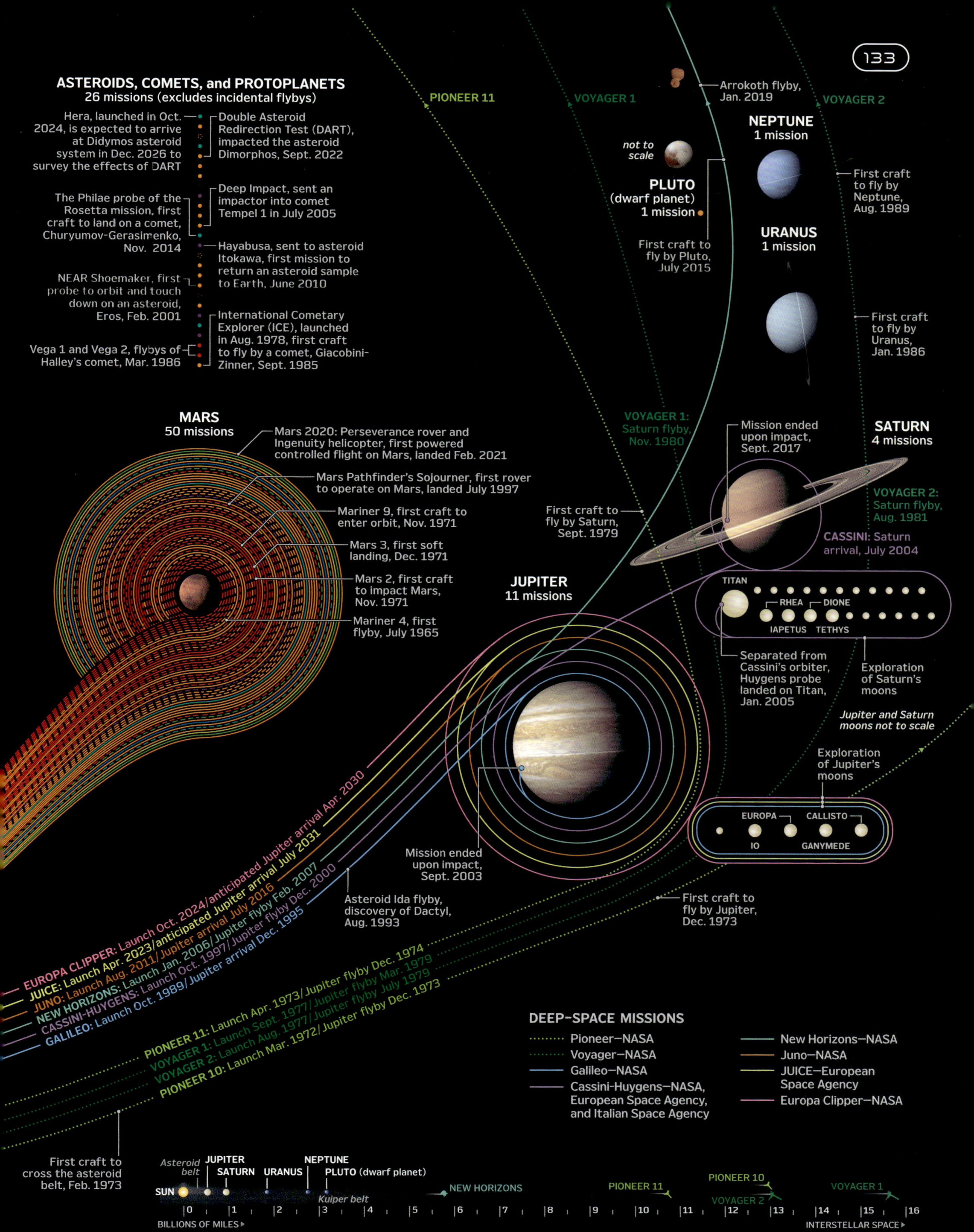
ASTEROIDS, COMETS, and PROTOPLANETS
26 missions (excludes incidental flybys)
Hera, launched in Oct. 2024, is expected to arrive at Didymos asteroid system in Dec. 2026 to survey the effects of DART
Double Asteroid Redirection Test (DART), impacted the asteroid Dimorphos, Sept. 2022
The Philae probe of the Rosetta mission, first craft to land on a comet, Churyumov-Gerasimenko, Nov. 2014
Deep Impact, sent an impactor into comet Tempel 1 in July 2005
Hayabusa, sent to asteroid Itokawa, first mission to return an asteroid sample to Earth, June 2010
NEAR Shoemaker, first probe to orbit and touch down on an asteroid, Eros, Feb. 2001
International Cometary Explorer (ICE), launched in Aug. 1978, first craft to fly by a comet, Giacobini-Zinner, Sept. 1985
Vega 1 and Vega 2, flybys of Halley's comet, Mar. 1986
PIONEER 11
VOYAGER 1
VOYAGER 2
Arrokoth flyby, Jan. 2019
NEPTUNE
1 mission
not to scale
PLUTO
(dwarf planet)
1 mission
First craft to fly by Pluto, July 2015
First craft to fly by Neptune, Aug. 1989
URANUS
1 mission
First craft to fly by Uranus, Jan. 1986
MARS
50 missions
Mars 2020: Perseverance rover and Ingenuity helicopter, first powered controlled flight on Mars, landed Feb. 2021
Mars Pathfinder's Sojourner, first rover to operate on Mars, landed July 1997
Mariner 9, first craft to enter orbit, Nov. 1971
Mars 3, first soft landing, Dec. 1971
Mars 2, first craft to impact Mars, Nov. 1971
Mariner 4, first flyby, July 1965
VOYAGER 1: Saturn flyby, Nov. 1980
Mission ended upon impact, Sept. 2017
SATURN
4 missions
VOYAGER 2: Saturn flyby, Aug. 1981
First craft to fly by Saturn, Sept. 1979
CASSINI: Saturn arrival, July 2004
TITAN
RHEA
DIONE
IAPETUS
TETHYS
JUPITER
11 missions
Separated from Cassini's orbiter, Huygens probe landed on Titan, Jan. 2005
Exploration of Saturn's moons
Jupiter and Saturn moons not to scale
Exploration of Jupiter's moons
EUROPA
CALLISTO
IO
GANYMEDE
Mission ended upon impact, Sept. 2003
Asteroid Ida flyby, discovery of Dactyl, Aug. 1993
First craft to fly by Jupiter, Dec. 1973
EUROPA CLIPPER: Launch Oct. 2024/anticipated Jupiter arrival Apr. 2030
JUICE: Launch Apr. 2023/anticipated Jupiter arrival July 2031
JUNO: Launch Aug. 2011/Jupiter arrival July 2016
NEW HORIZONS: Launch Jan. 2006/Jupiter flyby Feb. 2007
CASSINI-HUYGENS: Launch Oct. 1997/Jupiter flyby Dec. 2000
GALILEO: Launch Oct. 1989/Jupiter arrival Dec. 1995
PIONEER 11: Launch Apr. 1973/Jupiter flyby Dec. 1974
VOYAGER 1: Launch Sept. 1977/Jupiter flyby Mar. 1979
VOYAGER 2: Launch Aug. 1977/Jupiter flyby July 1979
PIONEER 10: Launch Mar. 1972/Jupiter flyby Dec. 1973
DEEP-SPACE MISSIONS
Pioneer—NASA
Voyager—NASA
Galileo—NASA
Cassini-Huygens—NASA, European Space Agency, and Italian Space Agency
New Horizons—NASA
Juno—NASA
JUICE—European Space Agency
Europa Clipper—NASA
First craft to cross the asteroid belt, Feb. 1973
Asteroid belt
JUPITER
SATURN
URANUS
NEPTUNE
PLUTO (dwarf planet)
SUN
Kuiper belt
NEW HORIZONS
PIONEER 11
PIONEER 10
VOYAGER 2
VOYAGER 1
0 1 2 3 4 5 6 7 8 9 10 11 12 13 14 15 16
BILLIONS OF MILES ▸
INTERSTELLAR SPACE ▸

CURRENT FACT COUNT 1 8 9 1

ROCKETS

Rockets are defined by their ability to function in the vacuum of space. To do so, they must carry and burn all their own fuel—a conundrum that required decades for scientists to conquer.

✦ TAKING FLIGHT

- While the fuel an airplane burns relies on the oxygen molecules of Earth's atmosphere, a rocket must carry its own oxygen to help burn fuel beyond the atmosphere.

- **WITHOUT FILLING STATIONS IN SPACE, A ROCKET MUST LAUNCH WITH ALL THE FUEL IT WILL EVER USE ON ITS JOURNEY.**

- Most of the fuel a rocket carries is needed just to launch the fuel it will need later.

- This is the perplexing rocket equation: If you want to punch something through Earth's gravitational grasp, you need a lot of fuel. But that fuel has weight to it. So, you must add more fuel to launch the weight of that fuel. The weight of that additional fuel now requires even more fuel for launch. And on, and on, and on.

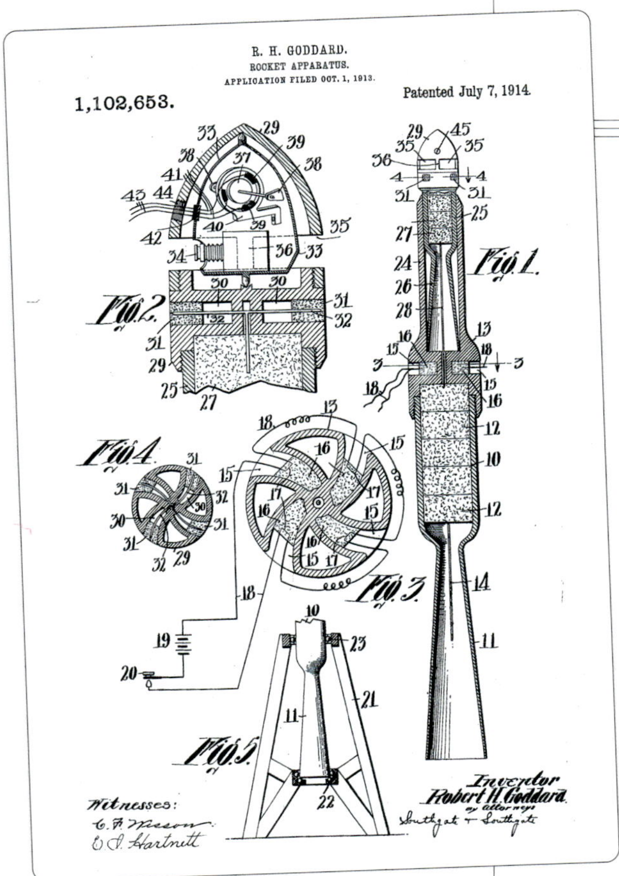

Robert H. Goddard's 1914 patent for his rocket apparatus—a design still used today

✦ EVOLUTION OF ROCKETRY

- Rocket propulsion technology dates back to 13th-century China, whose scientists adapted fireworks fueled by gunpowder into weapons.

- This is an example of solid-fueled rocketry.

- **IN 1903, KONSTANTIN TSIOLKOVSKY POSITED THAT A LIQUID-FUELED ROCKET COULD PROVIDE ENOUGH THRUST TO ESCAPE EARTH GRAVITY AND EXPLORE SPACE.**

- Robert Goddard successfully launched the first liquid-fueled rocket in 1926.

- Goddard's rocket was 10 feet tall.

- It flew for 2.5 seconds.

- And it rose 41 feet into the air and landed 184 feet away from where it launched.

- The first rocket to reach the Kármán line was a ballistic missile launched by Nazi Germany in 1944.

- Saturn V stood 35 stories tall—60 feet taller than the Statue of Liberty.

- It weighed 6.2 million pounds before launch.

A 1967 illustration comparing Saturn V's height to the Statue of Liberty

- Ninety-five percent of its total mass consisted of the fuel required to launch its tiny payload.
- Saturn V generated 7.6 million pounds worth of thrust during blastoff.
- Its rocket guzzled about 40,000 pounds of fuel per second.
- **SPACEX'S 400-FOOT-TALL STARSHIP ROCKET, DESIGNED TO CARRY HUMANS TO MARS, HAS 33 ENGINES AND MORE THAN TWICE AS MUCH THRUST AS SATURN V.**
- Where Saturn V was made of fully expendable parts, Starship's launch systems can be reused over and over, dramatically reducing the cost per pound of payload.

✦ FANTASTIC PROPULSION

- The Epstein drives featured in the hit TV series *The Expanse* (based on the novels by James S. A. Corey) utilize highly efficient nuclear fusion reactions to propel the spaceships that ferry humans through the colonized solar system.
- Though capable of high-g acceleration, they most often travel at a comfortable ⅓ g (one-third of Earth gravity).
- If our spaceships could travel at ⅓ g, we could reach the Moon in about six hours.
- **AT THAT SPEED, WE COULD REACH MARS IN ABOUT SIX DAYS.**
- And we could reach Jupiter in about 11 days at that speed.
- We could go all the way to Neptune in less than a month at that speed. (Remember, Voyager 2 took 12 years to achieve that.)

CURRENT FACT COUNT 1922

IN ORBIT

Much of modern society now relies on the ever growing constellations of satellites in orbit around Earth—from GPS and communications to weather tracking and climate modeling.

- From the onset of the first space race until around 2015, the entire world collectively launched about 200 total satellites a year.
- Since the onset of SpaceX's Starlink enterprise in 2019, the total satellites in orbit has jumped exponentially to multiple thousands per year.
- **THE VAST MAJORITY OF ALL SATELLITES IN ORBIT NOW BELONG TO STARLINK.**
- Every launch, a rocket goes up a ways before it arcs, ultimately becoming parallel to Earth's surface.
- **ONCE ORBIT IS ACHIEVED, WITH VERY HIGH SIDEWAYS VELOCITY, NO FURTHER ROCKET SUPPORT IS NECESSARY TO MAINTAIN IT.**
- If the orbit is not high enough above the outermost reaches of Earth's atmosphere, air molecules will exert a drag on the satellite, requiring a periodic boost, lest it deorbit all by itself.
- Blue Origin and Virgin Galactic space launches are suborbital, where the ship ascends above the Kármán line and falls back to Earth.

The paths of satellites above Alberta, Canada, appear in this blended image as dashes.

✦ LOW EARTH ORBIT

- Around 90 percent of all satellites are positioned in low Earth orbit.
- **LOW EARTH ORBIT (LEO) MEANS 1,200 MILES OR LESS ABOVE EARTH.**
- The International Space Station orbits in LEO.
- To remain in low Earth orbit, an object must maintain a speed of 17,500 miles an hour to counteract Earth's gravitational pull.
- **AT THAT SPEED, IT TAKES 90 MINUTES TO CIRCLE THE PLANET.**
- The higher the orbit, the longer an object takes to circle the planet.
- The higher the orbital altitude, the less speed an object requires to maintain that altitude, because Earth's gravitational force decreases.

✦ GEOSTATIONARY ORBIT

- There is a precise altitude and speed that keeps an orbiting object in perfect lockstep with Earth's rotation. We call it geostationary orbit (GEO).
- GEO is 23,000 miles above Earth's surface. Here, it takes 24 hours to circle the planet.
- Objects in GEO hover over a chosen spot on the Equator for as long as desired—as long as they maintain the necessary speed and altitude.

✦ MEDIUM EARTH ORBIT

- Between LEO and GEO is medium Earth orbit (MEO).
- GPS satellites orbit in MEO at an altitude of 12,500 miles.

- Here, a GPS satellite orbits the Earth twice a day.

- **MEO IS FAR ENOUGH FROM EARTH FOR TIMEKEEPING TO BE MATERIALLY AFFECTED BY GENERAL RELATIVITY.**

- In lower gravity, clocks run a smidge faster than clocks on Earth's surface due to special relativity.

- So GPS time is corrected before it displays on your smartphone and any other device.

- The U.S. government's Space Force operates a constellation of at least 24 GPS satellites in MEO constantly.

- Weather satellites tend to orbit far from Earth—at MEO or beyond—which grants them a wider view with which to track storms and developing weather systems.

✦ REENTRY

- While in free fall, every thing—and body—in the capsule is weightless.

- The reentry heat produced by a vessel returning to Earth is a feature, not a problem, of its return to Earth.

- **IT'S A FORM OF AEROBRAKING.**

- Reentry heat is the dissipation of the ship's orbital kinetic energy. It needs to go from 18,000 miles an hour to zero. The atmosphere actively slows down the vessel.

- Heat shields during reentry ensure the dissipated energy doesn't enter the capsule itself, protecting its occupants.

- Without aerobraking, the craft would need to carry fuel its entire trip, just to use it to slow during reentry.

The distances at which a few notable objects orbit, including the ISS, Hubble, and GPS satellites

CURRENT FACT COUNT 1950

THE OUTER SPACE TREATY

The Outer Space Treaty was originally conceived to prevent the militarization of space. But in the wake of the new commercial space race, we now look to this foundational framework to address questions nobody had conceived at the time.

- The 1967 Outer Space Treaty laid the foundation for international space law.

- Amid the nuclear arms race, the U.S., Soviet Union, and U.K. agreed upon, signed, and ratified a treaty set forth by the United Nations, binding themselves to the peaceful exploration of outer space.

- **UNDER THE OUTER SPACE TREATY, GOVERNMENTS AGREED TO:**
  - Pursue space exploration in a peaceful way and for the benefit of all humanity
  - Never place nuclear weapons or any other weapons of mass destruction into orbit or atop any other celestial body
  - Never claim ownership over space
  - Take responsibility for damage caused by their spacecraft
  - Avoid harmful contamination of space or other bodies

- The official name of the 1967 Outer Space Treaty is Treaty on Principles Governing the Activities of States in the Exploration and Use of Outer Space, Including the Moon and Other Celestial Bodies.

- More than 100 countries, including every major spacefaring nation, has signed on to the treaty, and not one has withdrawn from the original document since its inception.

- Of course, there's no international outer space police, or *Star Trek*-ian Starfleet of the United Federation of Planets, to keep each country in check. Like with all treaties, governments rely on diplomacy and alliances to maintain adherence.

In 2019, Israel's Beresheet lunar lander crashed while attempting to land on the Moon, scattering tardigrades onto the surface in the Sea of Serenity region.

U.S. State Department representative Jennifer R. Littlejohn at the 2024 Artemis Accords signings

- The treaty formed the first—and still foundational—legal framework around international cooperation in the final frontier. But specifics remain murky, as it was primarily spurred by the looming threat of nuclear war.

- What does the space treaty allow on the Moon or in space? Pretty much anything but nuclear weapons or other weapons of mass destruction. Technically, any country or private company can put whatever they wish into space, so long as it is done with peaceable intentions.

- **IT WAS NOT WRITTEN TO INCLUDE SUCH UNFORESEEN TECHNOLOGIES AS SATELLITE CONSTELLATIONS, PRIVATE TOURISM ENTERPRISES, OR ASTEROID MINING OPERATIONS.**

- A company or government cannot own an asteroid, but they can mine, own, and resell all of the resources they harvest from it.

- So, could this be a possible loophole in the treaty? Can a private company erect a water-mining operation on the Moon, establish a monopoly of fuel depots, and effectively govern spaceship traffic control?

- **THE TREATY MAKES NO DECLARATIONS AROUND WHAT'S PROHIBITED TO BRING BACK TO EARTH. SO, IF YOU PROMISE TO GIVE SOMEONE THE MOON AS A DECLARATION OF LOVE, THERE'S NO LAW TO STOP YOU—EXCEPT THOSE OF PHYSICS ITSELF.**

- In 2019, an Israeli spacecraft violently crash-landed on the Moon, spilling thousands of tardigrades onto the otherwise uncolonized Lunar surface.

- Tardigrades are a species of microscopic animal famous for their ability to survive extreme conditions.

- The mission intended to achieve a soft landing but instead scattered the contents, including the tardigrades, across an expanse of hundreds of feet in all directions (and effectively violating *Star Trek*'s Prime Directive).

- The tardigrades aboard the spacecraft were desiccated prior to launch. So, if any survived the impact, they're likely still in hibernation and unable to multiply.

- **THE MISSION RAISED ETHICAL CONCERNS ABOUT LAWS AND REGULATIONS REGARDING CONTAMINATION OF THE MOON AND CALLED FOR NEW AMENDMENTS TO THE OUTER SPACE TREATY TO PROTECT THE LUNAR ENVIRONMENT.**

- Yes, the Outer Space Treaty prohibits harmful contamination of outer space. But a smattering of desiccated tardigrades hardly poses a threat to humanity.

✦ THE ARTEMIS ACCORDS

- The latest international outer space treaty is called the Artemis Accords, drafted by NASA and the U.S. State Department.

- They were created in 2020 as an updated version of the 1967 Outer Space Treaty.

- They establish principles for a peaceful, sustainable, and cooperative future for humanity in space.

- The Artemis Accords require strict accounting of all objects a country puts into space.

- **THEY ESCHEW SCIENTIFIC SECRETS, REQUIRING ALL DISCOVERIES BE SHARED WITH ALL SIGNATORIES FOR THE SAFETY AND BENEFIT OF THEIR FUTURE MISSIONS.**

- If we can live peacefully together in space, then why can't we live peacefully together on Earth's surface?

CURRENT FACT COUNT 1981

WATCH OUT FOR THAT JUNK

Human-made space junk is accumulating at a dizzying pace that, if not curtailed, could pose fatal hazards to space exploration.

- Rogue particles of satellite debris, dubbed "space junk," pose a real threat to the International Space Station (ISS), other satellites within the same orbit, and spacewalking astronauts.
- This threat served as the main plot device in the 2013 film *Gravity*. Unlike in the film (and many others depicting collisions in low Earth orbit), you wouldn't catch a glimpse of a small hunk of junk whizzing past you at 17,000 miles an hour. At those speeds—faster than a speeding bullet—it would pass the entire ISS within half a second.
- **IT'S ESTIMATED THAT SOME HALF A MILLION PIECES OF SPACE DEBRIS BETWEEN 0.4 AND 4 INCHES IN SIZE ARE CURRENTLY ORBITING EARTH.**

In 2023, a tool bag became an accidental satellite after being lost during a spacewalk outside the ISS.

- It's also estimated that more than 100 million pieces of space debris smaller than 0.4 inch are orbiting Earth.
- The estimate of such tiny bits of space debris is calculated based on examination of surface damage done to returning spacecraft.
- About 10,000 tons of space debris orbit Earth.
- **FROM 1998 TO NOVEMBER 2024, THE ISS HAD TO TWEAK ITS COURSE 39 TIMES IN ORDER TO AVOID POSSIBLY COLLIDING WITH SPACE DEBRIS.**
- The velocity of space debris is so high that bits as small as a paint fleck can cause problems.
- In 2016, a tiny piece of space debris chipped the ISS cupola window.
- Low-orbit debris can fall back to Earth in several years, while high-orbit debris can circle our planet for centuries.
- Although most low Earth–orbiting satellites will simply burn up in the planet's atmosphere, their incineration releases toxic particulates known to harm the ozone.
- Thousands of tons of satellite debris have burned up in Earth's upper atmosphere.
- **EACH DAY, ONE KNOWN PIECE OF SPACE JUNK PLUNGES BACK TO EARTH.**
- In 1978, a Soviet satellite scattered radioactive debris over northwest Canada as it plunged through the atmosphere.
- The incident likely caused high rates of cancer among the Dene people, who live near where the debris landed.
- In 2009, a U.S. iridium satellite and a defunct Russian Cosmos satellite collided, creating thousands of pieces of space junk.
- In 2025, a 1,100-pound, eight-foot-wide ring of metal fell to Earth, landing in a Kenyan village. It's presumed to be part of a launch vehicle that now orbits as space junk.

- **SOME RANDOM PIECES OF SPACE JUNK (MOST HAVE LIKELY FALLEN FROM ORBIT BY NOW): A SPATULA, TOOL BAG, PLIERS, GLOVE, AND NUTS AND BOLTS**
- The Kessler effect describes a catastrophic hypothetical scenario where the debris from a destroyed satellite collides with another satellite, multiplying the debris, which continues colliding and multiplying until Earth is swarmed by an untraversable sea of junk.

✦ SHOWS OF FORCE

- The 2020 Artemis Accords specify the responsible mitigation and, where possible, removal of space debris—including satellites—created by the signatories.
- This removal can take place using an anti-satellite weapon (ASAT), typically a high-speed ballistic projectile that destroys with its kinetic energy rather than an explosive device.
- In 2007, China used a ballistic projectile to destroy one of its own nonoperational weather satellites.
- The satellite China destroyed orbited at more than 500 miles above Earth's surface, too high for debris to deorbit on its own, remaining a hazard forever.
- **IF YOU HAVE THE PRECISION WEAPONRY TO DESTROY ONE OF YOUR OWN SATELLITES IN ORBIT, YOU HAVE THE WEAPONRY TO DESTROY THE SATELLITE OF AN ADVERSARY.**
- Not to be shown up, in 2008, the U.S. decided it needed to destroy one of *its* satellites.
- U.S. Operation Burnt Frost obliterated a nonfunctioning reconnaissance satellite in February 2008.
- The satellite, destroyed by an ASAT missile, was in low Earth orbit.

The metallic ring that crash-landed in a Mukuku village in Kenya in 2025

- By June 2008, nearly all debris from that kinetic kill had reentered Earth's atmosphere, harmlessly burning up on descent.
- In 2021, Russia decided it needed to destroy one of its own satellites using a kinetic missile.
- The satellite orbited comfortably in low Earth orbit, ensuring nearly all the debris will, in due time, reenter Earth's atmosphere and burn up en route.
- In 2022, India destroyed one of its own defunct satellites. Like the others, it burned up on reentry.

CURRENT FACT COUNT 2026

THE INTERNATIONAL SPACE STATION

The first-of-its-kind cooperative research station 250 miles above Earth

- The International Space Station (ISS) is a cooperative venture of five space agencies representing 15 countries.
- **$150 BILLION: THE ESTIMATED CUMULATIVE VALUE OF THE ISS, EARNING IT THE AWARD FOR THE MOST EXPENSIVE STRUCTURE—OR OBJECT OF ANY KIND—EVER BUILT**
- The first orbiting live-in spacecraft was Russia's Mir orbital station.
- After 1998, Mir also served as a test bed for the design and operation of parts that would be used on the ISS.
- Primary ISS Mission Control stations are located in the U.S. and Russia, with other control centers in Canada, Japan, and Europe.
- On October 31, 2000, the first crew of ISS astronauts launched from Earth.
- This date marks the last time every human was on Earth together as an exclusively terrestrial species. Since then, at least one Earthling has been living off planet, caught somewhere in our orbit.
- Astronauts aboard the ISS witness meteor shower displays below the station as they peer down from above.
- **BECAUSE THE ISS REFLECTS SUNLIGHT, IT CAN BE SEEN DEEP INTO TWILIGHT WITH THE UNAIDED EYE.**
- Using NASA's Spot the Station mobile app or website *(spotthestation.nasa.gov)*, you can track the ISS's position in the sky from Earth.
- You can sign up for alerts for when the station is passing overhead in your part of the world, giving you a chance to watch it orbit from your backyard.

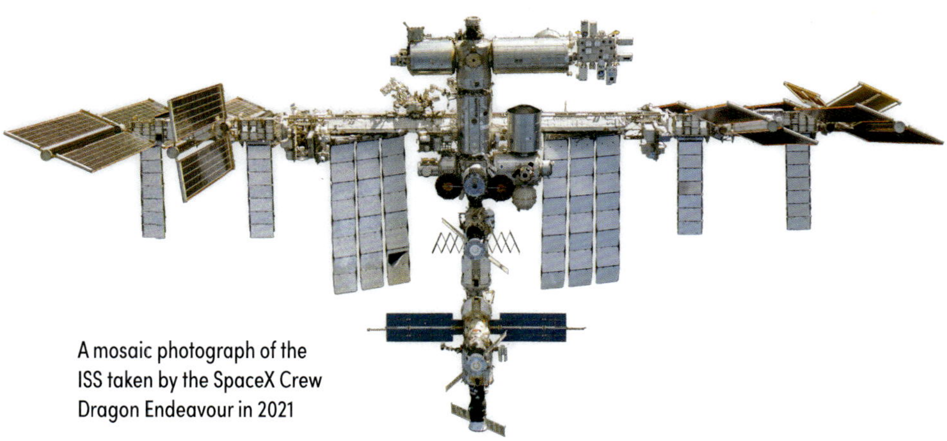

A mosaic photograph of the ISS taken by the SpaceX Crew Dragon Endeavour in 2021

✦ THE STATS

- The ISS orbits Earth roughly every 90 minutes or so.
- It travels at a speed of five miles per second.
- The station orbits 230 to 286 miles above Earth.
- Astronauts aboard the ISS experience 16 sunrises and sunsets every 24 hours.
- **IN A SINGLE DAY, THE ISS TRAVELS TWICE THE DISTANCE FROM EARTH TO THE MOON.**
- 356: the width, in feet, of the ISS solar array
- The habitable living quarters occupy about the same volume as a Boeing 747.
- The ISS has six separate sleeping areas, two bathrooms, and a gym.
- The first extravehicular activity (EVA) from the ISS occurred in 1988.
- In 1999, Tamara E. Jernigan was the first woman to perform an EVA from the ISS.
- **NASA ASTRONAUT PEGGY WHITSON HOLDS THE RECORD FOR THE MOST ACCUMULATED TIME IN SPACE, WITH MORE THAN 695 DAYS.**
- NASA astronaut Frank Rubio holds the NASA record for the longest uninterrupted time in the ISS at 371 days.
- In 2023, Frank Rubio was the first to grow tomatoes in space—which he then lost.
- Many accused Rubio of eating the two tiny tomatoes, but eight months later the rogue fruits were recovered, intact within a plastic bag on board the ISS.
- Astronauts on board the ISS tend to develop calluses on the top of their feet from stabilizing themselves by hooking their feet under railings.

- When the ISS falls through Earth's atmosphere, much of its hardware and external coating will burn up or melt.

The ISS's Tranquility module, home to life support systems, the observation cupola, and more

- With fully extended solar panels, the ISS—assembled in parts delivered by space shuttle—is as large as a football field.
- ISS electrical systems require eight miles of wire.
- **NASA ASTRONAUT SUNI WILLIAMS RAN THE BOSTON MARATHON IN SPACE IN 2007.**
- As Williams ran those 26.2 miles, the ISS traveled 77,880 miles around Earth.
- The Russian space program has depended on the Soyuz spacecraft since 1966, and until SpaceX entered the space field in 2020, the Soyuz was the only spacecraft capable of traveling to the International Space Station at all.

✦ RETIREMENT, ISS STYLE

- The ISS is scheduled to retire from orbit in 2031 after more than three decades of use, having housed hundreds of astronauts who conducted thousands of experiments.
- A deorbit vehicle, designed and built by SpaceX, will coast the massive ISS into a watery grave on Earth.
- **IT WILL CRASH-LAND IN THE "SPACECRAFT CEMETERY" OF POINT NEMO.**
- Point Nemo is in the middle of the Pacific Ocean and the most remote region on the planet.
- There, the ISS will join hundreds of other defunct spacecraft, including the Mir and Skylab space stations.

✦ OBJECTS IN MOTION

- You may have heard the space-age fable that NASA developed a high-tech pen that could work even in the vacuum of space, while the Soviets simply used a pencil. A fun story, but neither astronaut nor cosmonaut ever used pencils in space for fear of the flammable property of graphite. If the tip broke off, it could pose a serious risk to onboard electronics.
- The ISS is equipped with specialized armor called a Whipple bumper, named for astronomer Fred Whipple, who invented the technology to protect and deflect oncoming meteorites.
- In 2022, a micrometeorite less than one millimeter wide impacted and damaged the Soyuz capsule meant to carry two cosmonauts and a NASA astronaut back to Earth from the ISS. Russia had to launch a new uncrewed capsule to rescue the crew.
- The largest threat to astronauts, however, isn't from meteorites but rogue space junk.
- **IN NOVEMBER 2024, ASTRONAUTS ABOARD THE ISS PERFORMED AVOIDANCE MANEUVERS TO EVADE DEBRIS FROM A BROKEN-UP SATELLITE.**
- One dodge maneuver costs an estimated $1 million in fuel costs alone—and that's if the dodge is successful.
- Occasionally, astronauts have been required to "shelter in place" during an unpredictable onslaught of space junk collision.
- In these scenarios, astronauts don their space suits and board their escape capsule to prepare for potential emergency escape measures.

"EARTH IS THE CRADLE OF HUMANITY, BUT ONE CANNOT LIVE IN A CRADLE FOREVER."

—KONSTANTIN E. TSIOLKOVSKY, RUSSIAN SCIENTIST

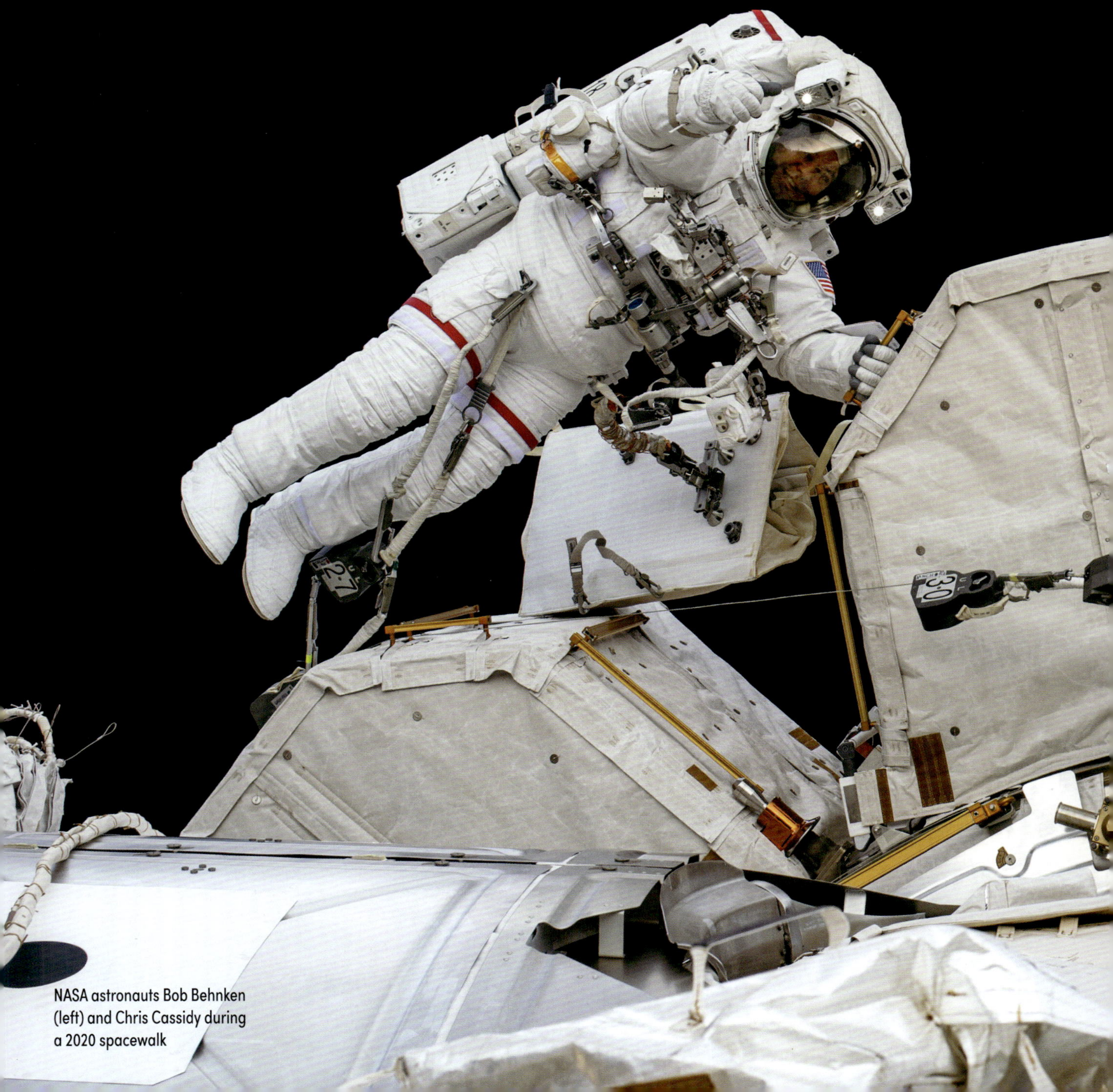

NASA astronauts Bob Behnken (left) and Chris Cassidy during a 2020 spacewalk

15

CURRENT FACT COUNT 2079

WHAT'S IT LIKE IN SPACE?

The human body did not evolve to withstand long stints of microgravity, so astronauts do what humans do best: adapt.

- You can float free and tumble in zero gravity, but your body's fluids slosh around.
- About half a gallon of fluid shifts from the lower body to the upper body before reaching a new equilibrium.
- After Scott Kelly's 340 days in space, enough fluid had traveled from his legs to his head to fill a two-liter soda bottle.
- The shifting fluids in their bodies make astronauts feel constantly congested.
- The fluids also cause long-term space travelers to develop puffy faces.
- **YOUR BRAIN IN SPACE IS LIKE AN INFLATED BALLOON, THANKS TO AN INCREASE IN CEREBROSPINAL FLUID.**
- The excess fluid may compress the brain's gray matter, but this effect largely reverses upon return to Earth.
- The excess fluid bulges the optic nerves and blurs your vision.
- If you open a soda can in space, bubbles form, but they don't rise to the top.
- So don't drink soda in space, because most of the carbonation remains in the liquid, causing a stomachache.
- Astronauts can lose up to 20 percent of their muscle mass in a single week while in space.
- To prevent muscle loss, they vigorously exercise about 2.5 hours a day.
- Without gravity, sweat doesn't drip. It just sticks to your face and body.

✦ THE SMELL TEST

- Nobody can smell anything in a vacuum, so astronauts smell nothing while in space.
- But after spacewalks, astronauts have reported smelling metal, seared steak, or welding fumes.
- Apollo astronauts returning from the lunar surface detected the scent of gunmetal.
- **THE DISTINCT EAU DE OUTER SPACE COULD BE ATTRIBUTED TO PARTICLES ADHERING TO SPACE SUITS.**
- Or it could be the lingering scent of ozone from repressurization.
- Or a combination of both.
- What would it be like to go swimming in zero g? It's accurately portrayed in the sci-fi film *Passengers*. When the interstellar space station loses artificial gravity, the swimming pool becomes a giant spherical blob, consistent with the way water behaves on the ISS. An astronaut in the pool nearly drowns, unable to escape the blob.
- **LIKE SCUBA DIVERS, ASTRONAUTS MUST TAKE PRECAUTIONS TO PREVENT GETTING DECOMPRESSION SICKNESS, ALSO KNOWN AS "THE BENDS."**
- The bends can happen during an abrupt transition from a place of high pressure to a place of low pressure.
- The air we breathe contains mostly nitrogen, which bops around harmlessly within our lungs, blood, and tissues.
- At higher pressures, however, the nitrogen molecules compress and move from our lungs into our bloodstream and tissues at much higher quantities than they'd otherwise carry.
- If a diver stays too long at great depths, tissues become saturated with compressed nitrogen. As the diver ascends, trapped molecules outgas from the tissues. Ascend too quickly and nitrogen will inflate rapidly within the blood vessels and tissues—to a dangerous and even deadly effect.
- That's why you should never fly within 24 hours of scuba diving. That nitrogen may still be working its way out of your system.

- The atmospheric pressure inside a space suit is only one-third the atmospheric pressure of the ISS, which is maintained at 1 atmosphere, or the air pressure at sea level, at all times.

- **TO PREVENT GETTING "BENT" THE WAY A SCUBA DIVER RISING FROM DEPTH MIGHT, ASTRONAUTS BREATHE PURE OXYGEN FOR A FEW MINUTES PRIOR TO A SPACEWALK.**

- This helps eliminate any nitrogen particles in their blood or tissues.

✦ TWINS STUDY

- NASA astronauts Scott Kelly and his identical twin, Mark, were the subjects of a groundbreaking study on how prolonged spaceflight impacts the human body.

- **BORN SIX MINUTES APART AND WITH NEARLY THE SAME GENETIC MATERIAL, SCOTT AND MARK BECAME THE FIRST IDENTICAL TWINS IN SPACE HISTORY.**

- Scott spent one year aboard the ISS while Mark, who was a former astronaut himself, remained on Earth, providing humanity with the first opportunity to study the impact of long-duration spaceflight on the human body, right down to our DNA.

- The NASA Twins Study examined how long-term exposure to weightlessness, radiation, stress, and other factors of life in space affects our bodies.

- The length of telomeres, the protective endcaps of our chromosomes, is a strong indicator of aging and stress.

- **AS HUMANS AGE, OUR TELOMERES SHORTEN. THIS PROCESS CONTINUES UNTIL THE DNA CAN NO LONGER REPLICATE.**

- While in space, Scott Kelly's telomeres actually lengthened. But the effect was short-lived. They rapidly shrunk and even broke off within days of landing back on Earth.

- Kelly describes the sensation of shifting body fluids as feeling like you're standing on your head all day.

- The phenomenon has also been described as the appearance of a "puffy face" and "bird legs."

- **THE ABSENCE OF GRAVITY TAKES ITS TOLL ON THE BODY'S MUSCLES, HEART, BONES—AND, RESEARCHERS BELIEVE, CHANGES IN GENE EXPRESSION.**

- Many of these effects resolve after the space traveler returns to Earth—at least in this study—but changes to vision and eye structures can last longer, likely because of fluid shifts in zero g.

- Are women astronauts more resilient to space than their male astronaut counterparts? Some studies suggest yes.

- The Space Omics and Medical Atlas (SOMA), a collaborative project of NASA and other scientific and academic institutions, tracks and compiles health biometrics of ISS and commercial astronauts.

- **SOMA'S RESEARCH REVEALED THAT ALL STUDIED CELLS FROM FEMALE ASTRONAUTS TEND TO RETURN TO BASELINE FASTER THAN THE CELLS OF MALE ASTRONAUTS.**

- Couple this with the fact that, on average, female astronauts tend to require less oxygen and resources, and all-female missions could make the most strategic sense for long-term space exploration.

- The 2021 Inspiration4 mission was the first spaceflight crewed exclusively by civilians, heralding a new era of a commercialized space, and threw open the gates for space physiology research.

- Prior to the commercialization and privatization of space, all data on the impacts of zero g on humans came from a carefully selected pool of mostly healthy, mostly white, and mostly male government-employed professional astronauts.

- **SHORT COMMERCIAL SPACEFLIGHTS ENABLE RESEARCHERS TO CONDUCT SCIENTIFIC STUDIES MUCH MORE QUICKLY AND ON A MUCH MORE DIVERSE POOL OF SPACEFARERS.**

- Their growing trove of data helps us better understand how the average human responds to microgravity.

✦ STELLAR ODDITIES

- **HUMANS HAVE SENT SOME STRANGE STUFF INTO SPACE.**
  - Luke Skywalker's lightsaber: The actual prop wielded by actor Mark Hamill flew to space aboard the space shuttle *Discovery*.
  - A gorilla suit: In 2015, during Scott Kelly's one-year mission aboard the ISS, he received a furry gorilla costume as a surprise birthday gift from his twin brother down on Earth (retired NASA astronaut Mark Kelly). Scott donned the ape suit to prank his fellow ISS crew members.
  - An acoustic guitar: Chris Hadfield sang David Bowie's "Space Oddity" as the first music video from space.
  - Tesla and Starman: In 2018, SpaceX sent a cherry red Tesla Roadster into orbit on the first launch of the Falcon Heavy rocket. A mannequin donning a SpaceX space suit nicknamed "Starman" occupied the driver's seat while David Bowie's "Life on Mars" played on loop.
  - A six-iron golf club: American astronaut Alan Shepard used it to play golf on the Moon during the Apollo 14 mission. His first shot landed in a nearby crater: a lunar hole in one.

CURRENT FACT COUNT 2112

COSMIC CUISINE

When you think of astronaut food, freeze-dried ice cream and orange soda may come to mind. Space cuisine has evolved quite a bit since those Apollo-era snacks, but zero gravity still poses unique gastronomical challenges.

- In 2019, International Space Station (ISS) astronauts baked cookies in space for the first time—indeed, the first time *anything* was baked in microgravity—using a contraption not unlike the Easy-Bake Ovens of our childhoods.
- The feat seems trivial, but it demonstrates the peculiar and profound implications of the behaviors of air and heat without gravity.

Martha Stewart at Baikonur Cosmodrome before her then boyfriend Charles Simonyi flew to the ISS

- Your traditional convection oven won't work in zero g. Convection relies on gravity to tug down on continuously rising hot air so it can collide with cooler food molecules. That circulation—or exchange—of cool and hot air transforms doughy blobs into crunchy cookies within minutes.
- Without gravity, hot air can neither rise nor sink to form that crucial temperature exchange with uncooked food particles. Instead, it forms a stagnant pocket around the food.
- NASA astronauts heat their prepared food inside a forced convection oven. Instead of relying on gravity to circulate hot air, it uses a small fan to blow air around the chamber. These ovens do not get hot enough to cook raw food, only to slowly warm a meal.
- The zero g oven used to bake cookies in space created pockets of hot air around pieces of dough strapped in the center of a cylindrical tube.
- Without convection, the cookies required a baking time of more than an hour at 300°F.
- **DEVASTATINGLY, THE ASTRONAUTS NEVER GOT TO TASTE THEIR "HOME-MADE" TREATS. THEY SHIPPED ALL FIVE SPACE COOKIES BACK TO EARTH FOR SAFETY INSPECTIONS.**
- One of the cookies is now on display at the Smithsonian National Air and Space Museum in Washington, D.C.

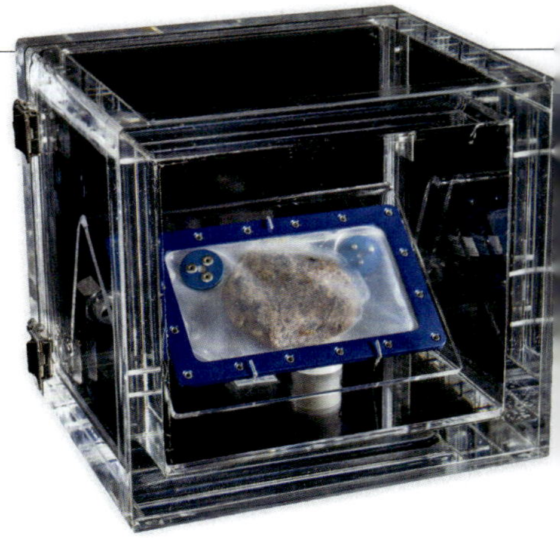

A cookie baked aboard the ISS

✦ INTERSTELLAR TOP CHEF

- Thermodynamics isn't the only obstacle hindering aspiring space chefs. Imagine crafting chocolate chip cookies from scratch aboard the ISS. A standard recipe requires flour, sugar, baking soda, salt, chocolate chips, raw eggs, and butter. The last two ingredients require refrigeration except in powdered form, so let's assume all ingredients are fine grains of some kind that can escape into the air. At best, granules in space are a pain to contain and an annoyance to encounter. At worst, they can damage electronics or pose serious health and safety risks to the human passengers.
- Crusty bread and other crunchy or crumbly foods are banned on the space station for this reason.
- **THE BEST FOODS TO CONSUME IN ZERO G ARE STICKY AND DON'T MAKE CRUMBS THAT CAN FLOAT AWAY. SOFT TORTILLAS, YES. BREAD AND CRACKERS, NO. PEANUT BUTTER, YES. POTATO CHIPS, NO.**
- Raw ingredients cost more to transport due to increased weight and storage requirements. It's far more economical and fuel efficient to send premade meals than individual ingredients housed in separate containers.
- Freeze-dried foods are frozen and, while in that state, have all their moisture evaporated from them, greatly reducing their launch weight.

- On long missions, it's best to take freeze-dried foods, which are shelf-stable for decades before being reconstituted with ordinary water.

- **RESEARCH HAS SHOWN THAT, WHILE IN SPACE, ASTRONAUTS PREFER COMFORT FOODS WHEN POSSIBLE.**

- Astronauts on board the ISS commonly trade foods of their nationality with astronauts from other countries, depending on who else happens to be on the space station at the time.

- Two things destroy food over time: microbes and the simple decay of large food molecules, especially proteins.

- Microbes make the food spoil. To ensure they don't bring spoiled food into space, NASA irradiates food before it's shrink-wrapped, rendering it shelf-stable for years.

- The decay of molecules can change the flavor and texture of food, even if it has no microbes crawling on it.

- Astronauts report that on space missions their taste buds become mildly desensitized, requiring a steady supply of hot sauce to render otherwise tasty foods palatable.

- Although astronauts miss the crunchy and spicy flavors of home, NASA works with renowned professional chefs to concoct the most delicious and nutritious meals possible for spacefarers.

- **MICROSOFT EXECUTIVE AND BILLIONAIRE CHARLES SIMONYI PAID MORE THAN $20 MILLION TO FLY IN A RUSSIAN SOYUZ TO THE ISS IN 2007.**

- At the time, Simonyi was dating lifestyle icon Martha Stewart, who advised on the meals he would bring to space.

- Stewart supplied Simonyi with an enviable stellar feast, some of which he shared with the other astronauts: roasted quail in Madeiran wine, duck breast with capers, chicken Parmentier, apple fondant pieces, rice pudding with candied fruit, and semolina cake with dried apricots.

Astronaut Leland Melvin floats amid some treats on the space shuttle *Atlantis*.

- When considering future long-term space missions to the Moon, Mars, or beyond, food surpasses its role as a mere allotment of calories and vitamins. It's a critical component to moral and psychological resilience.

- **DEVELOPING NEW TECHNOLOGIES TO HELP INCORPORATE TREATS THAT TASTE OF HOME INTO AN ASTRONAUT'S OVERALL DIET IS NOT A FRIVOLOUS EXPENSE, BUT A NECESSARY ADVANCEMENT FOR FUTURE CREWED MISSIONS.**

✦ WHAT'S ON THE MARTIAN MENU?

- In the future, when Moon colonies are common, restaurants would surely feature unusual menus. But they'd have no atmosphere.

- If you don't bring all the food you plan to eat on a space voyage, then you will need a way to create it or grow it at your destination.

- Moon and Mars colonies, if and when they exist, will not likely house farm animals anytime soon, requiring a full vegetarian menu to provide all long-term nourishment.

- Astronauts on board the ISS drink water recycled and purified from all their own body excretions: feces, urine, spit, sweat, and exhaled moisture.

- **FORTUNATELY, THE HUMAN DIGESTIVE SYSTEM DOES NOT REQUIRE GRAVITY TO MOVE FOOD THROUGH YOUR BODY.**

- In a scene from the film *2001: A Space Odyssey*, the character Heywood Floyd, while weightless in a transport vessel, sips liquid from a transparent straw. But when he stops sipping, the liquid slides back down the straw, revealing a force of gravity that wouldn't have existed in reality.

CURRENT FACT COUNT 2142

ANIMAL ASTRONAUTS

Laika may be the most famous animal astronaut (and the most famous dog, period). But she was far from the only animal sent by humans into space.

- The first animals in space were fruit flies, sent to space with a few kernels of corn.
- Aboard a V-2 rocket, the flies reached an altitude of 106 miles and survived their return trip to Earth.
- The first attempt to send a mammal into space was with Albert I, a rhesus monkey, launched from New Mexico in 1948.
- Albert I failed to reach space, making it 39 miles above Earth.
- **IN 1949, ALBERT II, ALSO A RHESUS MONKEY, BECAME THE FIRST MAMMAL IN SPACE, REACHING A HEIGHT OF 83 MILES.**
- Tragically, Albert II died on impact after his capsule's parachute failed.
- After the fatal flights of his monkey predecessors, Albert VI was cynically nicknamed Yorick before his launch (after the dead jester in *Hamlet*).
- Despite his nickname, Yorick became the first monkey to live through a launch and reentry, though he fell short of reaching the Kármán line.
- The first animal to orbit Earth was Laika, a husky mix who was launched in 1957.
- **LAIKA WAS A MUTT FOUND WANDERING MOSCOW, ALONE AND HUNGRY, BEFORE SHE BECAME THE FIRST LIVING CREATURE IN ORBIT ABOARD SPUTNIK 2.**
- The thinking was that if a stray dog can survive Moscow winters with hardly any food, the dog would be a good test case for an orbital mission.

Laika perches in the Sputnik 2 capsule before her history-making trip into space.

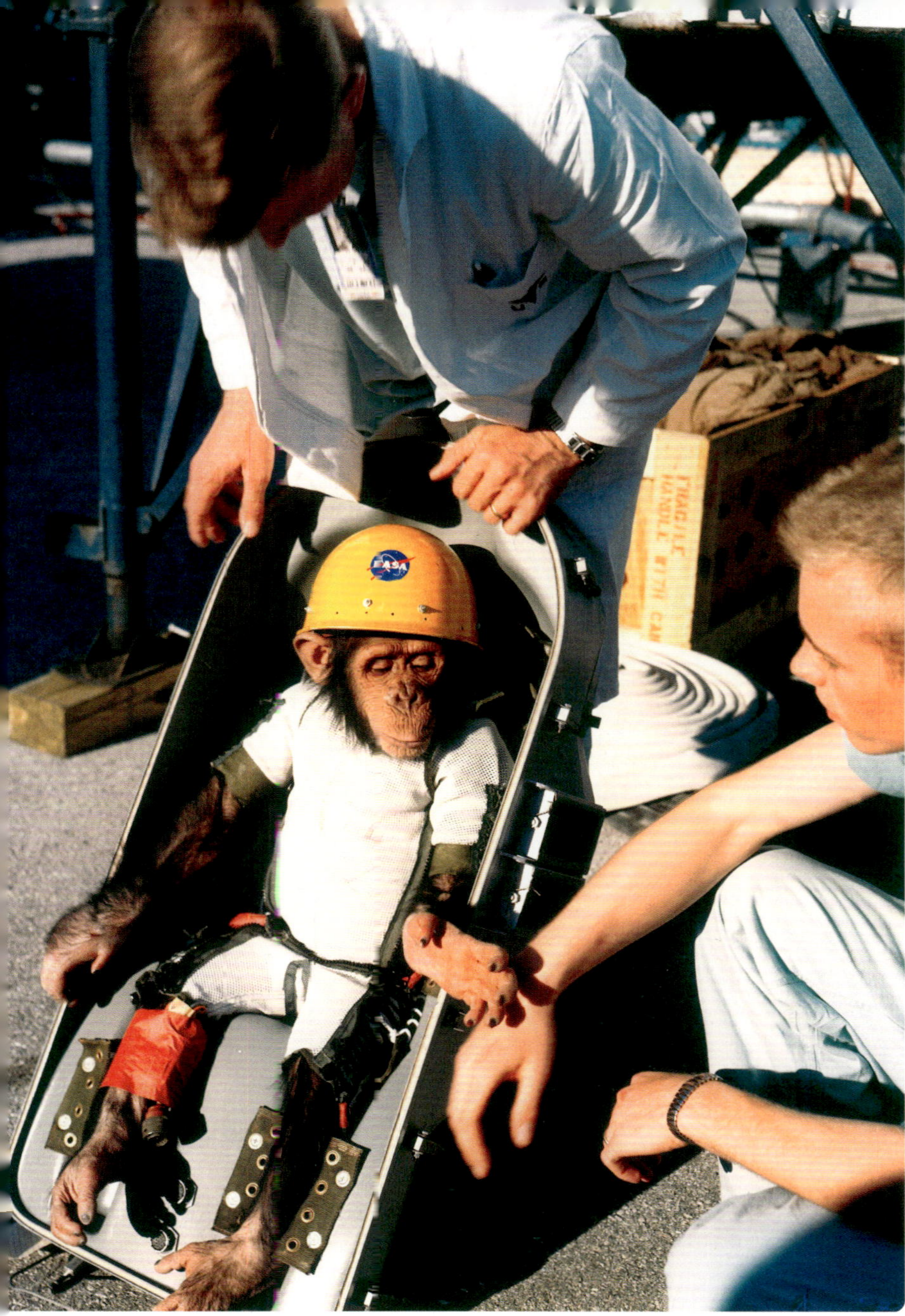

The chimpanzee Ham, the "Astrochimp," was the first hominin to travel to space.

- The name Laika is derived from the Russian verb *layat*, meaning "to bark."
- Due to technological limitations of the time, the Russian space program did not have any plans to bring Laika back alive.
- **SHE DIED IN SPACE A FEW HOURS AFTER LIFTOFF DUE TO OVERHEATING WHEN THE HEAT SHIELD FAILED. HER ONE-WAY MISSION AND TRAGIC FATE CEMENTED HER LEGACY ATOP THE RANKS OF THE MOST WELL-KNOWN DOG HEROES OF ALL TIME.**
- Only female dogs were considered for the first spaceflights. Their anatomy made it easier to design space suits accommodating their waste.
- Laika was followed by four other mammal species before John Glenn went into orbit.
- In 1960, Russian scientists successfully launched a spacecraft containing two dogs, a rabbit, two rats, 40 mice, and 15 containers with plants and fruit flies.
- Once back on Earth, one of the dogs on that 1960 launch had puppies, and one was given to U.S. President John F. Kennedy.
- In January 1961, Ham, a chimpanzee, launched into space and reached an altitude of 157 miles and a speed of 5,857 miles an hour.
- **TESTS ON HAM CLEARED THE WAY FOR ALAN SHEPARD'S FIRST U.S. SPACE LAUNCH LATER THAT YEAR.**
- Ham lived at the National Zoo until 1980 and then was moved to the North Carolina Zoo, where he died in 1983. He was buried in front of the International Space Hall of Fame in Alamogordo, New Mexico.
- The only chimpanzee to reach orbit was Enos, launched by the U.S. in November 1961.
- Enos was the third hominin to achieve orbit, after cosmonauts Yuri Gagarin and Gherman Titov, both in 1961.
- He was the first and only nonhuman hominin to achieve orbit.
- The first cat in space was Félicette, launched by France in 1963.
- Scientists monitored Félicette through electrodes implanted in her head.
- Félicette survived five minutes of weightlessness and returned safely to Earth.
- **IN 1972, FIVE MICE—NICKNAMED FE, FI, FO, FUM, AND PHOOEY—WENT ALONG FOR THE RIDE ABOARD THE APOLLO 17 COMMAND MODULE.**
- Two spiders were launched with the crew of Skylab 3 in July 1973 to test how well they could weave webs in space.
- Upon seeing a chimpanzee returned successfully from space, comedian Dick Gregory remarked of a photo he'd seen: "The caption read, 'Chimp returns.' That was a lie. That was a man that we sent up there, and that's how he looked when he came back."

CURRENT FACT COUNT 2165

EXTREMOPHILES

Life flourishes everywhere on Earth, even in the most hostile of environments. Those life-forms that can endure—and indeed thrive—in the most hellish of hellscapes are called extremophiles. They teach us about the limits of life—as we know it, anyway.

✦ ENTER EXTREMOPHILES

- Scientists have identified Earthbound species that can withstand extremes in temperature, salinity, atmospheric pressure, and radiation. They're called extremophiles.
- **EXTREMOPHILES ARE ANIMALS THAT NOT ONLY SURVIVE IN EXTREME ENVIRONMENTS, BUT PREFER AND THRIVE IN THEM.**
- The existence of extremophiles suggests the possibility of life-forms that could adapt to extreme conditions on other planets.
- To aliens who live comfortably in baths of acid, humans might be considered extremophiles.
- Panspermia is the idea that life can start on one planet, get cast into space on a rock from an asteroid collision, and land on another planet, seeding it with life.
- In evolution, no organism adapts to anything. It either dies or has random, built-in features that the new environment cannot kill and lives another day to propagate its variations.
- To survive a trip though interplanetary space, stowaway life would need native resistance to high and low temperatures, high and low pressure, high radiation, drought, and long intervals of time with no nourishment.
- Any life-form that survives such an interplanetary trip would have to be an extremophile.

✦ THE HUMBLE TARDIGRADE

- One such creature is the tardigrade, also known as a moss piglet or water bear.
- There are some 1,300 species of tardigrades worldwide.
- They are eight-legged creatures about half a millimeter long.

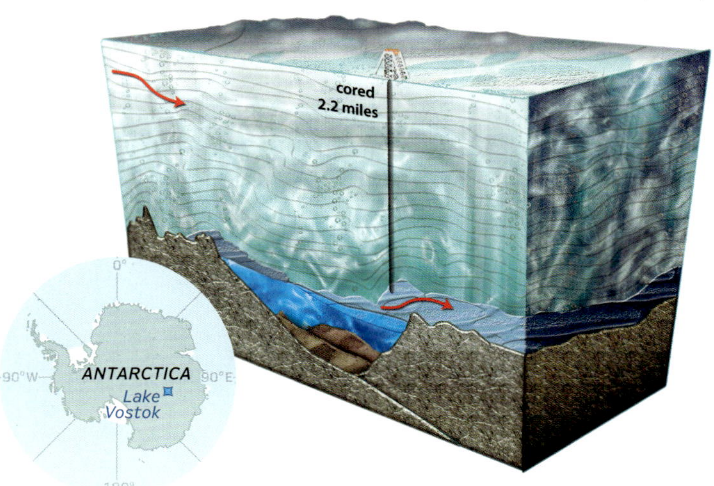

An artist's depiction of Lake Vostok, the largest known subglacial lake in Antarctica

- Tardigrades can withstand extremes of temperature and pressure.
- They can survive as long as 30 years without food.
- In 2007, a Russian spacecraft carried 3,000 living tardigrades into low Earth orbit for 10 days.
- Most of them survived the voyage and went on to birth normal offspring.
- **THUS, THE TINY BUT MIGHTY MOSS PIGLET BECAME THE FIRST ANIMAL TO SURVIVE THE HARD VACUUM OF SPACE.**
- When exposed to radiation, tardigrades excrete a special antioxidant pigment called betalain that protects their DNA from damage.
- Scientists are exploring ways to use this pigment in human cells to protect future long-duration astronauts from radiation.

✦ LAKE VOSTOK

- Lake Vostok, a pristine body of water buried and trapped beneath an Antarctic glacier, has not interacted with Earth's biosphere for at least 15 million years.
- The lake sits 2.5 miles below the surface ice of a glacier in the middle of the Antarctic continent.
- Sunlight, a fundamental source of energy for nearly all life on Earth, does not reach Lake Vostok.
- Stable life in Lake Vostok lives under thermal and chemical conditions found nowhere else on Earth.
- Lake Vostok is an Earth analog to the ice-covered, subsurface oceans of Jupiter's moon Europa and Saturn's moon Enceladus.

Tardigrades—photographed here at a magnification of 300:1—range from 0.1 to 1.5 millimeters.

CURRENT FACT COUNT 2184

ONWARD: BY THE NUMBERS

6

the number of rhesus monkeys named Albert sent up in rockets by the U.S. in the 1950s

848

the number of double stars cataloged by Caroline and William Herschel

9

the number of dogs sent up in Sputnik rockets by the U.S.S.R. in the 1950s

932,000

the orbital distance, in miles, of the James Webb Space Telescope from Earth

463

the mass of the International Space Station in tons

1,000,000

the SPF rating of the James Webb Space Telescope's sun shield

1,554

pounds: the weight of the James Webb Space Telescope's primary mirror

25,000

the number, at minimum, of known pieces of space junk larger than 10 centimeters, or about the size of a softball

17,999,124

total miles traveled by the Apollo missions

24

the number of U.S. astronauts who have flown to the Moon

12

the number of astronauts who have walked on the Moon

47

the weight, in pounds, of Moon rock brought back to Earth by Apollo 11

28,000,000,000

light-years: the distance to Earendel, the farthest star observed by the James Webb Space Telescope

$10,800,000,000

the estimated lifetime cost to NASA for the James Webb Space Telescope

17,000

the speed, in miles an hour, that the Hubble Space Telescope orbits Earth

180

the weight, in pounds, of an Apollo astronaut's suit on Earth. On the Moon, it would weigh 30 pounds.

23

the number of countries that have sent astronauts to the International Space Station

280

the number of people who have traveled to the International Space Station

9

extravehicular activity (EVA) hours spent by two astronauts in 2024 during China's Shenzhou 19 mission—the longest time yet

3

About 4.6 billion years ago, a star exploded, unleashing a shock wave through a nearby interstellar cloud rich with molecules of dust and gas. The cloud collapsed, and in its wake, a new star formed alongside a cohort of planetoid companions. Behold: the birth of our solar system.

Mercury, Venus, Earth, and Mars align in a planetary parade in this composite image of the inner solar system.

PLAN

ETARY

CURRENT FACT COUNT 2192

YOU ARE HERE

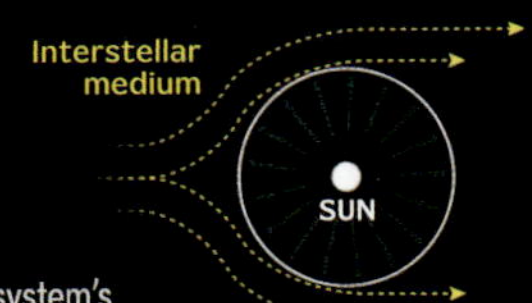

The Sun engulfs the solar system's planets in a bubble of charged particles that repel dangerous radiation and harmful materials originating in interstellar space.

The Milky Way's arms are filled with hazards to habitability, including radioactive clouds, areas of active star formation, and sterilizing blasts from dying stars.

A hidden black hole four million times the mass of the Sun makes the galaxy's heart a dangerous place, with intense bursts of radiation hostile to life.

PROFILE VIEW

Loose stars and roughly 150 dense stellar clusters orbit within the Milky Way's halo. Life-sustaining planets are unlikely here, with the heavy elements too sparse to birth Earthlike worlds.

A 10,000-light-year-wide bulge of dust, gas, and old stars surrounds the core.

Small, rocky planets like ours can't form without elements heavier than hydrogen and helium, which become less common at the far edges of the galaxy.

The solar system is comfortably nestled in a safe harbor between major spiral arms, and its nearly circular orbit helps it avoid the galaxy's perilous inner regions.

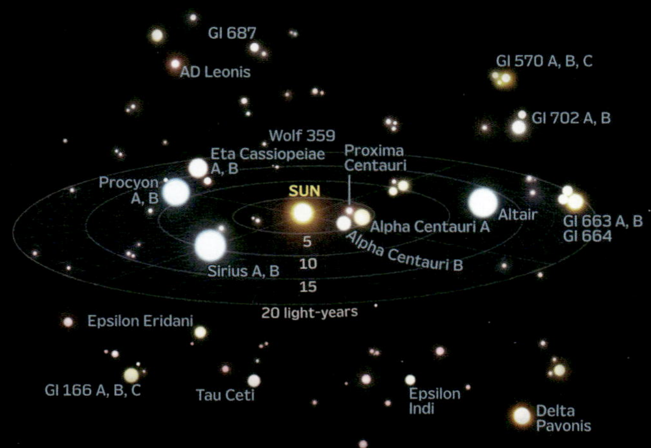

There are relatively few stars near the Sun, which reduces the risks to Earth from gravitational tugs, gamma-ray bursts, or collapsing supernovae.

CURRENT FACT COUNT 2230

THE SOLAR SYSTEM

Our cosmic backyard is home to planets, dwarf planets, hundreds of moons, and likely trillions of asteroids, comets, and other space rocks, all gravitationally bound to one star—and to one another.

✦ OUR INTERSTELLAR BACKYARD

- The word "planet" comes from the ancient Greek word *planetes,* meaning "wanderer," because the orbs appear to move across a fixed starry backdrop.

- **YOU CAN EASILY SPOT FIVE PLANETS WITH THE NAKED EYE: MERCURY, VENUS, MARS, JUPITER, AND SATURN.**

- The ancient Greeks saw and named seven wanderers, lumping the Sun and Moon in with the five visible planets named above.

- You can spot a sixth, Uranus, under pristine viewing conditions without any light pollution (and if you know exactly where to look).

- Although the ancient Greeks likely spotted Uranus, they would have mistaken it for a star, since it is so distant, dim, and moves so slowly across our sky that it appears fixed.

- **MORE THAN 90 PERCENT OF THE KNOWN OBJECTS IN THE SOLAR SYSTEM HAVE BEEN DISCOVERED SINCE THE YEAR 2000.**

- Looking for a day job? Don't even think about working on these planets: On Mercury, one day is almost six Earth months; on Venus, one day is about eight Earth months.

- On the other hand, you might want to consider a day job on these four fast-rotating planets with super-short days: Jupiter (10 hours); Saturn (11); Uranus (17); and Neptune (16).

- But you won't be able to get your feet on the ground on any of those planets, because they have no surface to stand on.

- In our solar system, Venus, Earth, Mars, and Jupiter's moons Ganymede and Europa all have free oxygen in their atmospheres.

- All our planets orbit the Sun in the same direction and on roughly the same plane because they were all birthed by the same rotating disk of gas and dust.

- **OUR SOLAR SYSTEM IS MADE UP OF FOUR INNER ROCKY PLANETS AND FOUR OUTER GIANT PLANETS.**

- Saturn and Jupiter, composed mostly of gaseous hydrogen and helium, are referred to as the gas giants.

- Uranus and Neptune, composed mostly of slushy concoctions of water, ammonia, and other heavier molecules, are called the ice giants.

- Uranus and Neptune were also considered gas giants until the 1990s, when Voyager 2 data revealed their heavier, icy composition hidden beneath gaseous atmospheres.

- The four inner planets and the four outer planets are divided by the asteroid belt.

- **ALL FOUR OUTER PLANETS, ALONG WITH SOME ASTEROIDS, HAVE RINGS.**

- Only Saturn was believed to have a ring until data from Voyager 1 and 2 revealed the ring systems of the other three giants in the late 1970s and '80s.

- Jupiter's rings are so faint that even our high-powered telescopes couldn't discern them from Earth. We needed Voyager 1 and the Hubble Space Telescope to help.

A to-scale artistic rendering of the Sun compared to the planets and moons of the solar system

✦ THE DEETS

- **A 20-YEAR-OLD HUMAN ON EARTH IS THE EQUIVALENT OF:**
  - 83 Mercury years old
  - 33 Venus years old
  - 11 Mars years old
  - 2 Jupiter years old
  - 1 Saturn year old
  - 0.2 Uranus year old
  - 0.1 Neptune year old
- When it comes to planetary size, Earth is about average. Planets from smallest to largest: Mercury, Mars, Venus, Earth, Neptune, Uranus, Saturn, and Jupiter.
- But if we're talking about density, Earth wins. Planets ranked from most to least dense: Earth, Mercury, Venus, Mars, Neptune, Jupiter, Uranus, and Saturn.
- **SOMEONE WHO WEIGHS 150 POUNDS ON EARTH WOULD WEIGH:**
  - 57 pounds on Mercury
  - 136 pounds on Venus
  - 57 pounds on Mars
  - 380 pounds on Jupiter
  - 160 pounds on Saturn
  - 133 pounds on Uranus
  - 169 pounds on Neptune
  - and only 25 pounds on the Moon
- To describe distances within the solar system, astronomers commonly use the "astronomical unit" (AU), defined as the distance between Earth and the Sun, equal to about 93 million miles. For example, Jupiter is 5.2 AUs from the Sun, while Neptune is 30.
- **WE MAY HAVE A HIGHER CHANCE AT FINDING LIFE BEYOND EARTH ON MOONS RATHER THAN PLANETS. TOP CONTENDERS INCLUDE JUPITER'S EUROPA, GANYMEDE, AND CALLISTO, ALONG WITH SATURN'S ENCELADUS AND TITAN.**

CURRENT FACT COUNT 2354

THE SUN

The pulsing heart of our solar system. Cataclysmic and catalytic, lethal and life-giving, our star feeds nearly all life as we know it.

✦ GLOW UP

- The Sun—the first body to form in our solar system—likely started life with dusty rings that condensed to form the planets.

- The Sun is a giant ball of plasma: the fourth state of matter beyond solid, liquid, and gas.

- Plasma can form when a gas becomes so hot, electrons are stripped away from their atoms and float amid the soup of charged particles and now neutral atoms.

- **EVEN IF YOU COULD TAKE THE HEAT, THERE'S NO WAY YOU COULD STAND ON THE SUN. IT HAS NO SOLID SURFACE, BUT RATHER A VIBRATING OUTER LAYER CALLED THE PHOTOSPHERE.**

- The Sun's photosphere writhes slowly in a chaotic pattern. But close-up images make it look scaly, almost reptilian, with burbling pockets called convection cells, each one about the size of Texas.

- The Sun increases its brightness by about 10 percent every billion years, gradually cranking up the heat on Earth (and every other planet). Yet human-induced climate change drastically outpaces what our star is capable of on its own.

- A billion years from now—if anthropogenic climate change hasn't already wiped out most life on our planet—increased solar brightness will begin to boil Earth's oceans.

- While Earth's magnetic field protects us from the Sun's assailment of charged particles, the Sun's own magnetic field—the heliosphere—cocoons our entire solar system, protecting all of the system's orbital bodies from interstellar radiation.

- In a sense, we all live inside the Sun's atmosphere.

- What would happen without the Sun's protective bubble? Two million years ago, that bubble may have shrunk, exposing Earth to a bombardment of heavy radioactive elements. That collision with interstellar radiation might have caused a global cooling event.

- **ASTRONOMERS STILL AREN'T SURE THE EXACT SIZE AND SHAPE OF THE HELIOSPHERE, BUT THEY KNOW IT EXTENDS AT LEAST 11 BILLION MILES BEYOND EARTH.**

- Interstellar space begins at the edge of the heliosphere. Only two probes have successfully reached interstellar space: Voyager 1 and 2.

- The New Horizons spacecraft will cross the interstellar barrier sometime before 2040 and with enough function and power remaining to relay interesting data about the size, scope, shape, and contents of the space beyond our star's protective bubble.

- Pioneer 10 and 11 are both on interstellar journeys, but neither are operational.

- At the Sun's core, energy roils with the same force as the creation and detonation of atomic bombs.

- Meanwhile, forces of mass and gravity threaten to force the Sun to collapse in on itself—a solar balancing act of magnificent proportions.

- **A SINGLE PHOTON OF LIGHT TAKES HUNDREDS OF THOUSANDS OF YEARS TO TRAVEL FROM THE SUN'S CORE TO ITS SURFACE.**

- The sunlight warming your face today started as photons in the Sun's core when Neanderthals were still around.

- The Sun peaks in the visible spectrum—specifically in the blue-green frequency corresponding with the color teal.

- You think the Sun is yellow? Think (but don't look) again. It emits all visible colors, a combination that yields white light.

- Half the energy emitted by the Sun is visible light, a narrow portion of the infinitely long electromagnetic (EM) spectrum and the only range our eyes are able to see.

- No wonder our eyeballs are exquisitely attuned to perceiving this oddly specific range of frequencies. Perhaps aliens around different stars might perceive different wavelengths of the EM, depending on what energy their host stars prefer to emit.

- The Sun looks yellow for the same reason the sky looks blue. As sunlight enters Earth's atmosphere, atmospheric particles scatter high-frequency blue photons more than lower-frequency colors.

> "FLYING THIS CLOSE TO THE SUN IS A HISTORIC MOMENT IN HUMANITY'S FIRST MISSION TO A STAR."
>
> —NASA ADMINISTRATOR NICKY FOX, REGARDING THE PARKER SOLAR PROBE

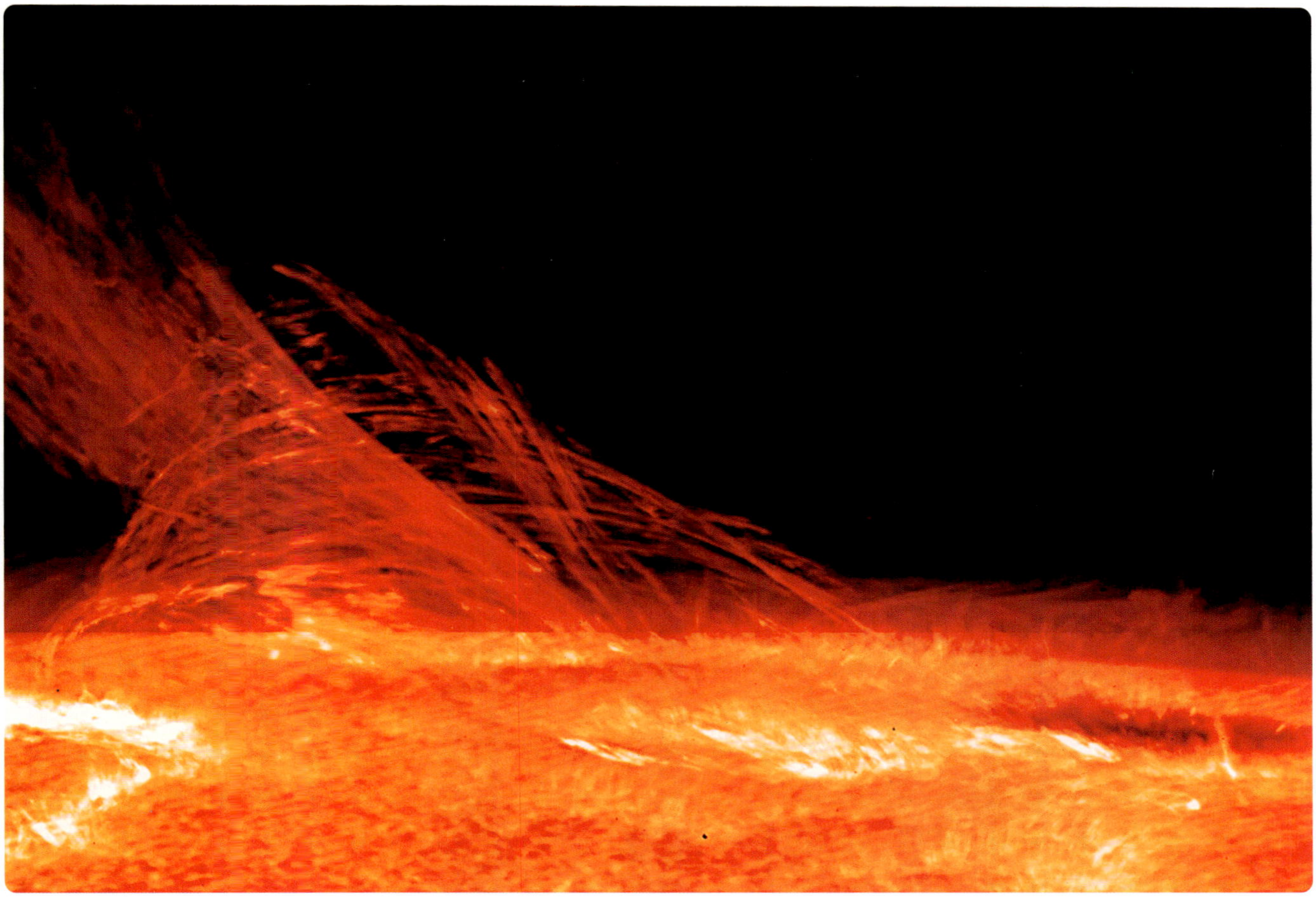

The Hinode (Solar-B) optical telescope took this photograph of the Sun's chromosphere in 2007.

- **THE INTERNATIONAL SPACE STATION ORBITS ABOVE THE ATMOSPHERE, SO TO ITS CREW, THE SUN APPEARS WHITE.**
- When the Sun crests the horizon, its light must traverse 10 times more atmosphere than when it is at its zenith.
- At sunrise and sunset, blue light is scattered away before it can reach our eyes, leaving the sky awash with the gorgeous red and orange hues.
- There are white stars, yellow stars, orange, red, and blue stars, but never green stars. Why? Any star emitting the perfect amount of energy to peak in the narrow visible spectrum containing green light is also emitting as much red light and blue light.
- These colors blend and, like our own Sun, the amalgam of colors is emitted as white light—impossible to pick apart with human eyes alone. Thus, a green star simply appears white.
- Twice a year, the Sun is positioned directly above the Equator. Those two days are the equinoxes.
- The vernal equinox marks the start of spring; the autumnal equinox marks the start of fall.

✦ IT'S 12 O'CLOCK SOMEWHERE

- Could Gary Cooper really meet the outlaws at *High Noon*? "High noon" refers to the precise moment when the Sun is directly overhead.
- But on nearly every day, everywhere on Earth, high noon can't and won't happen.
- **TO MEET AT LITERAL HIGH NOON, YOU WOULD HAVE TO BE AT ONE OF FOUR PLACES ON ONE OF FOUR DAYS OF THE YEAR:**
  - on the Tropic of Cancer (23.5° N) on the summer solstice
  - on the Tropic of Capricorn (23.5° S) on the winter solstice
  - on the Equator on the vernal equinox
  - on the Equator on the autumnal equinox
- During true high noon, the Sun is at its zenith. It briefly shines directly overhead and casts no shadows.

Sunlight glints off tidal creeks at Cape Romain National Wildlife Refuge in South Carolina.

✦ BIG IFS

- The sunrise you witness hasn't happened yet, and the sunset you witness happened a few minutes earlier. That's because the Sun's light refracts when it passes from the vacuum of space into Earth's atmosphere, creating an illusion. Like viewing a straw through a glass of water—the straw submerged in water appears to be where it isn't.

- If the Sun were the size of a hot tub, Earth would be the size of a sesame seed.

- You could pour a million Earths into the Sun.

- **THE SUN IS 1,000 TIMES MORE MASSIVE THAN ALL THE OTHER PLANETS IN OUR SOLAR SYSTEM *COMBINED.***

- If the Sun were the size of a beach ball, Neptune, the most distant planet, would still sit a mile away.

- **IF THE SUN WERE AN EXERCISE BALL, THEN:**
  - Jupiter would be a softball.
  - Saturn would be a baseball.
  - Earth and Venus would be peas.
  - Uranus and Neptune would be walnuts.
  - Mars would be a peppercorn.
  - Mercury would be a sesame seed.
  - Pluto would be a poppy seed.

- The Sun has occasional blemishes known as sunspots: evidence of magnetic eruptions.

- Sunspots look dark only because they are cooler than the surrounding area of the Sun's surface. If you plucked them off the Sun and plopped them in empty space, they would visibly glow unto themselves.

- The Sun rotates in the same direction as our home planet: counterclockwise.

- **BECAUSE THE SUN IS MADE OF PLASMA, ITS POLES ROTATE FASTER THAN ITS EQUATOR. IT TAKES 25 DAYS TO ROTATE AT THE EQUATOR BUT 35 DAYS AT THE POLES.**

- As the gaseous layers revolve at different rates, the magnetic fields get stretched, twisted, and disturbed.

- The expulsion of energy from thermonuclear fusion at the Sun's core forcibly ejects some charged particles into space.

- The steady, continuous outflowing of the Sun's charged particles is called solar wind.

- Sometimes solar wind flows out of the Sun, not in a steady, casual stream but as a spontaneous, explosive gasp.

- Solar flares are bursts of energy equivalent to millions of hydrogen bombs traveling at the speed of light in all directions.

- Fortunately, Earth's magnetic field absorbs most solar flare impacts.

- A massive solar flare, however, can trigger a much more dangerous solar event called a coronal mass ejection (CME).

- CMEs are gigantic blobs of charged plasma particles that are flung into space at a more leisurely pace than solar flares, reaching Earth in one to three days.

- CMEs can contain a billion or more tons of matter and are capable of disrupting Earth's magnetic field.

- **WHAT IF A CME FROM THE SUN WERE TO HIT EARTH?**
  - A powerful CME could throw society into chaos. It could short-circuit entire power grids and communication satellites, fry electronics, melt transformers, and plunge cities into total blackouts.
  - A storm could last hours or days. The repercussions could last years.
  - In 1859 the strongest geomagnetic storm recorded caused telegraphs around the world to spark, and auroras (the northern lights) were seen as far south as Cuba.
  - In August 1972, a dozen sea mines near North Vietnam mysteriously detonated. The suspect? A powerful solar storm.

- Solar prominences, also called solar filaments, are elongated loops of plasma extending hundreds of thousands of miles beyond the Sun's photosphere.

- Scientists are still unsure what causes them.

- All the violence on the Sun's surface—prominences, CMEs, flares, sunspots—are by-products of magnetic fields gone awry.

- Earth's magnetic field captures the Sun's errant charged particles and funnels them toward the North and South Poles. There, the particles collide with charged particles in our atmosphere and render them aglow. Behold, the auroras borealis and australis.

- **THE SUN'S MAGNETIC POLES REVERSE EVERY 11 YEARS OR SO. WHEN'S NEXT? 2035. AND 2046 AFTER THAT.**

- Before a solar magnetic pole flips, the Sun's magnetic activity kicks into high gear, creating intense auroras on Earth.

- Helium—an element named after Helios, Greek god of the Sun—is the second most abundant element in the universe.

- Helium was the first element discovered in space, via astronomical spectroscopy, by which the nature of distant objects is studied thanks to the light waves they emit.

A first-century B.C. bust of Helios

✦ FEELING HOT, HOT, HOT

- The Sun's surface is a scorching 10,000°F.
- **IT'S HOT ENOUGH TO NOT JUST MELT, BUT TO BOIL, DIAMONDS.**
- The Sun's corona—its outermost gaseous layer—is hotter than its surface. Scientists still wonder why.
- The word "corona" is Latin for "crown." During a total solar eclipse, the bright, jagged edges of this otherwise invisible outer layer are on full display.
- **THE CORONA IS SO HOT THAT IT STRIPS ELECTRONS FROM ITS GASES, LEAVING THE ELEMENTS AS BARE NUCLEI.**
- Solar flares are also hotter than the Sun's surface, reaching up to 3.5 million degrees Fahrenheit.
- As the Sun ages, its luminosity slowly rises, so that in a billion years, Mars could become as warm as Earth.
- **TWO 19TH-CENTURY THEORIES FOR WHAT FUELS THE SUN:**
    - First, that meteorites continually pelted the Sun and this mechanical energy was converted into heat.
    - This was quickly dismissed, however, after calculating that the Sun's output would require a total mass of meteors equal to the mass of Earth to bombard the Sun every 100 years.
    - Second, that the Sun was a huge, actively burning chunk of coal.
    - This theory, too, was discarded after geologists accurately calculated the age of Earth, which far exceeded the time required for the Sun to burn through the amount of coal required to maintain its heat output.
- **THE SUN TURNS 4,250,000 TONS OF MATTER (NEARLY ONE GREAT PYRAMID) INTO ENERGY PER SECOND.**
- Half of the Sun's mass is in its core. But its core only makes up about 1.5 percent of its volume.
- The Sun's core is gaseous, but it's so tightly compressed that it is 13 times denser than lead.
- The core can reach nearly 30 million degrees Fahrenheit. To take the Sun's temperature, NASA built a probe using materials resistant to extreme heat: niobium wires encased in sapphire crystal.

✦ THE PARKER SOLAR PROBE

- Launched in 2018, NASA's Parker Solar Probe currently orbits the Sun.
- The probe brushed our star in 2021 when it entered the corona, the first probe to do so.
- It was named for Eugene N. Parker, who first hypothesized solar winds in the 1950s.
- With the launch of the probe, Parker became the first person to witness a spacecraft named after him. He died in 2022.
- The Parker Solar Probe's heat shield is a carbon-composite shield 4.5 inches thick coated in Sun-reflecting white ceramic paint.
- Its instruments are made of tungsten, a metal that can withstand temperatures above 6000°F.
- The probe will count electrons, protons, and helium ions in solar winds, measuring their velocity, density, and temperature.
- **THE PARKER PROBE COMPLETED ITS PRIMARY SCIENCE MISSIONS IN 2025 AFTER ITS 25TH CLOSE APPROACH AROUND THE SUN.**
- It is the fastest artificial object, traveling at 120 miles a second, or about 430,000 miles an hour. That's still less than one-twentieth of one percent the speed of light.
- Although the Sun will incinerate most of the probe's components, the heat shield will remain intact, left to orbit the star for millions—or even billions—of years.
- **IF WE COULD TRAVEL AT THE SAME SPEED AS THE PARKER SOLAR PROBE, A TRIP TO OUR CLOSEST STAR SYSTEM, ALPHA CENTAURI, WOULD TAKE HUMANS 67,000 YEARS.**

✦ IF EVIL ALIENS SNATCHED AWAY OUR SUN ...

- Earthlings wouldn't know for 8 minutes 20 seconds (the time it takes for a photon from the Sun to reach us as well as any consequences of gravity).

- Earth and all other planets would fly off at a tangent into interstellar space (of course, with the Sun gone, the entire solar system would now be interstellar space!).

- **WHAT WOULD HAPPEN ON EARTH?**
  - Earth would be plunged into cold and darkness with nothing but the distant lights of the Milky Way galaxy to illuminate our world.
  - Photosynthesis would cease, but our atmosphere would remain intact.
  - Earth's core would continue to churn, warming the planet's crust with geysers and volcanoes.
  - Eventually, even the oceans would ice over, leaving frigid realms ruled by the extremophiles who thrive on hydrothermal vents.

- What if evil aliens turned the Sun into a black hole? Earth would still be plunged into darkness and cold, but it would remain in orbit exactly as it was before.

- Keep in mind: Oxygen liquefies at minus 300°F and freezes at minus 350°F.

- So, if our atmosphere reached such chilling temperatures, it's conceivable that populations would need to scoop up oxygen from the ground and heat it into a breathable gas. (A 1951 short story, "A Pail of Air" by Fritz Leiber, explored this very concept.)

✦ PROPULSION BY STARLIGHT

- In 2010, a 4.5-pound spacecraft unfurled a 2,000-square-foot sheet of highly reflective and extremely thin plastic in the vacuum of space. The JAXA spacecraft, called IKAROS (Interplanetary Kite-craft Accelerated by Radiation Of the Sun), became the first spacecraft propelled primarily by solar sails.

- **THE NAME IKAROS IS PRONOUNCED LIKE ICARUS, AN HOMAGE TO THE ANCIENT GREEK MYTH.**

- When light hits any object, it transfers momentum onto that object according to Newton's third law of motion: For every action (or force applied), there is an equal and opposite reaction.

- Small spacecraft can take advantage of this force to propel themselves through space with solar sails.

- Solar sail membranes catch photons like sailboat sails catch wind.

- **THE TRANSFER OF ENERGY FROM A PHOTON TO AN OBJECT IS CALLED RADIATION PRESSURE, OR THE FORCE OF LIGHT.**

- In the vacuum of space, radiation pressure affects objects more than any force except gravity.

- Over time, the radiation pressure will affect the trajectory of spacecraft and satellites, so orbital engineers must factor it into their calculations.

- The force imparted is so teensy—about two-millionths of a pound per 1,000 square feet of sail—that you won't notice right away.

- But in the vacuum of space, there are minimal forces acting to slow it down. Newton's first law tells us that an object in motion stays in motion unless acted upon by an outside force. So, every photon that hits a solar sail imparts more and more energy with nothing to stop it.

- **BECAUSE STAR POWER IS FREE, A SPACECRAFT NAVIGATING WITH SOLAR SAILS NEEDN'T CARRY ANY FUEL.**

- Over time, a solar sail could hypothetically zip through space at nearly 70 million miles an hour.

- Johannes Kepler first proposed this concept after noticing that the tails of comets always point away from the Sun, concluding that the Sun itself must impart that force. In a 1610 letter to Galileo, Kepler wrote, "Given ships or sails adapted to the breezes of heaven, there will be those who will not shrink from even that vast expanse."

CURRENT FACT COUNT 2376

THE SUN: BY THE NUMBERS

99.8

the percentage of the solar system's mass taken up by the Sun

93

million: the distance into space, in miles, you peer across if you look at the Sun. (But you'll also burn your retina, so don't do that.)

1

million miles an hour: the approximate speed at which the solar wind billows from the Sun. It could get from New York to China in less than 30 seconds.

20

years old: the technical age of the Sun, because it has only completed about 20 orbits around the Milky Way since its birthday

153

years: how long it would take to drive a straight path from Earth to the Sun at your typical highway speed of 70 miles an hour

330,000

the number of Earths it would take to equal the mass of the Sun

230,000,000

Earth years: how long it takes the Sun to orbit the center of the Milky Way

4×10^{26}

the wattage of the Sun if it were a light bulb

400,000

how many times brighter the Sun is than a full moon

100

the magnitude greater the Sun's diameter is compared to Earth's. (It's still just a medium-size star.)

250

the thickness of the photosphere, or the Sun's "surface," in miles. Earth's surface crust, by contrast, averages about 25 miles.

11

Earth days: the length of one day at the Sun's poles

4.57

billion years: the age of the Sun, making it a middle-aged star

100,000

miles across: the span of the largest sunspots, or 12 Earth diameters (but they're typically just the size of Earth itself)

140

miles a second: the speed at which the Sun orbits the center of the Milky Way

25

Earth days: the length of one day on the Sun compared to the Earth at its Equator

170,000

light-years: roughly the distance of the Sun's single orbit around the Milky Way

10

billion tons: the total mass of particles spewed into space by a single coronal mass ejection

1

million years: the length of time it can take for a single photon to escape from the Sun's core. Afterward, it takes just 500 seconds to reach Earth

3.8

million miles: the closest approach an artificial object has ever made to the Sun. The Parker Solar Probe made its closest approach in December 2024.

4

million tons: the amount of matter the Sun converts into energy per second

10

the percentage of the Sun's inner core in which thermonuclear conversion takes place

CURRENT FACT COUNT 2444

MERCURY

A world of extremes and eccentricities

✦ THE CLOSEST PLANET TO THE SUN

- Barely larger than Earth's Moon, Mercury is the smallest planet in the solar system.
- It's smaller than Jupiter's moon Ganymede and Saturn's moon Titan.
- **MERCURY IS THE LEAST EXPLORED OF THE FOUR INNER PLANETS.**
- Due to its proximity to the Sun, it is more fuel-efficient for a spacecraft to escape the solar system entirely than to visit Mercury.
- Only three spacecraft have ever attempted a visit: NASA's Mariner 10 flybys, NASA's MESSENGER orbital mission, and the European Space Agency and Japan Aerospace Exploration Agency's dual orbital probes called BepiColombo.
- No mission has ever attempted to land on the planet—although the MESSENGER probe crashed there after fulfilling its multiyear orbital mission and collecting hundreds of thousands of images.
- **MESSENGER LEFT BEHIND A 52-FOOT-WIDE IMPACT CRATER ON MERCURY'S SURFACE.**
- The ancient Romans named the planet after their swift-footed messenger god because of how quickly the planet moved across the sky compared with any other.

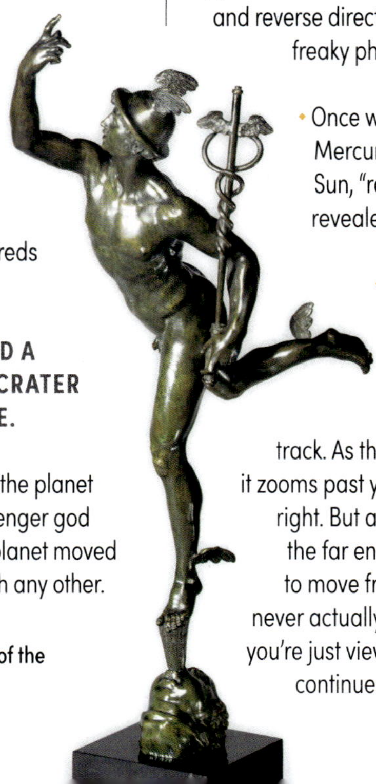

A circa 1800–1830 bronze statue of the god Mercury

- Babylonians called the planet Nabu after their god of learning and wisdom.
- Babylonian astronomers documented the retrograde motion of Mercury—when it appears to reverse directions across the sky—in the seventh century B.C.
- **IS RETROGRADE REAL? WELL, IT'S A VERY REAL OPTICAL ILLUSION. MERCURY DOESN'T ACTUALLY MOVE BACKWARD, EVER.**
- Our perspective of Mercury's motion from our orbit creates the illusion. When we see it reverse directions, it's simply continuing its orbit around the backside of the Sun.
- To ancient astronomers, who believed Earth was the center of all motion, Mercury appeared to move across the sky, slow down, and reverse direction—an understandably freaky phenomenon.
- Once we understood that Mercury (and Earth) orbit the Sun, "retrograde motion" was revealed for the illusion it is.
- Another way to understand retrograde: Imagine watching a race car drive around a circular track. As the car passes the stands, it zooms past you from your left to your right. But as it continues the loop at the far end of the track, it seems to move from right to left. The car never actually changed direction; you're just viewing its path as it continues around the track.

- Mercury follows the most elongated path among the eight planets, with almost 15 million miles between its closest and farthest points from the Sun.
- **MERCURY IS SO SMALL YOU COULD FIT 18 COPIES OF IT INSIDE EARTH.**
- But it's still twice Pluto's diameter.
- Mercury orbits so close to the Sun that you can only see it during dawn and dusk.
- For the best view of Mercury, look for it during a total solar eclipse.
- Ancient Greek and Roman astronomers thought Mercury's morning and evening appearances were the glow from two different stars.

✦ MERCURY'S EXOSPHERE

- Mercury has the thinnest atmosphere of all the planets—it's nearly nonexistent. Its thin smattering of particles is really an exosphere.
- An exosphere is the outermost portion of a planet's atmosphere, where particles escape easily into space. Exospheres are difficult to discern from space itself.
- For context, Earth's exosphere begins hundreds of miles above the Kármán line, the internationally accepted boundary of outer space.
- Mercury's thin atmosphere means that despite its proximity to the Sun, nighttime temperatures can plunge to minus 290°F.
- With little air through which to scatter sunlight, Mercury's daytime sky would not look blue, but rather more like our night sky: black and star-studded.
- Mercury's daytime temperatures are hot enough to melt lead (it's still not the hottest planet in our solar system; that's Venus).

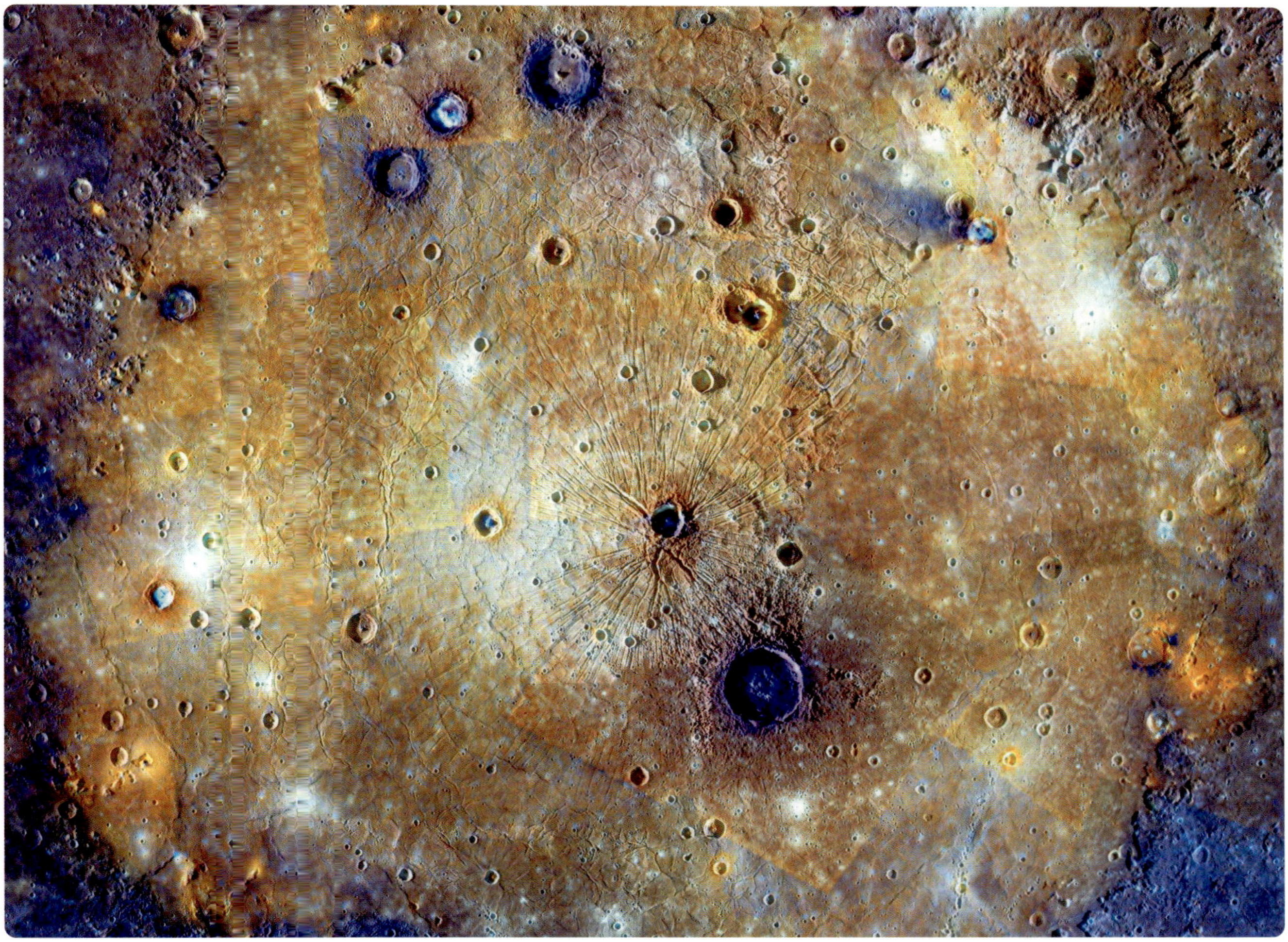

Mercury's youngest large-impact basin, Caloris, with its lava floods in orange and material dredged up from below the surface in blue

- Mercury's exosphere—mostly oxygen, sodium, hydrogen, helium, and potassium—is continuously blasted off its surface by solar winds and meteor strikes.

- While Earth has a strong magnetosphere (a protective bubble that deflects charged particles), Mercury has a weak one, so solar winds constantly bombard it.

- **MERCURY'S MAGNETIC FIELD IS LOP-SIDED: IT'S ALMOST THREE TIMES AS STRONG IN THE NORTH AS THE SOUTH.**

- With little atmosphere to protect it, the planet's surface has been pockmarked by the impacts of space rocks.

- On Mercury, the Sun would look three times larger than it does from Earth.

- **MERCURY IS ALL HEART: ITS METALLIC CORE MAKES UP MORE THAN HALF ITS VOLUME.**

- Mercury's surface might look a bit like our Moon's: gray, rocky, and cratered.

- Cliffs up to a mile high and hundreds of miles long formed on the planet as it cooled and contracted while forming.

- Mercury's colossal impact crater, Caloris Basin, is so large it would reach from the tip of Florida to Washington, D.C.

- The asteroid that made Caloris Basin slammed into Mercury so hard, the seismic waves created hills and fractures on the opposite side of the planet.

- The rugged landscape made by an impact's seismic waves is called "weird terrain."

- Mercury can experience magnetic tornadoes, when searing hot solar wind plasma funnels down to its surface.

- **TO EXPLAIN ANOMALIES IN MERCURY'S ORBIT, 19TH-CENTURY ASTRONOMERS INVENTED THE PLANET VULCAN, WHOSE GRAVITY WOULD ACCOUNT FOR DEVIATIONS FROM NEWTON'S LAWS.**

- In 1916, Albert Einstein published his general theory of relativity, which accounted for Mercury's anomalous orbit naturally, without the help of planet Vulcan.

- Vulcan was discarded immediately.

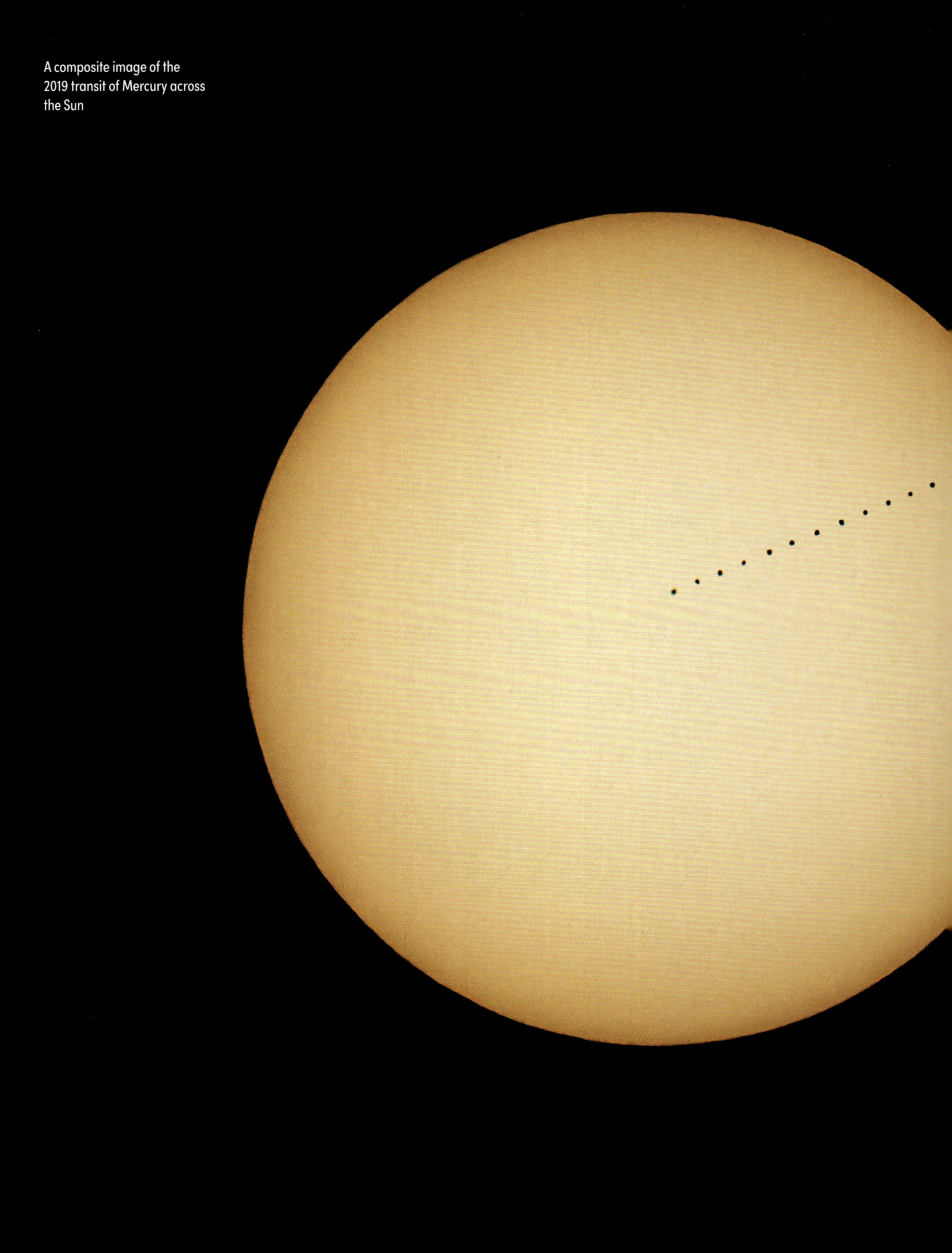

A composite image of the 2019 transit of Mercury across the Sun

✦ ALL YOUR SENSES

- On Mercury, just like in outer space, no one can hear you scream. There's not enough atmosphere for sound waves to propagate.

- On Mercury, sunlight shines 11 times brighter on the planet's surface compared to Earth's.

- Mercury only rotates three times for every two times it orbits around the Sun.

- In other words, one day on Mercury is more than twice as long as its year.

- Mercury is moving so fast that it could travel from New York City to Los Angeles in less than 85 seconds.

- If the Sun were to vanish, Mercury would be the first to notice, just 3.2 minutes later.

- If you lived on the planet, you'd see just one sunrise every two Mercury years. And if you lived on some parts of Mercury's poles, you'd be in permanent darkness.

- **MERCURY HAS NEARLY NO AXIAL TILT—IT SPINS PERFECTLY UPRIGHT—AND SO IT DOES NOT EXPERIENCE SEASONS THE WAY EARTH DOES.**

- Instead, it has thermal seasons: fluctuations of its global temperature dictated by its proximity to the Sun throughout its orbital year.

- Mercury's "summer" occurs during its perihelion (closest approach) and "winter" during aphelion (farthest distance).

- For comparison, Earth's perihelion occurs in early January. That's peak winter in the Northern Hemisphere.

- But because our orbit is nearly a perfect circle, there's no significant difference in temperatures during our closest point of approach or farthest point of retreat from the Sun.

- Mercury's orbit is more elliptical than Earth's, and therefore more vulnerable to temperature fluctuations throughout its orbit.

- At its aphelion, Mercury is 50 percent farther from the Sun than at its perihelion. Earth is a mere 3.4 percent farther at aphelion than at perihelion.

- Mercury's highly elliptical orbit doesn't just result in temperature fluctuations, but also orbital speed fluctuations that result in fascinating and bizarre phenomena.

- For example, at its perihelion, Mercury moves nearly twice as fast through its orbit than it does at aphelion, while its rotation speed remains steady.

- **IF YOU STOOD ON MERCURY'S EQUATOR DURING ITS PERIHELION:**
  - The Sun would rise, set, and rise again within the same day.
  - The Sun itself would go into retrograde during the day. Mercury is the only planet in the solar system where one could witness this illusion.
  - The Sun would appear to hover at the highest point of the sky for weeks on end (according to Earth time) before moving eastward, then reversing direction once again before setting in the west.
  - The Sun would appear three times larger in the sky than it would at its aphelion.
  - Temperatures on the equator would reach 800°F, while the same region only reaches to just above 500°F during aphelion.
  - Meanwhile, Mercury's north and south poles, where the Sun can never reach, are perpetually shadowed in darkness.

- **THE ONLY PLANETS WE CAN SEE IN SILHOUETTE AS THEY MOVE ACROSS THE SUN ARE MERCURY AND VENUS.**

- This is known as a transit.

- Mercury transits are rare, happening only about 13 times each century.

- Deep below Mercury's surface may be a layer of diamond 10 miles thick.

CURRENT FACT COUNT 2455

MERCURY: BY THE NUMBERS

29

miles a second: the speed of Mercury in orbit around the Sun

88

Earth days: the length of one year on Mercury

800°F

the daytime temperature on Mercury

2

the degrees from vertical of Mercury's axial tilt

70

the percent of Mercury's mass composed of the element iron, earning it the nickname "Cannonball"

-290°F

the nighttime temperature on Mercury

176

Earth days: the length of one day on Mercury

0

the number of moons orbiting Mercury

13

the average number of times per century that Mercury transits, passing directly between Earth and the Sun

29,000,000

miles: the closest distance that Mercury gets to the Sun during its elliptical orbit

43,000,000

miles: the farthest distance that Mercury journeys from the Sun during its orbit

Viewers experience a total solar eclipse at the Novosibirsk Reservoir in Siberia, with Mercury visible just above the Sun.

CURRENT FACT COUNT 2518

VENUS

A hellish world with a cautionary tale

✦ EARTH'S SISTER PLANET: A WORLD OF EXTREMES

- Venus is the brightest natural object in the night sky after the Moon, thanks to its highly reflective, thick blanket of clouds.

- **VENUS IS ALSO THE HOTTEST PLANET IN OUR SOLAR SYSTEM.**

- The temperature on the surface is hot enough to melt lead and zinc.

- Venus's clouds rain sulfuric acid, but the planet's hot temperatures vaporize the drops before they reach its surface.

- While Venus is a hellish world, it's often called Earth's twin, because the two are remarkably similar sizes and densities.

- **VENUS MAY BE A CAUTIONARY TALE FOR EARTH: IT ONCE ENJOYED AN EARTHLIKE CLIMATE, BUT A RUNAWAY GREENHOUSE EFFECT LED TO SOARING TEMPERATURES.**

- What kick-started Venus's runaway greenhouse effect? One hypothesis proposes a series of cataclysmic volcanic eruptions pushed Earth's twin down a path toward the scalding hellscape it is now.

- It has more than 1,600 major volcanoes—the most of any planet in the solar system.

- Some of Venus's volcanoes might still be fizzing with activity. In 2024, scientists combed through 30-year-old imagery and noted bright spots that might be fresh lava.

- Venus's thick, carbon dioxide–rich atmosphere holds so much heat that its surface temperatures stay hot both day and night.

- Even the most advanced protective gear available today couldn't keep you safe in the searing temperatures.

- **TEMPERATURES ARE SO HIGH THAT EVEN ROBOTS STRUGGLE TO FUNCTION FOR EXTENDED PERIODS.**

- If you could stand on Venus's surface, you'd be instantly crushed to death under the pressure of the planet's thick atmosphere.

- This pressure is equal to that of a horse balancing on your big toe.

- The dense atmosphere allows sound to travel fast while lowering the pitch, so if you could talk on Venus, you'd sound like a duck.

- The longest any spacecraft has survived the high temps and crushing pressure is a little more than two hours.

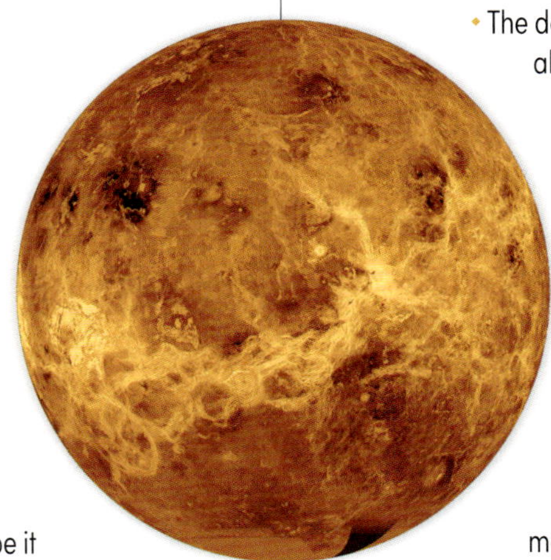

A view of the entire planet, mapped using a computer-simulated globe

- In 1982, Venera 13 visited the planet. It was only designed to withstand the conditions for 32 minutes.

- Venus has no seasons, thanks to a nearly circular orbit and the barely noticeable tilt of its rotation—just three degrees.

- Hurricane-force winds whip through its upper atmosphere, hurtling the clouds around the planet every four Earth days.

- **VENUS'S SUPERFAST WINDS ARE GETTING EVEN FASTER, INCREASING FROM 186 MILES AN HOUR TO 249 MILES AN HOUR IN 10 YEARS.**

- Lightning might flash on Venus, but whether the phenomenon is from atmospheric electricity or flashes from meteors is still debated.

- Heavy-metal snow may fall on Venus's mountaintops, creating frosty metallic snowcaps.

- Some of Venus's volcanoes look like "arachnoids," with weblike networks of cracks.

- Other odd volcanoes look like ticks, their domes crowned with wispy lava flows that evoke bug legs.

- Like Earth, Venus also has its own form of continents, called Tesserae, that may have formed in the same way as ours and resemble how we believe early Earth continents may have looked.

- Unlike Earth, Venus's Tesserae never experienced plate tectonics.

- If you filled a football to NFL regulations in the high-pressure Venusian atmosphere of 13 pounds per square inch and then brought it back to Earth, it would burst instantly.

- The correct Latin and English adjective for something or someone from Venus isn't really Venusian. It should be Venereal.

- As that word was already taken by the medical profession, astronomers had to agree on a new adjective.

A color-coded perspective view of Venus shows part of the lowlands to the north of Ovda Regio plateau.

- An artist once misinterpreted the name of Venus's quasi-moon 2002VE as Zoozve on a children's poster—and now Zoozve is its official name.

✦ LIFE IN THE CLOUDS

- If microbial life could survive the hellish Venus environment, it would probably live in the clouds around 30 miles up, where temperatures are about 140°F and pressures are similar to Earth's at sea level.
- In 2020, scientists announced the discovery of phosphine—a biomarker of life on Earth—within Venus's thick atmosphere.
- The finding bolstered a hypothesis that Venus's temperate clouds may harbor microbial life.
- **BUT THE EXCITEMENT WAS SHORT-LIVED. A REANALYSIS SUGGESTED THAT THE DATA WERE INITIALLY MISREAD, THOUGH THE POSSIBILITY OF LIFE COULDN'T BE RULED OUT.**
- If phosphine is present in Venus's atmosphere, does that mean there's alien life? Not necessarily. Chemical reactions from high pressures and significant volcanic activity could create phosphine. NASA's next mission to Venus, scheduled to arrive in the early 2030s, will seek answers as it further investigates the volcanic activity and hunts for phosphine in the atmosphere.
- Although Venus's clouds mimic pressures and temperatures on Earth's surface, that's about where the comparison ends.
- A microbe surviving in the clouds would need to endure extremes that might kill even the hardiest of extremophiles on Earth.
- The clouds are composed mostly of sulfuric acid with very little to no available water vapor.
- Because Venus lacks the protective ozone layer found in Earth's atmosphere, a microbe on the planet would have to withstand both extreme acidity in addition to high radiation levels.

The 2012 transit of Venus, which lasted 6 hours 40 minutes

✦ SHINE BRIGHT LIKE A DIAMOND

- Venus, the brightest object in our night sky, is often the first point of light you see at night—if you mistake it for the first star you see tonight, don't expect your wish to come true.

- **VENUS IS SO BRIGHT THAT UNDER THE RIGHT CIRCUMSTANCES, YOU CAN SEE IT DURING THE DAY.**

- At its faintest, it outshines the brightest star in the night sky, Sirius, by almost 10 times.

- At its brightest, it outshines Sirius by almost 25 times.

- Venus's brightness once led many people to mistake the planet for a UFO, including Jimmy Carter while he was governor of Georgia.

- During WWII, the U.S.S. *New York* spotted the glint of what they thought might be a Japanese balloon weapon and fired 300 rounds at the soaring object. Turns out they were shooting at Venus.

- What if you held a compass on Venus? It might only occasionally—and erratically—spin, as the planet has a weak induced magnetic field.

- Venus and Mercury are the only planets we're able to see in silhouette as they move across the Sun (called a transit). They look so small during solar transits that they're only visible using a telescope or camera with a proper filter.

- **VENUS'S TRANSITS HAPPEN IN A PAIR OF PAIRS: TWO TRANSITS ARE SEPARATED BY EIGHT YEARS, THEN 121 YEARS PASS BEFORE ANOTHER PAIR OF TRANSITS HAPPEN SEPARATED BY EIGHT YEARS.**

- This pattern then repeats every 243 years.

- The two inner planets, Venus and Mercury, are only seen just before sunrise or just after sunset.

- Venus and Mercury are alike in another way: They're the only planets in our solar system with no moons.

- Not all planets spin the same way. Venus and Uranus both turn clockwise, the opposite direction of others in our solar system.

- On these planets, the Sun rises in the west and sets in the east.

- Venus rotates so slowly that it takes upwards of eight hours for the Sun to rise over the planet's horizon.

- Long ago, a giant asteroid likely rammed into Venus so hard it tipped the planet upside down, causing its apparent backward spin.

- Venus waxes and wanes in the night sky like our Moon, but its phases come from our changing view of the planet as it orbits the Sun.

- When Venus appears "full" in our sky, the planet is tiny, as it's at the farthest position from Earth, opposite the Sun.

- But during Venus's closest approach to Earth, the planet looks like a crescent shape.

- Because of its relative distance to Earth during its phases, Venus appears brighter as a crescent than it does when it's full.

✦ THE VENERA PROGRAM

- **THE SOVIETS' VENERA PROGRAM SHAPED OUR UNDERSTANDING OF VENUS WHILE PROPELLING PLANETARY EXPLORATION THROUGH A SERIES OF "FIRSTS":**
  - In 1967, Venera 4 was the first probe to enter the atmosphere of another planet.
  - In 1970, Venera 7 became the first probe to soft-land on another planet.
  - It also transmitted the first data from the surface of an alien planet.

CURRENT FACT COUNT 2524

VENUS: BY THE NUMBERS

584

Earth days: a full cycle of Venus's phases

21

years: the span of time across which the Babylonians logged Venus's appearances, one of the oldest surviving astronomical records

1962

the year of humanity's first successful space probe to fly by another planet: the Mariner 2 pass of Venus in December

93

how many times denser Venus's atmosphere is than Earth's, making it the densest among the solar system's rocky planets

243

Earth days: the length of one day on Venus; a year on Venus is only 225 Earth days, so a day lasts longer than a year there

867 DEGREES

the temperature, in Fahrenheit, after Venus's thick atmosphere of carbon dioxide caused a runaway greenhouse effect

Mount Wilson Observatory captures the conjunction of the Moon, Venus, and Jupiter

CURRENT FACT COUNT 2601

EARTH

The world where life found a way

✦ OUR HOME PLANET

- Earth is the only planet in our solar system not named after a Greek or Roman god or goddess.
- In most languages, the name of our planet comes from a word meaning "ground" or "soil."
- The English word "earth" comes from the Old English word *eorth*, meaning "ground."
- **EARTH HAS SEVERAL QUASI-MOONS: ASTEROIDS TUGGED BY EARTH'S GRAVITY AS THEY ORBIT THE SUN.**
- The famous Route 66 highway that crosses the U.S. is longer than the distance from Earth's surface to the outer edge of its core.
- Earth is not flat, but it's also not perfectly round, bulging around the Equator—a result of its rotation.
- Want evidence Earth isn't flat? If you lie on the ground to watch the Sun set below the horizon, and in that instant stand up, allowing you to see beyond your previous horizon, you will see the Sun rise backward for a few seconds before once again setting.
- Earth is 26 miles wider at the Equator than pole to pole. We call this slight pear shape an oblate spheroid.
- The thickness of Earth's bulge is akin to a piece of duct tape around a volleyball.
- Mount Chimborazo in Ecuador is the 39th highest peak in the Andes Mountains, but its location near Earth's oblate Equator means that its summit is the farthest spot on Earth from the planet's center.
- If we take Earth's equatorial bulge into account, then Mount Chimborazo is more than 6,889 feet higher than Everest.
- Earth is the only planet in the solar system with plate tectonics.
- While granite is common on Earth, it's extremely rare anywhere else in the solar system. Granite's celestial rarity is due to the rock type being created by plate tectonics recycling other rocks.
- Most minerals on Earth rely on water or life to form—with a third relying solely on life-forms—which is why few minerals exist on the Moon, Mercury, and Mars.
- Out of the thousands of minerals known today, only about 296 may have existed before Earth formed.
- **THE OLDEST MINERAL CRYSTALS FOUND ON EARTH ARE TINY BITS OF ZIRCON DATING BACK ALMOST 4.4 BILLION YEARS.**
- The formation of 72 minerals on Earth is tied to bat and bird guano and urine.
- After spending decades wandering northern Canada, Earth's magnetic North Pole is heading farther north, now passing Earth's geographic North Pole, as it drifts toward Siberia.
- Earth is smaller than the Sun's sunspots and almost any explosion on its surface.

"I HAD ALWAYS KNOWN THE SKY WAS FULL OF MYSTERIES—BUT NOT UNTIL NOW HAD I REALIZED HOW FULL OF THEM THE EARTH WAS."

—RANSOM RIGGS, *MISS PEREGRINE'S HOME FOR PECULIAR CHILDREN*

✦ EARTH'S BIG IFS

- **IF EARTH'S 4.5-BILLION-YEAR HISTORY WERE CONDENSED INTO A SINGLE DAY:**
  - Photosynthesis would start around 6 a.m.
  - Multicellular organisms would appear by 2 p.m.
  - The reign of dinosaurs would begin around 10:40 p.m.; they'd be extinct by 11:40 p.m.
  - Anatomically modern humans would appear in the final six seconds.
  - And all humans alive today would be born within the day's final 1/500th of a second.
- If Earth's core cooled completely, the continents would cease their drift. Volcanoes would grow cold, and new mountains would never form. Deep-sea hydrothermal vents would sputter out and succumb to the ocean's icy embrace. Our protective magnetic field would dissipate, leaving us vulnerable to the Sun's deadly onslaught of storms and radiation.

Exploradores Glacier in Aysén, Chile

- What if Earth spun the opposite way? Global ocean and atmospheric currents would reverse, altering climate worldwide.

- Earth's continual rotation means that you're constantly moving through space. If you're standing at the Equator, you're spinning at about 1,000 miles an hour.

- The closer you are to one of the poles, the slower you are rotating through space.

- **WHAT IF EARTH SUDDENLY STOPPED ROTATING?**
  - Everything would be launched eastward at hundreds of miles an hour. If you lived on the Equator, your body would be thrown at 1,000 miles an hour.
  - Residents of Oslo, Norway, would roll toward Sweden at about 500 miles an hour.
  - The closer you lived to the poles, the more likely you'd be to survive the tumble. Santa Claus and his elves would pirouette in place.
  - Winds stronger than any ever recorded would scour the surface of the planet.
  - All the oceans would crash on shores in massive global tsunamis.
  - One side of the planet would be roasted by sunlight for six months at a time.
  - Without the centrifugal effect from rotation, the oceans would shift to the poles, creating one giant landmass around the Equator and two polar oceans.

- If Earth increased its spin to 17,000 miles an hour instead of 1,000, residents at the Equator would be suddenly weightless, hovering above Earth's surface.

- In this scenario, as polar ice melted, water would collect around the Equator, making Earth more flattened and slowing its spin.

- If all the water in the atmosphere rained down on Earth at once, the entire planet's surface would be covered by about an inch.

- If we filled a giant balloon with all the water on Earth, it would stretch 860 miles in diameter.

- If Earth were a perfect sphere, and retained its surface water, the ocean depth would be 1.7 miles everywhere.

- The amount of water that evaporates, or transpires, from plants into the atmosphere each day is roughly enough to fill some 46 trillion Olympic swimming pools.

- **IF ALL OF EARTH'S GLACIERS AND ICE CAPS MELTED, SEA LEVELS WOULD RISE BY ROUGHLY 230 FEET, FLOODING COASTAL CITIES AROUND THE WORLD.**

- During the last glacial maximum, 20,000 years ago, so much water froze into ice sheets that sea level was more than 400 feet lower than it is today.

- At the time, ice covered approximately 25 percent of the land around the world.

- The amount of land on Earth currently covered by glaciers is roughly 11 percent—and rapidly shrinking.

A 60-foot explosion of lava in Hawaii Volcanoes National Park, Hawaii

✦ UNDER PRESSURE

- Earth has layers like a jawbreaker candy: a solid metal inner core, liquid metal outer core, mostly solid mantle, and solid crust.

- **EARTH'S MANTLE ISN'T MOLTEN, BUT A SOLID ROCK THAT SLOWLY FLOWS LIKE TAFFY OVER GEOLOGIC TIMESCALES.**

- Miles under the surface the rocks may be almost entirely sparkling green thanks to an abundance of the mineral olivine.

- The deepest point on Earth is nearly seven miles underwater.

- This is more than a mile deeper than Mount Everest is tall.

- The pressure is so intense near Earth's center that the liquid metal found in its outer core is compressed into a solid orb.

- Earth's gravity drops continually as you ascend away from the planet's surface, but extends outward to infinity.

- For this reason, you weigh slightly less on a building's rooftop than you do in its lobby.

- Every day, about 20 volcanoes erupt on Earth.

- While it might seem like more volcanoes erupt each year, the eruption rate hasn't changed. Instead, more interest and better communication have led to more reported eruptions.

- Some of the heat that drives Earth's volcanic eruptions and plate tectonics is a relic from our planet's birth.

- **VOLCANOES ACT LIKE A RELIEF VALVE FOR EARTH, HELPING RELEASE SOME OF THE PLANET'S INTERNAL HEAT.**

- Rather than heat Earth's surface, large explosive eruptions release ash and gas into the atmosphere that reflect solar radiation, creating a slightly cooling effect.

- Scientists aren't sure how plate tectonics first began, but one hypothesis suggests the impact that hypothetically birthed our Moon may have also birthed plate tectonics.

- The motions of tectonic plates continue to shape the topography of Earth.

- Every 200 million years, our surface is entirely remade with new crust.

- For millions of years, plate tectonics has acted like a thermostat on Earth, but human emissions are overriding the meter.

- **ALMOST 1.5 BILLION YEARS FROM NOW, THE RESTLESS TECTONIC PLATES THAT SCULPTED EARTH'S SURFACE MAY GRIND TO A HALT.**

- If plate tectonics stops, mountains will not only stop growing, they'll shrink, as erosion wears them down into hills. The oceans will gradually fill with sediment, increasing sea levels until continents eventually flood.

- Space rocks collided with Earth just as much as our heavily cratered Moon. But plate tectonics has erased most of the evidence by frequently resurfacing our planet.

- Earth has had at least three supercontinents, when all landmasses were smashed together.

- Two hundred to 250 million years from now, the continents may collide again, forming a new supercontinent.

- Days are not all precisely 24 hours; they naturally vary over time, but the differences are too small to notice. As of this writing, July 5, 2024, was the shortest day recorded since the adoption of atomic clocks, measuring 1.66 milliseconds faster than 24 hours.

- In 2022, the eruption of Hunga Tonga-Hunga Ha'apai shocked scientists when the blast sent a sonic boom that zipped around the world twice.

- The volcanic eruption unleashed so much water vapor that it could fill more than 58,000 Olympic swimming pools.

- The next time you say "light as air," know that air on Earth weighs about 15 pounds per square inch at sea level.

- A suction cup is hard to lift, not because of suction (which is not really a thing), but because of the weight of Earth's atmosphere on top of it. A suction cup with 10 square inches of area feels 150 pounds of atmospheric weight on top of it.

- Due to tidal forces between Earth and the Moon (and other factors such as earthquakes and glacier melting), Earth's rotation rate varies and must occasionally be corrected with leap seconds.

- Earth days are growing longer by 1.7 milliseconds every century due to these phenomena.

- By adding a leap day every four years, we make our calendar year 44 minutes longer than a true year.

- To compensate, every 100 years, we skip the leap year—unless the year is evenly divisible by 400.

- By that rule, we skip the addition of a leap day again in 2100, 2200, and 2300.

An illustration of Earth's inner core, outer core, mantle, and crust

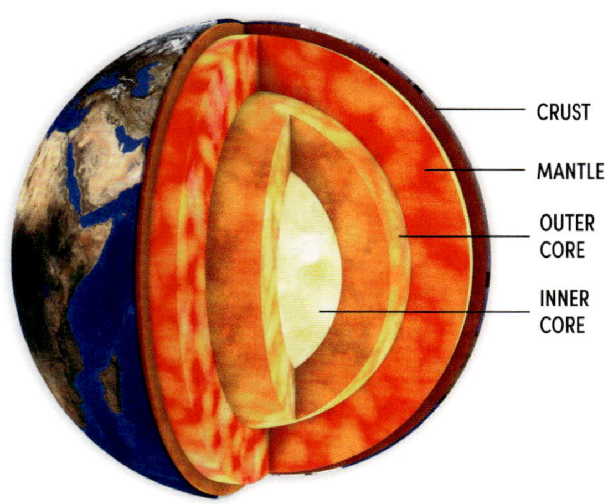

CURRENT FACT COUNT 2611

EARTH: BY THE NUMBERS

40

feet: how much Earth's tides can vary in some places around the world

6

hours: the length of Earth's day after it first formed

17,670

miles an hour: how fast Earth would have to spin for humans to be flung into space. That's about 17 times faster than our planet's current rotation.

1.5

inches: how far the Moon spirals away from Earth each year due to tidal forces between it and Earth

4

billion years: the age of Earth's oldest known intact rock. It was discovered on the Moon.

44

trillion watts: how much heat is estimated to radiate from Earth into space. Half of that comes from radioactive decay of elements in the crust and mantle.

9,800°F

the temperature of Earth's core—nearly as hot as the Sun's surface

332,500,000

the number of cubic miles of water on Earth

15,000

pounds per square inch: the amount of weight pushing on something at the Mariana Trench, the deepest point in the sea

81

how many times more you would weigh if Earth and the Moon were the same size

Low tide at Belliveau Cove in Bay of Fundy, Nova Scotia, Canada

"AS YOU PASS FROM SUNLIGHT INTO DARKNESS AND BACK AGAIN . . . FOR THE FIRST TIME IN YOUR LIFE YOU FEEL IN YOUR GUT THE PRECIOUS UNITY OF THE EARTH AND ALL THE LIVING THINGS IT SUPPORTS."

—RUSSELL "RUSTY" SCHWEICKART, APOLLO 9 ASTRONAUT

The autumn equinox as seen from a weather satellite 22,000 miles above Earth

CURRENT FACT COUNT 2692

THE MOON

A world likely birthed from Earth's very own crust

✦ OUR ONE AND ONLY

- Where did the Moon come from? It was likely hewn from Earth itself billions of years ago, according to the giant impact hypothesis, the leading proposed idea in answer to this question.

- **ACCORDING TO THE HYPOTHESIS, A PROTOPLANET THE SIZE OF MARS SMASHED INTO EARTH AND RIPPED CHUNKS FREE FROM OUR ROCKY INNARDS. THESE EVENTUALLY AMASSED INTO THE MOON.**

- That hypothetical protoplanet was named Theia after the mother of Selene, the ancient Greek goddess of the Moon.

- Before that debris coalesced into the Moon, it orbited Earth as a ring. Yes, like Saturn, Earth may have once had rings of its own.

- Rocky blobs deep near Earth's core may be remnants from Theia.

- Once the Moon formed, it was a molten magma ocean.

- The Moon is slowly shrinking as its molten core cools and contracts. It's lost about 150 feet in diameter over the last several hundred million years.

- The Moon's interior started hot but has slowly been cooling and shrinking.

- As the Moon's core cools, the entire orb contracts, puckering the surface like a raisin and triggering moonquakes that form ridges tens of meters high.

- **A SINGLE MOONQUAKE CAN LAST MORE THAN AN HOUR.**

- Even though we only ever see one side of the Moon, it does rotate on its axis as it orbits Earth. Its rotation rate matches its orbital rate (called synchronous rotation).

- The Moon is spiraling away from Earth at the same rate at which human fingernails grow (about 1.5 inches per year).

- If you could see the Moon the way it looked 4.4 billion years ago, just after its formation, it would look 20 times larger in the sky because it would be orbiting 20 times closer.

The Greek goddess Theia, one of Uranus's children

- Billions of years ago, the closer position of the Moon would have enabled many more lunar eclipses than we experience today.

- In addition, the Moon may have had a magnetic field connected with Earth's, perhaps helping retain our planet's atmosphere—a key to the development of life on our planet.

- Now, you could fit every planet in the solar system between Earth and the Moon when they are at their farthest distance apart.

✦ THE MOON'S SCARS

- The lunar features we've come to colloquially call "the man in the Moon" are the remnants of ancient floods of molten rock.

- **IN THE 17TH CENTURY, ASTRONOMERS MISTOOK THE DARK MARKS ON THE MOON'S SURFACE (CALLED MARIA) FOR OCEANS.**

- Lunar maria formed by ancient space rock impacts that melted the Moon's surface and triggered volcanic eruptions, creating vast pools of lava.

- Most of the Moon's dark spots were created more than three billion years ago, during a period when space rocks bombarded the inner solar system.

- We might call part of the material covering the surface "moondust," but it isn't powdery. It's made of sharp shards of shattered rock.

- Solar wind and ultraviolet light from the Sun electrostatically charge this moondust, making it hover several centimeters above the lunar surface.

- Some volcanic eruptions on the Moon may be recent, geologically speaking—as young as 50 million years old.

- Most impact craters on the Moon are perfect circles.

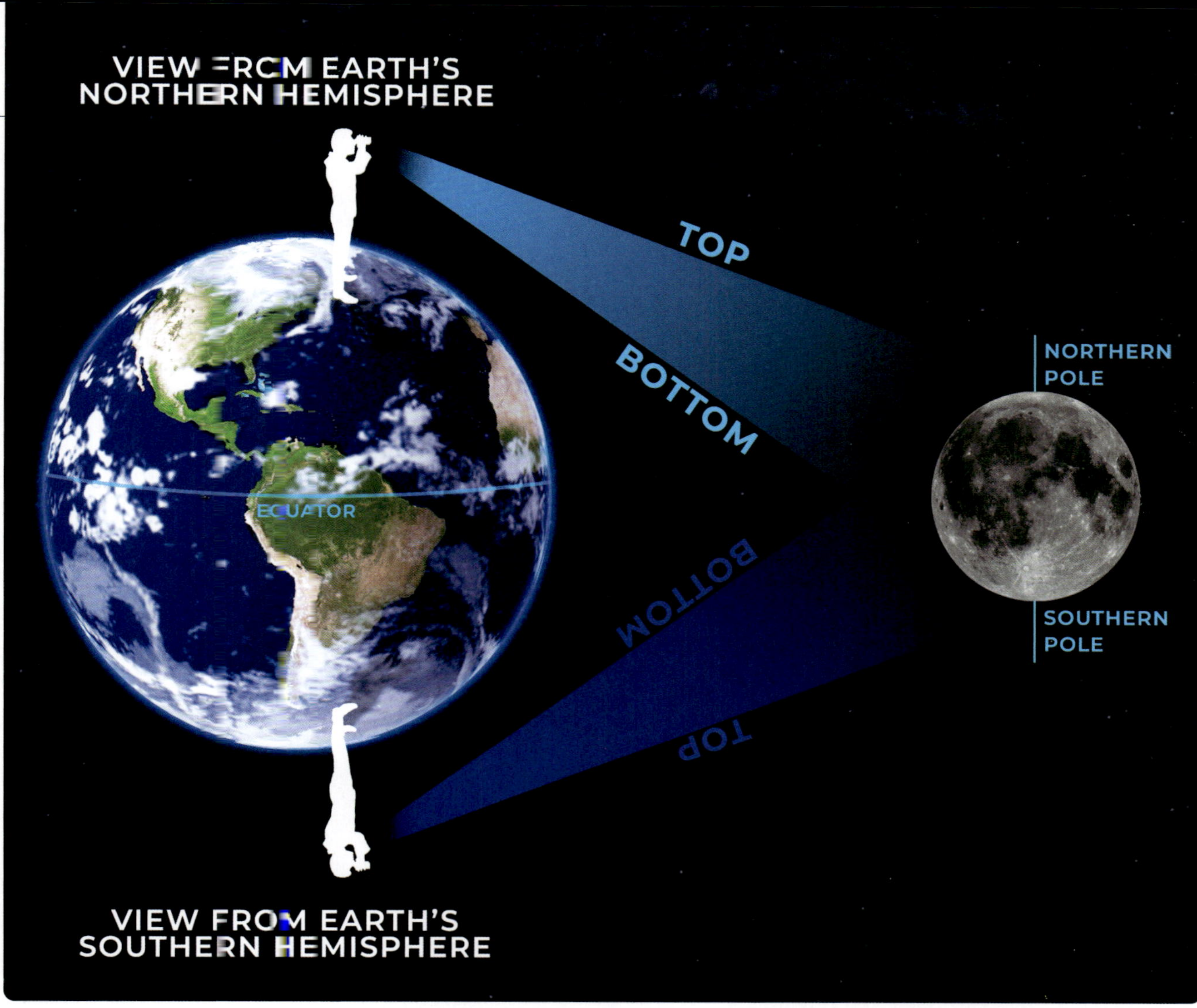

Perspective: From Earth's Northern Hemisphere, the Moon's north pole appears at the top. From Earth's Southern Hemisphere, that's reversed.

- Why? When impactors hit the surface, all that kinetic energy instantaneously converts most of that matter into heat, forming a shock wave and flinging out the surrounding debris equally in all directions.

- Only about 5 percent of Moon and Martian craters are elongated, which is caused by an impactor that arrived at a very low angle (like skipping a rock across a pond), creating multiple shock waves along its route.

- The Moon is unusually large for a moon. Although Earth is 81 times more massive than the Moon, the Moon is a quarter the size of Earth in diameter.

- By comparison, the largest moon in the solar system (Ganymede) is only 6 percent the total size of its host planet (Jupiter).

- What if the Moon were more massive? Earth would rotate more slowly, lengthening days and nights.

- An even *more* massive Moon could cause a double tidal lock with our planet. Sounds like a wrestling move, but it's a real celestial phenomenon.

- Under our synchronous tidal lock with the Moon, all of Earth sees one face of the Moon. Under a double tidal lock, only one side of Earth would ever see one face of the Moon.

- All large moons in the solar system are tidally locked to their planet.

- Though the same side of the Moon always faces Earth, the face appears differently across latitudes.

- Moon-gazers in the Northern Hemisphere see a Moon upside down relative to Moon-gazers in the Southern Hemisphere.

- At the Equator, the Moon appears to rotate 180 degrees over the course of the night.

- **THE FIRST GLIMPSE HUMANITY HAD OF THE MOON'S FAR SIDE WAS IN 1959 FROM THE SOVIET UNION'S LUNA 3 SPACECRAFT.**

- The Moon is in the sky for 12 hours every day, but we can't always see it because the Sun's brilliance blots it out.

- Due to the laws of reflection, the full moon is not twice as bright as a half moon. No, it shines *six* times brighter.

✦ SUPER MOONS AND MICROMOONS

- Super moons happen when the Moon's orbit comes closest to Earth, making it appear 14 percent bigger than when it's at its farthest.
- A micromoon occurs when a full moon coincides with its farthest distance from Earth.
- A super moon outshines a micromoon by 30 percent.
- Visually, it's exceedingly difficult to determine the size difference, but that significant difference in brightness is detectable to even untrained eyes.
- For comparison, the apparent size difference between a super moon and a micromoon is proportional to the size difference between a quarter and a nickel.

✦ THE TIDES

- Spring tides are the especially high tides during a full or new moon, when Earth, the Moon, and the Sun all align.
- They occur twice per month.
- **SPRING TIDES ARE ABOUT 20 PERCENT HIGHER THAN HIGH TIDES DURING WAXING OR WANING PHASES.**
- King tides are the high tides that coincide with either a super moon or a new moon.
- Yes, there's such a thing as a super new moon: a new moon at Earth's perigee, or the closest point. (They receive far less media hype, of course.)
- Despite the name, the difference between king tides and high tides during a standard full moon is nearly negligible, with only about two inches of difference.
- We blame the Moon for high tides, but it has a partner in tidal effects: the Sun. Gravity from each combine to pull the water on Earth in predictable patterns.
- Two to one: On average, the Moon accounts for two-thirds of the strength of tides and the Sun for one-third.
- **TIDE-GENERATING FORCES VARY ACCORDING TO THE DISTANCE BETWEEN COSMIC OBJECTS. THE MORE DISTANT THE OBJECT, THE LESS PULL, CALCULABLE AS THE CUBE OF THE DISTANCE BETWEEN THE OBJECTS.**
- The back-and-forth sloshing of tides is slowing Earth's rotation at a rate of about two-thousandths of a second per day per century.
- In *Bruce Almighty*, Jim Carrey's godlike character pulls the full moon closer to enhance a romantic moment, unwittingly causing ferociously high tides that pound the world's shorelines.

"WHENEVER I GAZE UP AT THE MOON . . . I AM BACK TO THAT PRECIOUS PINPOINT OF TIME, STANDING ON THE FOREBODING—YET BEAUTIFUL—SEA OF TRANQUILITY. I COULD SEE OUR SHINING BLUE PLANET EARTH POISED IN THE DARKNESS OF SPACE."

—BUZZ ALDRIN

✦ BUT IS IT *REALLY* RARE?

- Humans have a funny tendency to declare certain cosmic events as "rare" or "once in a lifetime." Examples include super moons (a few times a year), lunar eclipses (twice a year), partial solar eclipses (twice a year), total solar eclipses (every year or two), and retrograde motion of planets (constantly, depending on the planet).
- We also have a tendency to append meaning to these events. Friday the 13th, for example, is not an unusual or rare occurrence. In fact, the 13th of any month is more likely to land on a Friday than any other day of the week.
- **EVENTS THAT ARE ACTUALLY RARE:**
  - A nova, or cosmic explosion, that's visible to the naked eye. These typically occur once every 80 years.
  - Transit of Venus: The next one is due in the year 2117.

- Transit of Mercury: We get around 13 of these per century.
- A quintuple conjunction, or when Mercury, Venus, Mars, Saturn, and Jupiter (the only planets visible to the naked eye) can all be witnessed at the same time from Earth.
- Halley's comet, which only comes around every 76 years or so.
- A total solar eclipse in your home town. A total eclipse might occur somewhere on Earth every year or two, but a singular location will only witness one every 375 years on average.
- A full alignment of all eight planets: never. This is statistically impossible.

✦ WHAT IF EARTH HAD TWO MOONS?

- If Earth had two moons instead of our one, Earth-based stargazing would be twice as difficult. Stars are best observed on moonless nights without the lunar light pollution cast by the Moon's highly reflective surface. Two moons visible orbiting Earth would mean two times the natural light pollution.
- Tides would be chaotic, ranging from twice as high to nearly undetected.
- Evenings would shine twice as brightly, throwing off the instincts of nocturnal animals and rendering romantic moonlit walks rather unromantic.
- Our calendars would cease working, and we would require a new method of timekeeping.
- **THE TWIN MOONS WOULD GRAVITATE TOWARD ONE ANOTHER, EVENTUALLY CULMINATING IN AN INEVITABLE COLLISION.**

- Two large moons orbit the desert planet Arrakis in the film *Dune,* based on the novels by Frank Herbert.
- We know that they are large moons because they're both spherical (a shape that celestial bodies usually obtain through the inward compression of high gravity) and because they have strong magnetic fields that can disrupt Arrakis's compasses.
- If Arrakis's moons are that large, and if Arrakis is of a similar size to Earth, then their orbits are likely highly unstable. The larger the moons, the more unstable their orbits.
- **THIS IS THE THREE-BODY PROBLEM: ANY THREE OR MORE GRAVITATIONALLY BOUND CELESTIAL OBJECTS WILL ORBIT ONE ANOTHER IN COMPLEX AND UNPREDICTABLE PATTERNS, LEADING TO COLLISIONS AMONG THEMSELVES OR EJECTION OF THE OBJECTS FROM THE SYSTEM.**

✦ WHAT IF YOU COULD PLAY SPORTS ON THE MOON?

- The gravity on the Moon is one-sixth that of Earth, rendering you, your golf ball, javelin, or any other ballistic object airborne for six times longer.
- **THE CURRENT WORLD RECORD FOR A JAVELIN THROW ON EARTH IS A LITTLE MORE THAN 300 FEET; ON THE MOON, IT COULD REACH NEARLY 2,000 FEET.**
- Dimples on a golf ball exploit surrounding air molecules to remain airborne longer, enhancing performance. So, driving a golf ball off the top of Mount Everest, where air pressure is one-third of that at sea level, would demonstrate reduced air resistance—but also reduced effectiveness of the ball's dimples.
- On the Moon, which has no atmosphere, golf balls would not require any dimples, resulting in no hooked or sliced shots, only straight drives.

✦ LUNAR DIGS

- With real estate prices these days, would you rather live in a lava tube on the Moon than on Earth? In 2024, scientists found evidence of cave networks snaking their way beneath the lunar surface.
- These subsurface pits and tunnels likely formed the same way lava tubes form here on Earth: when rivers of molten lava crust over, leaving behind hollow subsurface channels.
- **A MOON CAVE WOULD OFFER NATURAL PROTECTION FROM THE SURFACE'S WILD TEMPERATURE SWINGS AND DEADLY RADIATION AND COULD PROVIDE AN IDEAL BASE FOR FUTURE LUNAR HABITATS.**
- Humans already feel relatively weightless underwater due to our similar density to the displaced water (called Bernoulli's principle). If your future lunar base boasted a swimming pool, the waves you made while swimming would be six times higher than those you'd make while swimming on Earth.
- You could jump in and out of the water like a dolphin.

CURRENT FACT COUNT 2714

THE MOON: BY THE NUMBERS

20

the number of times closer the Moon was to Earth when it formed compared to today

47

pounds: the weight of Moon rock samples that returned to Earth with Apollo 11 astronauts—the first lunar samples ever

8,000

the number of times stronger Earth's tides are now (20 x 20 x 20 = 8,000) compared to when the Moon first formed

842

pounds: the total weight of the rocks that Apollo astronauts brought back from the Moon

27

the number of astronauts who have orbited the Moon

24

the number of Apollo astronauts who have orbited the Moon

29.5

Earth days: the length of one day on the Moon

12

total number of Apollo missions

6

total successful crewed Apollo Moon landing missions

3

marshes, 12 bays, 20 lakes, 22 seas, and one ocean make up the Moon's maria

4

the number of moons in the solar system that are bigger than Earth's Moon

1,600

the number of miles across the Ocean of Storms, the Moon's largest mare, which is slightly larger than the Bering Sea that cleaves the continents of North America and Asia

1

the number of times the Moon rotates on its axis for every cycle of its phases

300

trillion: the number of times greater the sea level air pressure is on Earth compared to the air pressure on the Moon's surface.

30

the number of Earths that could line up between the Moon and us

0

the number of rings the Moon has

50

the number of times brighter the full Earth is when viewed from the Moon than when the full moon is viewed from Earth

40

yards: the distance astronaut Alan Shepard hit a golf ball on the Moon

1,550

miles: the diameter of the South Pole–Aitken basin, the largest crater on the Moon

18,046

feet: the elevation of Mons Huygens, the tallest mountain on the Moon

25

miles: the thickness of the Moon's crust on the side that faces Earth (the crust on the far side is 37 miles thick)

2,160

miles: the diameter of the Moon, which is less than a third the width of Earth

CURRENT FACT COUNT 2773

MAP OF THE MOON

NEAR SIDE

1 Luna 2 (U.S.S.R.) Crashed Sept. 14, 1959
2 Ranger 6 (U.S.) Crashed Feb. 2, 1964
3 Ranger 7 (U.S.) Crashed July 31, 1964
4 Ranger 8 (U.S.) Crashed Feb. 20, 1965
5 Ranger 9 (U.S.) Crashed Mar. 24, 1965
6 Luna 5 (U.S.S.R.) Crashed May 12, 1965
7 Luna 7 (U.S.S.R.) Crashed Oct. 7, 1965
8 Luna 8 (U.S.S.R.) Crashed Dec. 6, 1965
9 Luna 9 (U.S.S.R.) Landed Feb. 3, 1966
10 Surveyor 1 (U.S.) Landed June 2, 1966
11 Surveyor 2 (U.S.) Crashed Sept. 23, 1966
12 Luna 13 (U.S.S.R.) Landed Dec. 24, 1966
13 Surveyor 3 (U.S.) Landed Apr. 20, 1967
14 Surveyor 4 (U.S.) Crashed July 17, 1967
15 Surveyor 5 (U.S.) Landed Sept. 11, 1967
16 Surveyor 6 (U.S.) Landed Nov. 10, 1967

NEAR SIDE

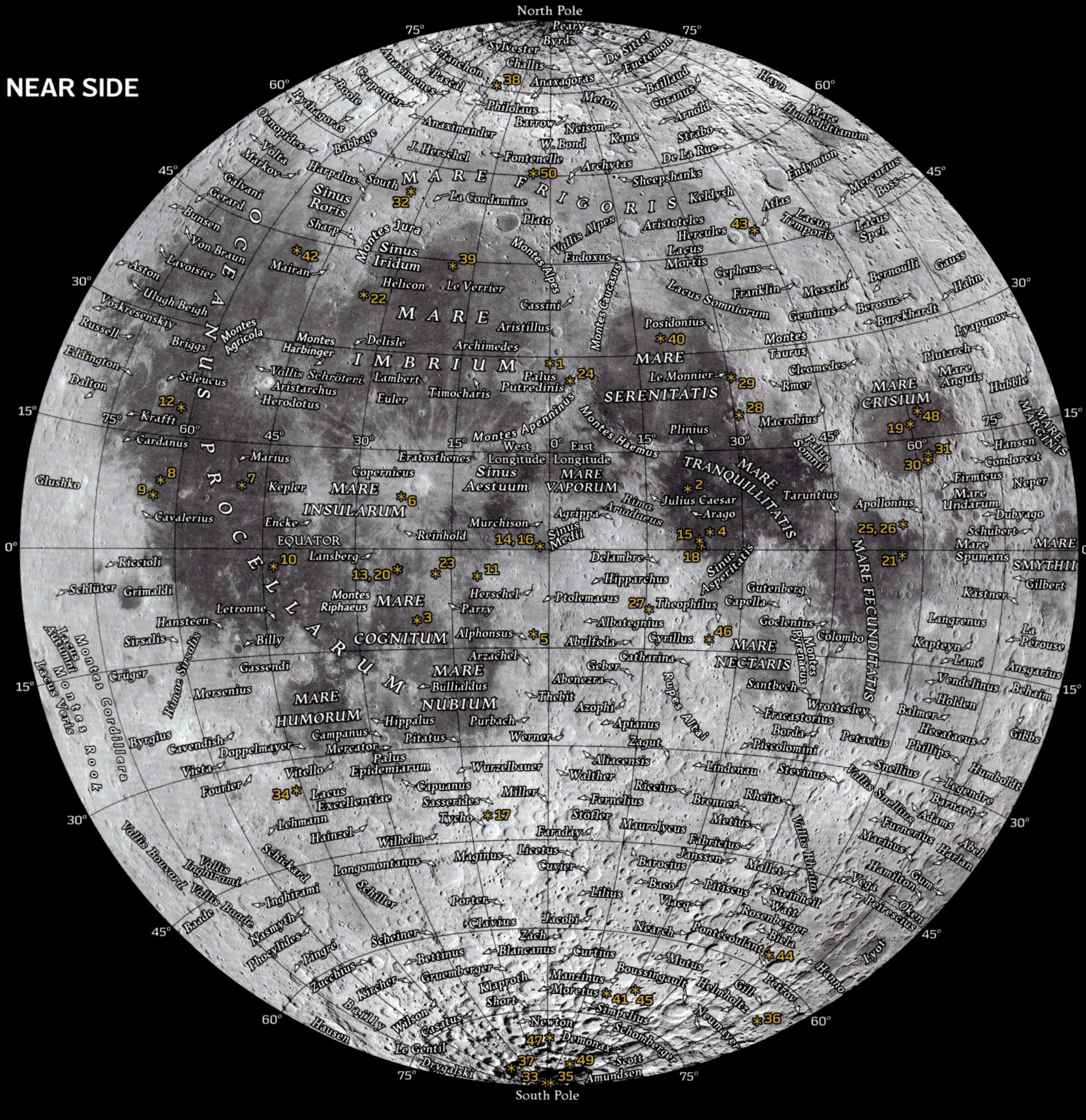

17 Surveyor 7 (U.S.) Landed Jan. 10, 1968
18 Apollo 11 (U.S.) Landed July 20, 1969
19 Luna 15 (U.S.S.R.) Crashed July 21, 1969
20 Apollo 12 (U.S.) Landed Nov. 19, 1969
21 Luna 16 (U.S.S.R.) Landed Sept. 20, 1970
22 Luna 17 (U.S.S.R.) Landed Nov. 17, 1970
23 Apollo 14 (U.S.) Landed Feb. 5, 1971
24 Apollo 15 (U.S.) Landed July 30, 1971
25 Luna 18 (U.S.S.R.) Crashed Sept. 11, 1971
26 Luna 20 (U.S.S.R.) Landed Feb. 21, 1972
27 Apollo 16 (U.S.) Landed Apr. 21, 1972
28 Apollo 17 (U.S.) Landed Dec. 11, 1972
29 Luna 21 (U.S.S.R.) Landed Jan. 15, 1973
30 Luna 23 (U.S.S.R.) Landed Nov. 6, 1974
31 Luna 24 (U.S.S.R.) Landed Aug. 18, 1976
32 Hiten (Japan) Crashed Apr. 10, 1993
33 Lunar Prospector (U.S.) Crashed July 31, 1999
34 SMART-1 (ESA) Crashed Sept. 3, 2006
35 Chandrayaan-1 Moon Impact Probe (India) Crashed Nov. 14, 2008
36 Kaguya/SELENE (Japan) Crashed June 10, 2009
37 LCROSS, Centaur Impactor (U.S.) Crashed Oct. 9, 2009
38 GRAIL A (Ebb), GRAIL B (Flow) (U.S.) Crashed Dec. 17, 2012
39 Chang'e 3 (China) Landed Dec. 14, 2013
40 Beresheet (Israel) Crashed Apr. 11, 2019
41 Chandrayaan-2 Vikram lander (India) Crashed Sept. 7, 2019
42 Chang'e 5 (China) Landed Dec. 1, 2020
43 Hakuto-R 1 (Japan) Crashed Apr. 25, 2023
44 Luna 25 (Russia) Crashed Aug. 19, 2023
45 Chandrayaan-3 Vikram lander (India) Landed Aug. 23, 2023
46 SLIM (Japan) Landed Jan. 19, 2024
47 IM-1 Odysseus lander (U.S.) Landed Feb. 22, 2024
48 Blue Ghost 1 (U.S.) Landed Mar. 2, 2025
49 IM-2 Athena lander (U.S.) Landed Mar. 6, 2025
50 Hakuto-R 2 RESILIENCE lander (Japan) Crashed June 5, 2025

FAR SIDE

1 Ranger 4 (U.S.) Crashed April 26, 1962
2 Lunar Orbiter 1 (U.S.) Crashed Oct. 29, 1966
3 Lunar Orbiter 3 (U.S.) Crashed Oct. 9, 1967
4 Lunar Orbiter 2 (U.S.) Crashed Oct. 11, 1967
5 Lunar Orbiter 5 (U.S.) Crashed Jan. 31, 1968
6 Okina, Selena Orbiter (Japan) Crashed Feb. 12, 2009
7 LADEE (U.S.) Crashed April 18, 2014
8 Chang'e 4 (China) Landed Jan. 3, 2019
9 Chang'e 6 (China) Landed June 1, 2024

FAR SIDE

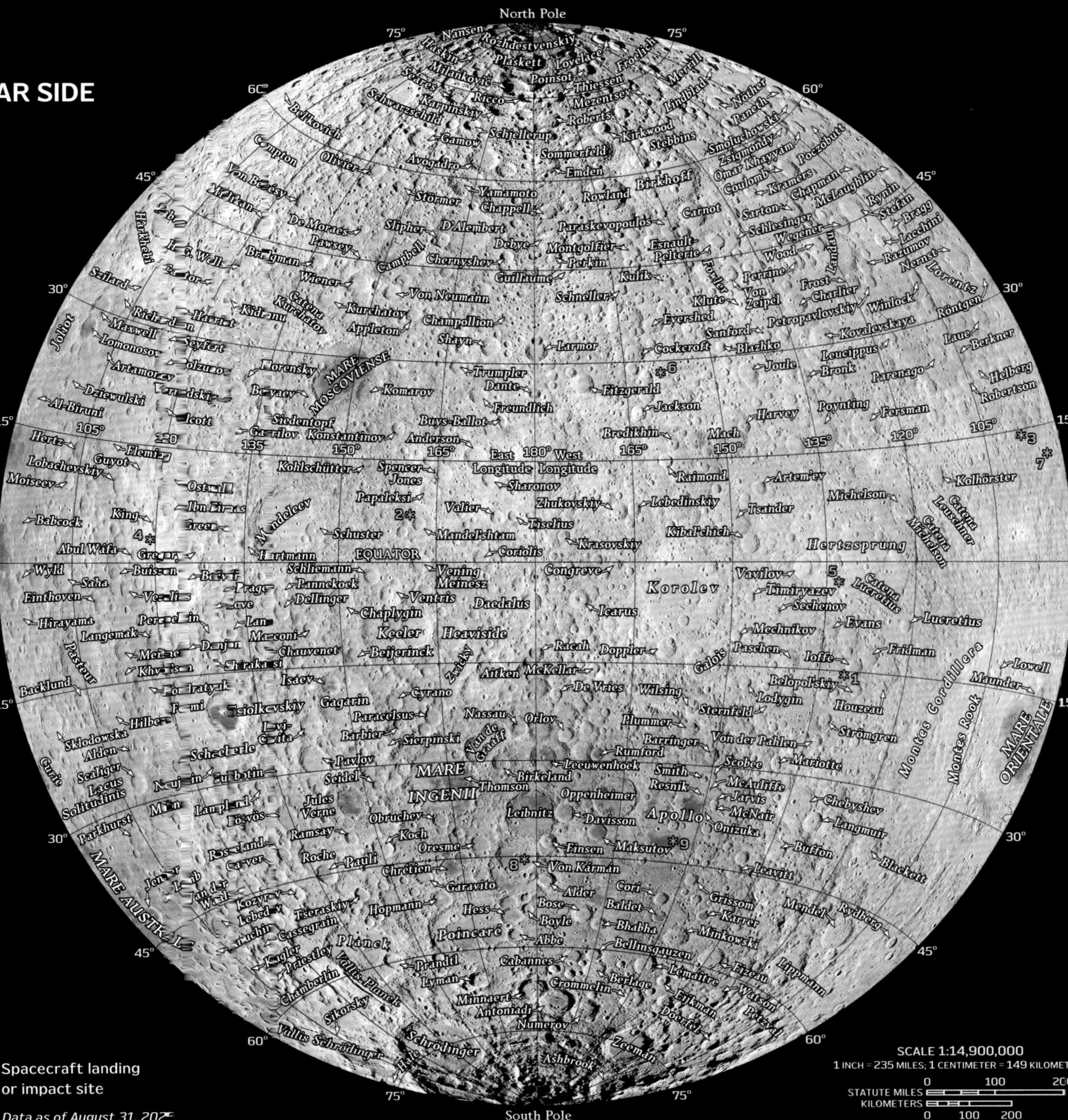

* Spacecraft landing or impact site

Data as of August 31, 2025

SCALE 1:14,900,000
1 INCH = 235 MILES; 1 CENTIMETER = 149 KILOMETERS
STATUTE MILES 0 100 200
KILOMETERS 0 100 200

CURRENT FACT COUNT 2956

MARS

A world teeming with possibilities and threaded with clues to our biggest questions about where life began and whether life could, or even did, exist beyond Earth

✦ OUR RED NEIGHBOR

- To the naked eye, Mars glows a rusty red. Oxidation of Mars's iron-rich rocks and regolith imbues its surface with the iconic red-orange coloring.

- The ancient Romans named the planet for their god of war because the reddish glow was reminiscent of blood.

- Human blood (and that of many other animals) runs red for a similar reason: oxygenated iron atoms within.

An original movie poster for *The War of the Worlds*

- Vulcans, such as Lieutenant Spock in the *Star Trek* universe, bleed green instead of red due to their copper-imbued blood.

- Copper-blooded creatures on Earth—such as octopuses, horseshoe crabs, and emperor scorpions—all bleed blue, not green.

- A common misconception is that human blood is red only when exposed to oxygen. But in the vacuum of space (or within vacuum-sealed blood bags and syringes at a hospital), our blood still runs scarlet, something nearly all Hollywood sci-fi directors get right—whether or not they know why.

- Mars appears red to us, but up close, swaths of its surface can look brown, tan, or even slightly green.

- Mars's egg-shaped orbit is increasingly squashed, so it's inching toward Earth on its closest pass.

- From Mars, the Sun looks about two-thirds smaller than it does from Earth.

- **MARS HAS SEASONS, CAUSED NOT ONLY BY ITS EARTHLIKE TILT BUT ALSO ITS VARYING DISTANCE FROM THE SUN ON ITS ELONGATED ORBIT.**

- Its atmosphere mysteriously changes composition seasonally, with methane and oxygen increasing in spring and summer.

- The Martian sky is the inverse of Earth's. During the day, a Martian would see a reddish orange sky, but sunrises and sunsets would appear blue.

A first- or second- century A.D. statue of Mars

- Martian dust particles scatter red and orange frequency light more readily than blue. Earth's atmospheric particles scatter blue frequency light more readily.

- **MARS-BOUND HUMAN EXPLORERS SHOULD PLAN FOR A 300-MILLION-MILE, SEVEN-MONTH-LONG JOURNEY.**

- Martian air density is only one percent that of our planet.

- The thin atmosphere can carry sound, but only across short distances, with higher frequencies fading more quickly than lower-pitched ones.

- For the same reason, flying a kite on Mars would require much stronger winds than it would on Earth.

- Mars sings with a constant low-frequency hum beyond human hearing. Scientists don't know why.

- **IN SPACE, NO ONE CAN HEAR YOU SCREAM. ON MARS, SOMEONE LESS THAN 50 FEET AWAY MIGHT HEAR YOU SCREAM FAINTLY.**

- The lack of atmosphere means any sunlight warming the planet quickly escapes into space.

- If you stood on the Martian equator, your feet would be a comfortable 75°F, while your head would endure freezing temperatures.

- **A DAY ON MARS IS CALLED A SOL. IT'S THE ONLY PLANET UPON WHICH WE'VE BESTOWED SUCH A DISTINCTION.**

- Timekeeping on Mars and in space: 1 sol = 24 hours 37 minutes.

- Some NASA Mars mission scientists wear watches with Martian sol time to help track rover mission agendas.

- The only samples we have of Mars are chunks blasted out by space rocks that eventually fell to Earth.

- NASA's Perseverance rover gathered and sealed 43 samples of the red planet, ready and awaiting a return to Earth.

Dust storms swirl in this image of Mars taken by Hubble in 2001.

✦ BIG REPUTATION

- In 1877, Italian astronomer Giovanni Schiaparelli drew a map of Mars, labeling some features *canali* (channels). English media mistranslated it as "canals," inspiring generations of myths about Martian engineering.

- In 1895, astronomer Percival Lowell sparked a frenzy with ardent yet unsubstantiated claims he could see a network of artificially crafted canals etched into Mars's surface.

- H. G. Wells, inspired by Lowell's claims, wrote *The War of the Worlds,* in which creatures from Mars landed on Earth. The work capitalized on the public fascination with the prospect of intelligent life on the red planet.

- Yes, Mars is famous for its dust storms, but they're likely not as strong as the one in the 2011 novel (and 2015 film) *The Martian*.

- Martian dust storms can reach gale speeds of 60 miles an hour.

- With Mars's meager atmospheric pressure, the storm would feel more like a gentle breeze, incapable of knocking over a human—or an entire spaceship.

- Still, future astronauts will need to deal with dust obscuring solar panels, as depicted in *The Martian*.

- Mars's dust storms can cover vast swaths of land as large as Earth's continents, or even zip around the entire planet.

- **THESE GLOBAL STORMS KICK UP ENOUGH PARTICULATE TO BLOCK THE SUN, BLANKETING THE SURFACE IN DARKNESS.**

- The storms can be self-sustaining: The dust absorbs heat, intensifying the temperature difference between ground and air, which drives the winds.

- Dust can stay aloft on Mars for weeks or months, thanks to the planet's thin atmosphere and weak gravity.

- The dust swirls in columns across Mars's surface, like Looney Tunes's Tasmanian Devil.

- Martian dust sticks to everything because of its slight electrostatic charge.

- Mars's dustiest season is summer in the southern hemisphere, when the planet orbits closest to the Sun.

- Mars has noctilucent clouds (high-altitude clouds that seem to shine at dusk and dawn, rarely seen on Earth, and even then, only at high latitudes).

- The red planet hosts the largest mountain, the largest canyon, and the largest impact crater in the entire solar system.

✦ 0/10 DO NOT RECOMMEND THE TOXIC SOIL

- **THE MOON AND MARS DO NOT HAVE SOIL BUT REGOLITH.**

- What's the difference? Regolith lacks microbes: the tiniest packages of life, in the form of bacteria, fungi, archaea, algae, protozoa, and viruses.

- Martian regolith is toxic thanks to high concentrations of the chlorine-containing compound perchlorate.

- "I'm going to have to science the shit out of this," says protagonist Mark Watney upon finding himself stranded on Mars with dwindling resources in the hit film *The Martian*.

- When he grew potato crops using his own feces as fertilizer, he took his catchphrase to heart. But would this really work? Earth's soil is teeming with microorganisms that do not exist in human feces. Without those microorganisms, there would be no plants—no life.

- In an experiment that would make Watney proud, NASA successfully grew potato plants in fertilized Mars-like regolith, although most of them died shortly afterward.

Grow potatoes on Mars? Exciting but not likely.

✦ OLYMPUS MONS

- The largest mountain in the entire solar system is the Martian volcano Olympus Mons.

- **IT'S NEARLY THREE TIMES AS TALL AS MOUNT EVEREST.**

- It's about 100 times the volume of Mauna Loa, Earth's largest volcano.

- Its area matches that of Arizona.

- It's so heavy it compressed the planet's crust around it into a moat-like depression.

- If you were to stand at the base of Olympus Mons, you couldn't see the peak. It's so far away, it's beyond the horizon.

- Olympus Mons is a shield volcano: It grew from successive layers of lava flow.

- Olympus Mons likely grew so large because Mars's gravity is only about a third of that on Earth.

✦ VALLES MARINERIS

- Mars's Valles Marineris is the solar system's largest canyon.

- It's known as the Grand Canyon of Mars, but that's quite the understatement. It stretches the distance between Los Angeles and New York City.

- The Valles Marineris probably started as a gigantic crack as the planet cooled and further widened from erosion.

✦ UTOPIA PLANITIA

- The largest impact crater in the entire solar system is found on Mars: Utopia Planitia.

- It is more than 2,000 miles across.

- That's more than 10 times larger than the largest crater on Earth: the Vredefort Dome in South Africa.

- **IN 2016, NASA DISCOVERED THIS GIANT BASIN CONTAINS UNDERGROUND ICE—THE EQUIVALENT OF ENOUGH WATER TO FILL LAKE SUPERIOR.**

✦ HUMAN-MADE MARTIANS—AND THEIR FATES

- The first spacecraft to land on the Martian surface, the Soviet Mars 3 lander, survived only 110 seconds before failing. It returned a single gray, featureless image.

The Perseverance rover took this photo of the Ingenuity helicopter during the first flight on another planet.

- Mars 3 carried along a small rover, but nobody knows if it ever deployed.

- NASA's Viking 1 and 2 landers became the first U.S. spacecraft to land safely on the surface of Mars.

- Viking 1 launched on August 20, 1975, and landed on July 20, 1976. Viking 2 launched on September 9, 1975 and landed on September 3, 1976.

- In 1997, NASA's Mars Pathfinder mission carried a base-station lander, now named the Carl Sagan Memorial Station, and the Sojourner rover to the red planet.

- The Pathfinder lander and Sojourner rover played pivotal roles in *The Martian,* the hit 2011 novel by Andy Weir that inspired the film. Mark Watney exhumes the dust-covered spacecraft and upcycles Pathfinder's radio camera into an Earth communication device.

- Sojourner traveled 330 feet and was operational for 95 Earth days.

- In 2004, twin rovers Spirit and Opportunity were operational.

- **SPIRIT DROVE 4.8 MILES AND REMAINED OPERATIONAL FOR 6 YEARS 77 DAYS—20 TIMES LONGER THAN EXPECTED—BEFORE IT BECAME CAUGHT IN A SAND TRAP.**

- The Opportunity rover traveled 28 miles and remained operational for 14 years 138 days—57 times longer than expected.

- By 2012, the Curiosity rover's planned two-year mission had exceeded its life span more than six times over. It is the longest operational rover yet.

- In 2021, the Perseverance rover launched with Mars's first helicopter, aptly named Ingenuity.

- **THE FIRST NON-AMERICAN ROVER SUCCESSFULLY EXPLORED MARS IN 2021 WITH THE CHINESE ZHURONG ROVER—CHINA'S FIRST SUCCESSFUL LANDING ON ANOTHER PLANET.**

- It operated nearly a year before going into hibernation to ride out an approaching dust storm.

- Zhurong was expected to awaken several months later but never did.

✦ INGENUITY

- Ingenuity was the first helicopter to take flight on another world.
- It served as a scout to its companion, the Perseverance rover, directing it to the safest routes toward interesting exploration targets.
- It reached a max ground speed of more than 22 miles an hour, while Perseverance can only achieve 0.1 mile an hour.
- The four-pound copter buzzed about the red planet for three years before crash-landing in a barren river valley.
- **IT FLEW MORE THAN TWO HOURS ACROSS 72 FLIGHTS, COVERING MORE THAN 11 MILES.**
- It rose nearly 80 feet above the surface.
- It was designed to fly merely 15 feet, if it could even take flight at all.
- The landing site of Perseverance (and Ingenuity with it) inside the Jezero crater was named Octavia E. Butler Landing, after the science fiction author.

✦ SEEKING ANCIENT LIFE

- Scientists conceived of the Perseverance rover with a return mission in mind—bringing Martian samples back to Earth.
- **PERSEVERANCE'S JOB WAS TO SEEK OUT SIGNS OF LIFE AROUND JEZERO CRATER AND SELECT AND STASH THE MOST INTERESTING FINDS FOR A FUTURE SAMPLE RETURN.**
- Billions of years ago, Jezero crater contained an ancient lake system more than 20 miles across flooded with liquid water and fed by multiple rivers.
- On Earth, highly abundant and diverse microbial ecosystems thrive within muddy-bottomed river mouths and estuaries, very similar to those found around Jezero crater.
- Sometimes, a river carries more sediment than it can erode, forming a vast delta of microbe-rich buildup.
- Each new layer quickly buries and preserves the layer beneath, creating a time capsule of climate conditions—and perhaps even microbial fossils.
- If microbes ever did call Mars home, and if they were anything like microbes here on Earth, then the vast deltas of Jezero should have been teeming with them.
- **ONE OF PERSEVERANCE'S MOST INTERESTING DISCOVERIES WAS A SMALL BOULDER WITHIN AN ANCIENT, SEDIMENT-RICH RIVERBED.**
- The rock contained organic molecules that could have formed the building blocks of life, as well as imprints that could have been made by flowing water.
- Scientists named it Cheyava Falls, after the highest waterfall of Earth's Grand Canyon.
- Cheyava means "intermittent waters" in the Native American Hopi language.
- Even on Earth, searching for signs of ancient microbes can be a challenge. Only careful analysis and contextual clues can verifiably differentiate a natural mineral deposit from a true biosignature. So, short of encountering a complex life-form like Marvin the Martian, we'll need to bring these rocks home to study.

✦ THE SAMPLE PLAN

- NASA and the European Space Agency hatched an ambitious plan to retrieve samples from Mars. The task would be a feat of engineering and orbital dynamics unlike any we've yet witnessed.
- The sample-return spacecraft will consist of three main bodies: the Earth Return Orbiter to be deployed around Mars, the Sample Retrieval Lander to obtain the samples from Perseverance, and the Mars Ascent Vehicle to launch from the surface.
- The Sample Retrieval craft must land extremely close to the Perseverance rover with an escape rocket in tow.
- Then, it must safely transfer the samples collected by Perseverance into the rocket without causing any Earth-Mars cross contamination.

- The Mars Ascent Vehicle, about the size of a volleyball, will then launch it into orbit, the first rocket ever launched from another planet.

- **WE MAY HAVE SUCCESSFULLY LAUNCHED FROM THE MOON MANY TIMES, BUT MARS HAS TWICE THE MOON'S GRAVITY AND IS 530 TIMES FARTHER FROM EARTH.**

- Finally, the Earth Return Orbiter will intercept the rocket and make its way back home. This would mark the first time any interplanetary spacecraft captured another orbiting object and the first round-trip mission to Mars.

✦ THE GODS OF FEAR AND PANIC

- The names of Mars's moons—Phobos and Deimos—come from the Greek gods of fear and panic.

- **HOW MARS GOT ITS MOONS IS A MYSTERY. THEY MAY BE ASTEROIDS CAPTURED BY THE RED PLANET'S GRAVITY.**

- Phobos orbits so fast around Mars you could see it wax or wane in a single night if you were on the planet.

- To someone on Mars, Deimos would look like Venus does from Earth. Phobos would look like a glowing potato about half the size of our Moon.

- Like Earth's Moon, Phobos and Deimos always show the same side to Mars.

- Solar winds electrically charge parts of Phobos's surface to hundreds of volts, enough to fry robotic explorers' sensitive electronics.

- Phobos has so little gravity that if a person weighing 150 pounds visited, they'd weigh 2.4 ounces.

- It orbits closer to Mars than any other known moon around a planet, circling just 3,700 miles away.

- It speeds around Mars three times per Martian day.

- It is getting closer to Mars by six feet every century.

- Phobos could be our launchpad to Mars: Some suggest we colonize its Mars-facing side as a base from which to explore the red planet.

- **A MARS-FACING BASE ON PHOBOS WOULD HAVE ITS ADVANTAGES:**
  - Explorers would be protected from solar radiation.
  - Communication between Phobos and Mars would incur nearly no delay, unlike the 20 minutes it takes for Martian signals to reach Earth.
  - Because of Phobos's weak gravity, landing a spacecraft there would be easier, and far more fuel-efficient, than on Mars.
  - If you stood on the side of Phobos that faces Mars, the red planet would practically fill the sky.
  - Within 50 million years, Phobos may get so close to Mars that it will either collide or shred apart under the planet's gravity.
  - If Phobos breaks apart, Mars will lose a moon but gain a ring.

- Deimos's orbit is slowly growing larger and larger. Eventually, the moon will be lost to space.

- Deimos is so small that if you ran around the entire moon, you'd be just shy of completing a marathon.

- Astronomer Asaph Hall discovered the moons of Mars in 1877.

- A century and a half before their discovery, author Jonathan Swift wrote about two small moons in orbit around Mars in his novel *Gulliver's Travels*. Swift was no prophet. He was simply astronomy-fluent. At the time, it was known that Venus has no moons, Earth has one, and Jupiter had (at least) four. So, he approximated that Mars should have about two.

- Using Kepler's laws, Swift even gave roughly accurate estimates of the moons' orbital speeds.

Swift hypothesized the moons of Mars in *Gulliver's Travels*

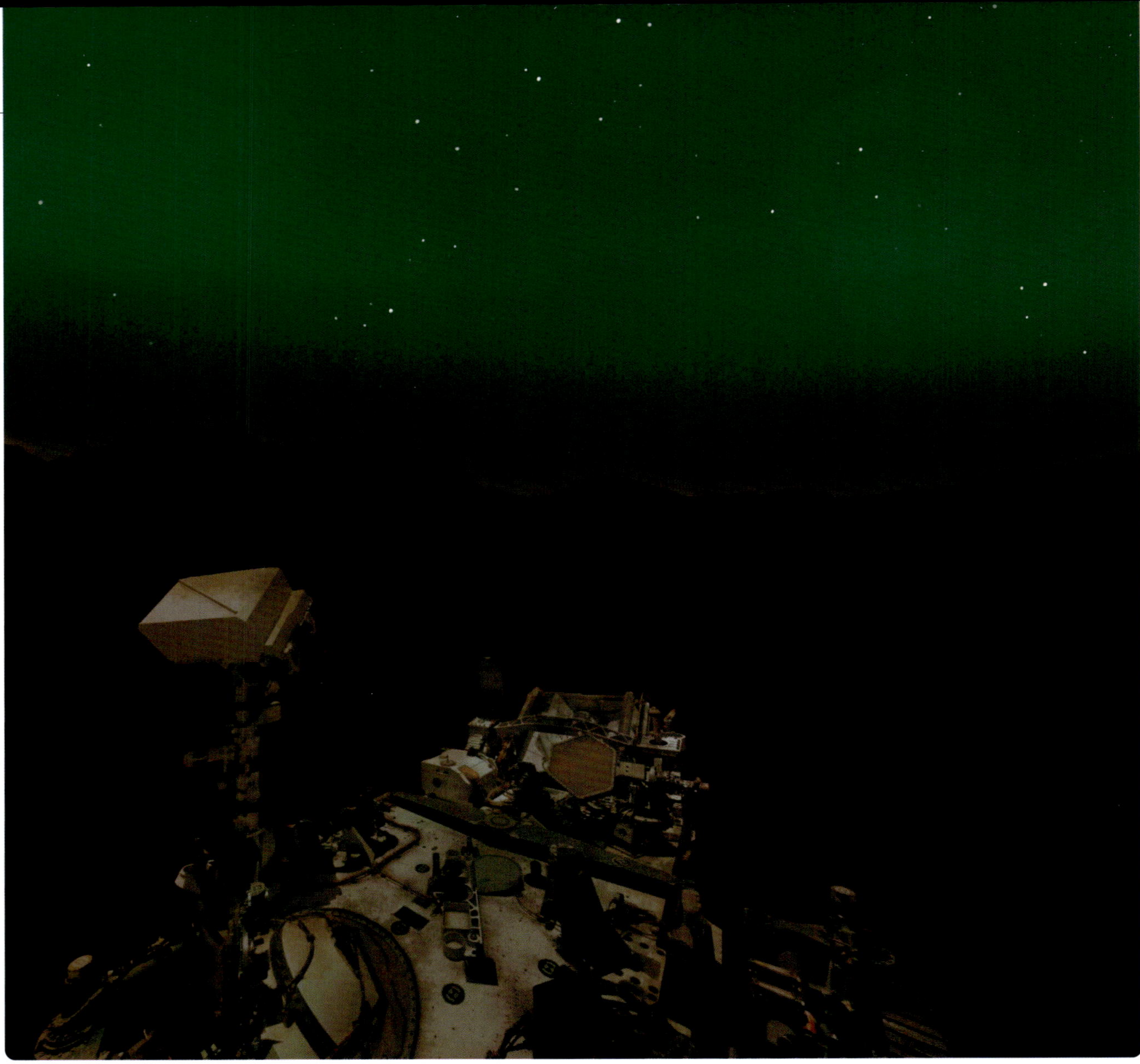

An artist's concept of the Perseverance under glowing green auroras above Mars

✦ BABY, IT'S COLD OUTSIDE

- Mars has polar ice caps like Earth's. But parts of Mars's caps are made from dry ice.
- **MARS WINTERS GET SO COLD THAT ATMOSPHERIC CARBON DIOXIDE FREEZES AND FALLS AS SNOW, GROWING THE PLANET'S POLAR ICE CAPS.**
- Carbon dioxide freezes in four-sided shapes, so Martian snowflakes would be snow cubes.
- A snow of water ice falls on Mars, but it never reaches the ground, turning to vapor in the thin air.
- In the spring, Mars's dry ice doesn't melt. It sublimates—it turns directly to gas.
- As Mars's dry ice sublimates in the spring, trapped gas bursts through the surface like geysers.
- The explosive effect carves strange spider-web-like patterns into the ice caps.
- Mars spins faster by fractions of a millisecond each year, perhaps as the planet's ice masses shift.
- Mars has a core of liquid iron that sloshes, causing the planet to wobble.
- Mars may have undergone some 20 ice ages since it formed, far ahead of Earth, which has known five.

✦ SHAKE, SHAKE, SHAKE

- Just like Earth, Mars gets quakes too: Marsquakes.
- Unlike Earth, Mars experiences no tectonic activity—no forces that move, shake, or re-form the crust, the cause of quakes on Earth.

- So, the source of Marsquakes remains a mystery.
- Some hypotheses speculate that Marsquakes are caused by contraction as the planet cools or by magma squishing through the subsurface.
- In 2019, scientists identified the first active fault zone on Mars by tracing Marsquakes to a line of deep fractures, known as the Cerberus Fossae.
- The Cerberus Fossae, 750 miles long, were named for the Latin for "trench" *(fossa)* plus Cerberus, the three-headed dog who kept the dead from leaving Hades in Greek mythology.

✦ THE MARTIAN MICROCLIMATE

- Mars once had vast oceans and lakes, but they evaporated after the planet's magnetosphere winked out and solar winds stripped its atmosphere.
- **ANCIENT FLOODS LIKELY CARVED DEEP VALLEYS ON MARS'S SURFACE.**
- Until 2024, scientists believed Martian water could only be found at its frozen poles. Now, we know some water nestles within the calderas of volcanoes scattered across the entire surface.
- The calderas form unique and surprising microclimates, perfect for frost formation during Mars's cold seasons.
- Humid winds drift up toward the volcano peaks, then condense and freeze, coming to rest within the calderas.
- **EACH DAY, THE AFTERNOON SUN EVAPORATES THE FROST POOLS. AND EACH NIGHT, THE CYCLE BEGINS ANEW.**
- While water atop Mars's crust exists as films of frost and solid patches of ice, liquid water may yet lurk beneath its desiccated surface.
- If stashes of ancient liquid water in subsurface aquifers exist, they're too deep to detect with the standard radar used by current Mars probes and rovers.
- **SCIENTISTS COULD ONE DAY LISTEN TO MARSQUAKES TO FIND THESE HYPOTHETICAL HIDDEN POCKETS OF WATER.**
- The proposed technique, called the seismo-electrical method, would listen for specific patterns of disturbances emitted when a seismic event passes through liquid water.
- The same technique might one day help us assess the exact depths and chemical compositions of icy ocean worlds and their potential habitability.

✦ MARTIAN AURORAS

- Mars, like many planets and moons in the solar system, experiences its own auroras.
- While Earth's auroras can typically only be seen in the polar regions, Mars's lack of magnetic field means a solar storm can envelop the entire planet in an aurora.
- **IN 2024, ALL OF MARS WAS ENGULFED IN AN AURORA WHEN A POWERFUL SOLAR STORM BOMBARDED ITS THIN ATMOSPHERE WITH CHARGED PARTICLES.**
- An astronaut exploring the Martian surface during that storm would have witnessed a jade-colored sky and endured the equivalent radiation dose of approximately 30 chest x-rays.
- Mars's weak magnetic field cannot protect the planet from onslaughts of solar radiation the way Earth's field does, so future exploration missions must consider protective measures for the astronauts' health as well as solar-powered equipment prone to degradation by radiation.
- Without a protective magnetic field, future Martians will be exposed to high levels of radiation and left vulnerable to the effects of solar flares and coronal mass ejections, especially during years of solar maximum.

✦ A MISHAP IN METRIC

- At least one space calamity can be attributed to a simple mistranslation between the metric and imperial systems.
- In 1999, the perfectly operational Mars Climate Orbiter (MCO) spent 10 months en route to Mars only to overshoot its orbital insertion window and smash into the planet.
- Analysis revealed that the NASA Jet Propulsion Lab's (JPL) spacecraft engineering team had programmed the orbiter to relay data in metric units, while the navigation engineering team at Lockheed Martin programmed their software in imperial units.
- The botched mission cost NASA hundreds of millions of dollars and caused enduring embarrassment.
- **THE INVESTIGATION BOARD REPORT CITED NASA'S OWN HISTORY OF SUCCESS AS A CONTRIBUTING FACTOR.**
- It stated, "In the case of MCO there was a widespread perception that 'Orbiting Mars is routine.' This perception resulted in inadequate attention to navigation risk mitigation." In other words, even scientists sometimes let confidence breed complacency.
- The board then recommended that "project personnel should question and challenge everything—even those things that have always worked. JPL top management should provide the necessary emphasis to bring about a cultural change."

Raindrops cling to tortula moss on a concrete wall in Wales.

✦ TERRAFORMING MARS

- Scientists and engineers have proposed myriad creative solutions to coax a breathable atmosphere back onto Mars.

- **ALL THE PROPOSED METHODS AIM TO KICK-START THE SAME GREENHOUSE EFFECT WE OBSERVE HERE ON EARTH.**

- One proposal: Introduce fluorine compounds (similar to the chemicals banned by the 1987 Montreal Protocol for creating a hole in Earth's ozone layer).

- These compounds are thousands of times more potent as greenhouse gases than carbon dioxide.

- To be effective, we would need to deliver, or create on planet, at least three times the amount we once produced on Earth prior to the Montreal Protocol.

- Another proposal: Nuke it. If we strategically bomb Mars's poles, we can quickly vaporize its vast stores of ice and release all the trapped carbon dioxide and other greenhouse gases.

- Even if nuclear detonation did not elicit a long list of ethical and legal conundrums, it still could not raise the temperature or atmospheric pressures any meaningful amount. Mars's poles simply lack the necessary mass.

- Another alluring suggestion is to just throw some glitter on it—literally. A team of researchers proposed introducing millions of tons of reflective aluminum and iron nanoparticles into the Martian atmosphere, where they would remain aloft, high above the dust particles, and scatter light back onto the surface like a planetary disco ball.

- Even if we temporarily create an atmosphere warm enough to sustain life, we'll need to maintain a magnetic field to contain it and prevent solar wind from blowing it back out to space.

- One suggestion is to create a faux magnetosphere (a magnetic shield) powered by an artificially crafted closed electric circuit.

- The circuit would rest at the Lagrangian point between Mars and the Sun, thereby deflecting oncoming solar wind. As long as the shield remained in place, the atmosphere might stand a chance.

- **ANOTHER IDEA: THE HUMBLE TORTULA MOSS MAY BE KEY TO FUTURE MARTIAN POST-TERRAFORMING EFFORTS.**

- Mosses are exceptional at carbon dioxide capture and sequestration and require little to no maintenance, so a mossy Mars may help expedite oxygen production.

- The unassuming but mighty tortula moss ranks among the most resilient plants on Earth—perfect for those who lack a green thumb as well as future Mars residents.

- Its resistance to drought, extreme heat, light fluctuations, and low-oxygen and high-UV environments make it an extremophile.

- It can tolerate desiccation beyond 98 percent water loss and anoxic (low-oxygen) conditions. It not only survived, but thrived, after enduring gamma blasts that would kill a human.

"THERE THEY GO, OFF TO MARS, JUST FOR THE RIDE."

—RAY BRADBURY, *RHODOMAGNETIC DIGEST*

CURRENT FACT COUNT 2970

MARS: BY THE NUMBERS

3

feet: nearly how much higher you could jump on Mars than on Earth, because of its weaker gravitational pull

26

the number of months our planet takes to pass between Mars and the Sun, making the red planet its brightest and easiest to see

79

years, 10 months, 8 days: how much time it would take to drive from Earth to Mars at 200 miles an hour

34,700,000

miles: the distance between Mars and Earth in 2003, the closest in nearly 60,000 years

2287

the next year Mars will come this close to Earth

63,000,000

miles: roughly the farthest Mars ever gets from Earth

20

degrees: the angle of Mars's axis, caused some three billion years ago when a trio of clustered volcanoes grew so large, they caused it to tip

5

the number of Grand Canyons that could fit top to bottom in Mars's Valles Marineris canyon, between four and six miles deep

7

Earth years: the average amount of time between global dust storms that encircle the entire planet of Mars

60

miles an hour: the top speed of dust storm winds on Mars

-100°F

the temperature below which tortula moss can survive for years, making it a candidate for terraforming Mars

2003

the year in which a particularly strong solar flare destroyed the radiation detector of the Odyssey orbital spacecraft

4.7

the strongest magnitude of a Marsquake ever recorded, much less intense than some of the quakes we've experienced on Earth

10

hours: the duration of the longest recorded Marsquake. (But who knows how long they lasted before we started recording?)

CURRENT FACT COUNT 2996

THE ASTEROID BELT

The domain separating the inner from the outer solar system. It carries the broken remnants of planets that never were, no longer are, or never will be.

- The rocky stretch of debris that orbits between Mars and Jupiter is collectively known as the asteroid belt.
- The belt is about 1 AU thick, or about the same width as the distance from Earth to the Sun.
- Millions of asteroids and one dwarf planet call the asteroid belt home.
- The belt holds about two million asteroids larger than 0.6 mile across and countless millions of smaller objects.
- The largest known asteroid in the belt is Vesta at more than 300 miles in diameter.
- If we collected the mass of the asteroid belt into a singular heap, it would add up to just 3 percent of the total mass of the Moon.

A travel poster from JPL's "Visions of the Future" series

- **THE TWO LARGEST OBJECTS IN THE ASTEROID BELT, THE DWARF PLANET CERES AND ASTEROID VESTA, ACCOUNT FOR NEARLY HALF OF THE ASTEROID BELT'S TOTAL MASS.**
- In *Star Wars:* Episode V *The Empire Strikes Back,* Han Solo and his crew aboard the *Millennium Falcon* narrowly escape an asteroid field as they dodge deadly chunks of careening rocks. It makes for nail-biting suspense and good use of a special effects department, but in reality, a trip through our solar system's asteroid belt would be a humdrum experience.
- Although the asteroid belt contains millions of asteroids, your spaceship is unlikely to come within a quarter million miles of one.
- Despite what Hollywood directors might have you believe, the chance of accidentally encountering an object in the asteroid belt is near zero. You wouldn't even know you've entered it.
- What about a more populated asteroid field in another solar system? A debris field containing enough matter to render itself dangerous to oncoming spacecraft would be an anomaly. The interacting, mutual gravitational pulls of closely situated debris will quickly coalesce into one single object—so quickly you won't need to worry about navigating such a field in the wilds of space.
- To create the dramatic debris-field scenario, a sci-fi director might instead consider navigating a planet's rings, like Saturn. But even Saturn's rings are mostly empty space, and the risk of collision is minimal.

An illustration of the solar system with the planets, asteroid belt, and Kuiper belt in order

✦ CERES

- The dwarf planet Ceres is the largest object and the only dwarf planet taking up residence within the asteroid belt.
- Ceres makes up about a third of the belt's total mass.
- **CERES IS THE CLOSEST DWARF PLANET TO EARTH AND THE FIRST OBJECT EVER DISCOVERED WITHIN THE ASTEROID BELT.**
- Ceres was discovered by the Italian astronomer Giuseppe Piazzi in Sicily in 1801.
- The dwarf planet has gone by many names: Hera, Juno, Piazzi, and Ceres Ferdinandea.
- Piazzi named it Ceres Ferdinandea: Ceres after the patron deity of Sicily and Roman goddess of the harvest and Ferdinandea after the king of Sicily.
- After heated debate, European astronomers rejected "Ferdinandea" but relented on the less political name of Ceres.
- Astronomer Wilhelm Olbers wrote on the matter, "I like the name Ceres since it reminds me of Sicily ... but the affix Ferdinandea will meet with as little luck as Herschel's George's planet" (referencing the original name of Uranus).
- Jupiter's strong gravity may have flung Ceres from the outer solar system into its current location while the planets were forming 4.5 billion years ago.
- Jupiter's gravity probably prevented Ceres from growing into a full planet.
- **CERES IS ABOUT 600 MILES ACROSS AND 80 PERCENT LARGER THAN THE ASTEROID VESTA.**
- Unlike Vesta, Ceres is round, with a rocky core and icy crust rife with geologic activity, including at least one three-mile-high cryovolcano (ice volcano).
- Ceres is so different from any of its surrounding neighbors that in 2006 it was promoted from asteroid to dwarf planet in the same decision that demoted Pluto to the same status.
- If Earth were the size of a nickel, Ceres would be the size of a poppy seed.

CURRENT FACT COUNT 3105

JUPITER

A massive world with a system all its own. Even some of Jupiter's many moons are entire worlds unto themselves.

✦ THE GIANT OF THE SOLAR SYSTEM

- Jupiter, the biggest planet in our solar system, has more than twice the mass of all the other planets in our solar system combined.
- It's considered a gas giant: a very large planet that is composed mainly of helium and hydrogen.
- Our solar system has two gas giants: Jupiter and Saturn.
- Unlike the rocky planets of the inner solar system, a gas giant doesn't have a hard surface to stand on.
- Jupiter may have once been as close to the Sun as Mars is now as it roamed throughout the solar system, influencing the formation of its celestial neighbors.
- These ancient wanderings scattered asteroids and water-rich comets that bombarded the inner planets. So, we may have Jupiter to thank for Earth's water.
- **JUPITER'S EARLY ROAMING MAY HAVE EJECTED A FIFTH GIANT PLANET FROM OUR SOLAR SYSTEM ENTIRELY.**
- Jupiter may have formed beyond Pluto's orbit before migrating to its current position.
- If you could stand on Jupiter (though there's no surface to do so), you might hear the powerful booms of intense thunder echoing between the planet's many cloud layers.
- With no solid land to break them up, Jupiter's storms last a long, long time.
- A searing ocean of liquid hydrogen likely lurks under Jupiter's gaseous exterior.
- The song "Fly Me to the Moon" mentions the desire to "see what spring is like on Jupiter and Mars." While Mars does experience seasons, including spring, thanks to its axial tilt, Jupiter's meager three-degree tilt makes for a seasonless world.

✦ A WORLD OF ONE'S OWN

- Jupiter is a mini system of its own: It harbors 95 known moons (and counting) and thousands of smaller asteroids, sometimes called moonlets.
- So many objects orbit Jupiter that in 2014 the International Astronomical Union had to make a rule reserving the naming of moons for objects larger than 1 kilometer or that carry significant scientific interest.
- Some of Jupiter's outermost moons orbit opposite the planet's rotation, hinting that they're actually captured asteroids.
- Thousands of asteroids, called "trojans," share Jupiter's orbit. They're classified into two swarms: those trailing the planet and those leading it.
- Stable gravity wells, called Lagrange points, keep the asteroids trapped in position without risk of colliding with Jupiter.
- Five Lagrange points exist around any two orbiting celestial bodies. They're where gravity and centrifugal forces balance, forming veritable parking spots for smaller objects such as spacecraft or asteroids.
- As remnants of our solar system's formation, these trojans have been trapped for billions of years.
- Think of Jupiter as the pied piper: It may have captured the trojans in the outer solar system before migrating to its current position.
- Its intense gravitational field influences the paths of comets and asteroids. It may help protect Earth from deadly impacts, but it could also nudge a space rock toward our planet instead.
- Jupiter's rings might be so puny because it has gigantic moons.
- Unlike Saturn's vivid rings, even the best ground-based telescopes can't image Jupiter's rings, because they're largely made of dust, not ice.
- In 1973, Pioneer 10 photographed Jupiter in its crescent phase—a first for planets in the outer solar system.
- **IN 1994, SCIENTISTS WATCHED 21 FRAGMENTS OF THE COMET SHOEMAKER-LEVY 9 CRASH INTO JUPITER WITH THE FORCE OF HUNDREDS OF MILLIONS OF ATOMIC BOMBS.**
- Shoemaker-Levy 9 is named for the discoverers Carolyn Shoemaker, Gene Shoemaker, and David Levy. The number 9 represents their ninth comet discovery together as a team.
- Before smashing into Jupiter, the array of Shoemaker-Levy 9 fragments stretched as wide as three times the distance between Earth and the Moon.

- The Shoemaker-Levy 9 fragments slammed into Jupiter's gaseous atmosphere at speeds so fast, the gases reached 53,000°F.
- **THE IMPACT CAUSED RIPPLES IN THE PLANET'S SURROUNDING CLOUD COVER AND SPLASHED GIANT PLUMES UP FROM THE LOWER ATMOSPHERE.**
- After the Shoemaker-Levy 9 impact, the dark clouds of debris in Jupiter's atmosphere took months to disperse.
- Shoemaker-Levy 9's collision with Jupiter sounded the global alarm: Smashups still happen in our solar system—and Earth could be next.

✦ IN VIVID COLOR

- Jupiter's stripes are superfast jet streams that swirl around the planet.
- The colorful bands striping Jupiter likely owe their vibrant hues to sulfur and phosphorus gases.
- **THE DARK ORANGE STRIPES AND THE WHITISH STRIPES ON JUPITER FLOW IN OPPOSITE DIRECTIONS.**
- Jupiter's Great Red Spot is a violent anti-cyclone: a storm with counterclockwise winds raging on the planet.
- Even in the 1600s, observers using the first telescopes on Earth saw a huge spot on Jupiter.
- **THE GREAT RED SPOT IS SHRINKING. SCIENTISTS AREN'T SURE WHY. IT MAY DISAPPEAR IN THE NEXT FEW DECADES.**
- The spot is changing over time, growing taller as it loses width, forming a more circular shape.
- In the 19th century, the Great Red Spot was about three times the width of Earth. Today, it's shrunk to about 1.3 times the width of Earth.
- The spot's winds swirl at up to 400 miles an hour—about 1.6 times the fastest winds ever measured on Earth.
- The winds at the edges of the spot are speeding up while those in the center are slowing.
- In 2019, the Juno spacecraft finally estimated the depth of the mysterious storm by measuring the effect of gravity as it flew over it.
- Juno transmits continuous radio signals to Earth that shift and warp in response to its motions. So, when Juno flew over the dense red spot, the frequencies shifted in direct response to the gravitational tug on the spacecraft over that specific area.
- We now know that the storm reaches nearly 300 miles deep into Jupiter's gaseous surface.
- That's more than 50 times deeper than Earth's oceans.
- **IF YOU COULD TRAVEL TO JUPITER'S CORE, 650 MILLION POUNDS OF PRESSURE WOULD CRUSH YOUR BODY IN ALL DIRECTIONS.**
- That would feel like more than 50,000 elephants piled on every square inch of your body.
- Jupiter's core likely isn't one dense mass, but a gassy cloud spread across half the planet's diameter.
- Vivid auroras at Jupiter's poles pulse like a galactic nightclub, hundreds of times more energetic than Earth's auroras.
- Jupiter's magnetic field bombards its moon Europa's icy surface, rendering it aglow like a night-light in Jupiter's sky.
- Its auroras come from not only solar wind but also particles from the moon Io's volcanic eruptions.

"[JUPITER] MAINTAINED ENOUGH HEAT AND MASS AND PRESSURE TO CRAM ATOMS VERY CLOSE TOGETHER, TO THE POINT THEY STOP BEHAVING LIKE THE ATOMS WE RECOGNIZE ON EARTH. INSIDE JUPITER, THEY ENTER A LIMBO OF POSSIBILITY."

—SAM KEAN, *THE DISAPPEARING SPOON: AND OTHER TRUE TALES OF MADNESS, LOVE, AND THE HISTORY OF THE WORLD FROM THE PERIODIC TABLE OF THE ELEMENTS*

✦ JUPITER'S MANY MOONS

- Jupiter has at least 95 known moons. Astronomers are still discovering more.

- **THANKS TO ITS MOONS, JUPITER EXPERIENCES DOUBLE, TRIPLE, OR EVEN MORE COMPLEX ECLIPSES.**

- Pairs of Jupiter's moons eclipsed the Sun almost every day from April to July 1991.

- From 2009 to 2010, the position of the moons of Jupiter, Saturn, and Uranus caused hundreds of solar eclipses for the gas giants.

- Jupiter LII is the planet's smallest known moon, measuring just a mile across.

- Jupiter's four largest moons—Io, Europa, Ganymede, and Callisto—are collectively called the Galilean moons, named after Italian astronomer Galileo Galilei, who first observed them in 1610.

- Jupiter's foursome of major moons is considered a mini-solar system, formed from rocky rubble left over from the planet's formation.

- Galileo initially mistook a trio of Jupiter's moons for stars.

- His recognition that the four objects were moons, and that the moons orbit Jupiter instead of Earth, overturned the idea of a geocentric universe.

- Galileo's discovery shook the very foundation of the Christian worldview, which held Earth as the center of everything. He witnessed the first evidence that anything could orbit something other than Earth.

- He spent the remainder of his life under house arrest after the findings earned him condemnation by the Roman Inquisition, which found him guilty of heresy in 1633.

- Astronomer Simon Marius discovered Jupiter's four largest moons around the same time as Galileo. Since he didn't publish his finding first, he's mostly forgotten.

- Simon Marius, not Galileo, named Jupiter's first four moons based on suggestions made by Johannes Kepler.

- **KEPLER SUGGESTED THEY TAKE NAMES FROM ZEUS'S MORTAL LOVERS IN GREEK MYTHOLOGY, AS ZEUS IS THE GREEK COUNTERPART TO THE ROMAN GOD JUPITER.**
  - Ganymede was a young man abducted by Zeus to become cupbearer to the gods.
  - Callisto was a huntress Zeus turned into a bear.
  - Io was a beautiful princess who Zeus turned into a heifer.
  - Europa was a princess Zeus pursued after turning himself into a white bull.

A third-century A.D. Roman floor mosaic of Zeus and Ganymede

✦ ALL ABOUT IO

- Io is the most volcanically active world in our solar system.

- It is teeming with hundreds of active volcanoes surrounded by lakes of molten lava and suffused in a toxic sulfuric atmosphere.

- Some eruptions spew dozens of miles high.

- **IO LOSES MORE THAN 2,000 POUNDS OF THE VOLCANICALLY OUTGASSED MOLECULES TO SPACE EVERY SECOND.**

- Io is so volcanically active that some eruptions can be seen from Earth by telescope.

- A gravitational tug-of-war between Jupiter and its moons heats Io's interior and causes the dramatic eruptions.

- Frequent eruptions resurface Io so fast that very few impact craters appear on its surface.

- A Voyager image taken in 1979 first revealed Io's active volcanic activities.

- Io's atmosphere is mostly made of sulfur dioxide.

- The abundance of sulfur on Io's surface gives it a vibrant yellow hue and likely causes its atmosphere to smell like rotten eggs.

- As Io's noxious atmosphere sheds into space, it casts auroras across not only itself, but also Europa, Callisto, and even Jupiter.

✦ ALL ABOUT EUROPA

- Europa is encased in an icy shell 10 to 15 miles thick that conceals an ocean nearly 100 miles deep.

- Europa and our Moon are roughly the same size.

- If Europa took the Moon's place, our nights would be much brighter, as its icy exterior reflects 5.5 times more light.

- Europa is a quarter of Earth's diameter, but it may hold an ocean twice as voluminous.

- **IT MAY HAVE STARTED AS A TINY GRAIN OF HAIL TRAPPED AROUND JUPITER DURING THE SOLAR SYSTEM'S FORMATION AND GRADUALLY GREW.**

- Jupiter's intense gravity, combined with Europa's elongated orbit, creates powerful tidal forces on Europa that constantly flex and crack its surface.

- As the surface bends and stretches, it generates continuous frictional heat, much like a paper clip grows warm as you bend it back and forth.

- The continuous tidal heating prevents the liquid subsurface ocean from freezing solid.

- **PLATE TECTONICS MAY SCULPT EUROPA'S SURFACE, BUT THE PLATES ARE MADE OF ICE, NOT ROCK.**

- Charged particles from Jupiter constantly bombard Europa's icy surface, splitting water into hydrogen and oxygen.

- Some of Europa's oxygen might migrate into its subsurface ocean, raising hopes for the potential for life.

- The squiggles across Europa's icy surface are actually cracks, perhaps colored by salts and sulfur rising from a subsurface ocean.

✦ ALL ABOUT GANYMEDE

- Jupiter's moon Ganymede is the largest moon in the solar system.

- Ganymede is bigger than both Mercury and Pluto.

- It is the solar system's only moon with a magnetosphere: It possesses its own magnetic field that dominates and surrounds it, protecting it from solar radiation.

- Ganymede's magnetosphere causes dancing auroras around its poles.

- The partial overlap between the magnetic fields of Ganymede and Jupiter causes the moon's auroras to sway.

- On Earth, auroras are mostly green. On the Galilean moons, they are red or orange.

- **GANYMEDE HAS A SALTY SUBSURFACE OCEAN THAT LIKELY HARBORS MORE WATER THAN ALL EARTH'S OCEANS COMBINED.**

- Irregular lumps seem to hide in Ganymede's ocean, hinting at giant rock formations and the potential for life-supporting ingredients.

- Ganymede retains a very thin atmosphere of oxygen, water vapor, and possibly bits of ozone.

✦ ALL ABOUT CALLISTO

- Jupiter's moon Callisto is the solar system's most heavily cratered object.

The Galilean moons (top to bottom): Io, Europa, Ganymede, and Callisto

- Any geologic activity that would otherwise erase the impact pockmarks has likely been dead for billions of years.

- Unlike the other three Galilean moons, Callisto orbits beyond the reach of Jupiter's radiation belt, further preserving every feature and facet of its surface.

- Astronomers think Callisto may have the oldest or most preserved surface in the entire solar system, offering interesting clues about our ancient beginnings.

- **CALLISTO MAY HARBOR A SUBSURFACE SALTWATER OCEAN.**

- Unlike Europa and Ganymede's oceans, what remains of Callisto's ocean isn't a likely candidate to harbor life. The moon lacks tidal forces, and thus the geologic energy, necessary to sustain life as we know it.

- Callisto is probably equal parts rock and water ice.

- Bright speckles on its surface are likely craters lined with water ice.

CURRENT FACT COUNT 3120

JUPITER: BY THE NUMBERS

10

the number of Jupiter widths it would take to equal the width of the Sun

10

hours: the length of a day on Jupiter

43

the rough number of minutes it takes sunlight to reach Jupiter

11

the number of Earths that could fit across Jupiter's diameter

1,300

the number of Earths that could fit inside Jupiter

26,400

miles an hour: the speed of a resident of Jupiter's equator—26 times faster than a resident of Earth's Equator

1,000,000

the number of years after the Sun's formation that Jupiter formed, suggesting it may be our solar system's oldest planet

50,000,000

the number of years after Jupiter's formation that Earth formed

600

miles: the width of each storm swirling across both the north and south poles of Jupiter

10

how many times deeper the subsurface ocean of Jupiter's moon Ganymede may be than the deepest of Earth's oceans

24

the number of hours it would take on Jupiter's moon Europa to generate enough oxygen to keep one million humans breathing

95

miles: the thickness of Ganymede's icy crust—two to three times thicker than Earth's rocky crust.

100

miles: the height of geysers erupting on Europa's surface

75

years: how long it would take to reach Jupiter in the same airliner that could get to the Moon in 18 days

90,000,000

years: the oldest estimated age of Europa's surface. Callisto's may be a few billion years old.

CURRENT FACT COUNT 3 1 2 6

EUROPA AND ITS OCEAN

Scientists suspect that hiding beneath Europa's frozen surface is a saltwater ocean with twice the volume of all Earth's oceans combined. Flying as close as 16 miles above the moon's surface, NASA's Europa Clipper spacecraft, launched in 2024, will use ice-penetrating radar and other sensors to seek out signs of liquid water within Europa's ice shell and study organic molecules on its surface.

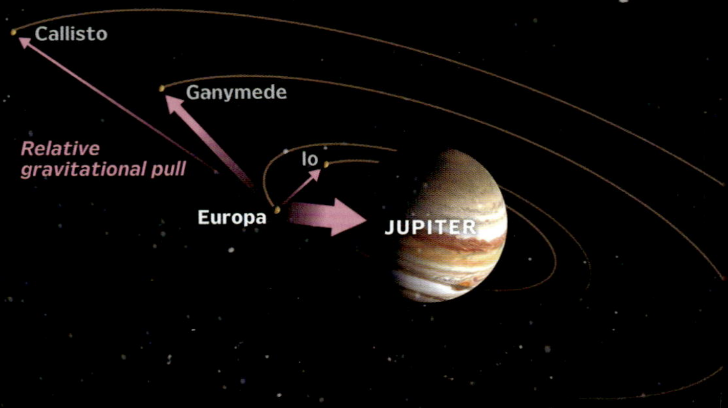

TIDAL FORCES

The orbits of three of Jupiter's four inner moons regularly align, amplifying tidal forces from the planet's immense gravity. Europa is pulled and stretched by both Jupiter's gravity and that of the other larger moons.

GRAVITATIONAL HEATING

Scientists believe that tidal forces create friction—and heat—within Europa's rocky mantle. This heat could power seafloor hydrothermal vents that spew organic chemicals and create warm upwellings that interact with the frozen, free-floating crust.

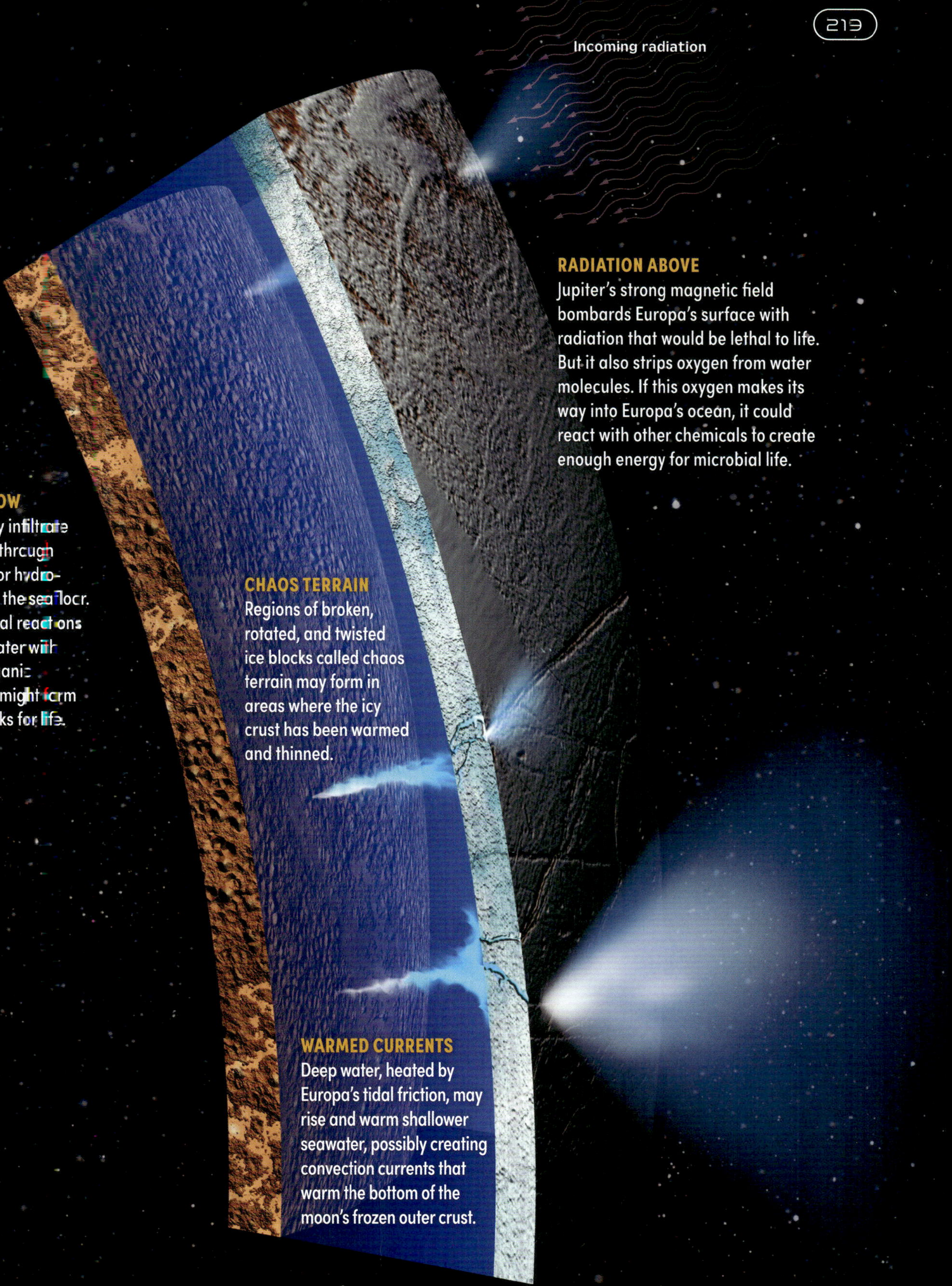

RADIATION ABOVE

Jupiter's strong magnetic field bombards Europa's surface with radiation that would be lethal to life. But it also strips oxygen from water molecules. If this oxygen makes its way into Europa's ocean, it could react with other chemicals to create enough energy for microbial life.

DEEP HEAT BELOW

Ocean water may infiltrate Europa's mantle through cracks, fissures, or hydro-thermal vents on the sea floor. Heat and chemical reactions could load the water with minerals and organic compounds that might form the building blocks for life.

CHAOS TERRAIN

Regions of broken, rotated, and twisted ice blocks called chaos terrain may form in areas where the icy crust has been warmed and thinned.

WARMED CURRENTS

Deep water, heated by Europa's tidal friction, may rise and warm shallower seawater, possibly creating convection currents that warm the bottom of the moon's frozen outer crust.

CURRENT FACT COUNT 3199

SATURN

A world festooned with seven spectacular rings, Saturn hoards more moons than any other planet—some may even harbor life beneath their ice-crusted oceans.

✦ THE RINGED PLANET

- In the night sky, Saturn glows a yellowish tan hue visible to the naked eye.
- Saturn was the most distant of the seven wanderers and, therefore, traced the slowest arc across the sky.
- Its sluggish pace may have inspired the ancient Romans to name the planet for their god of time and agriculture.
- Perhaps if the ancients could have seen Saturn's dazzling rings, they would have named it Venus instead, for their goddess of beauty.
- **SATURN IS SO LIGHT IT WOULD FLOAT ON WATER—IF YOU COULD FIND A BIG ENOUGH POOL.**
- Saturn's rings are composed of thousands of narrow ringlets, separated into seven primary rings.
- Saturn's rings likely formed around 300 million years ago.
- That means that sharks roamed Earth's oceans for 100 million years before the first rings appeared.
- The rings are so big and bright that you can see them from Earth even with a small telescope.
- **BUT THEY ARE ALSO SO SKINNY THAT VIEWED EDGE ON, THEY SEEM TO DISAPPEAR—SOME ARE ONLY 30 FEET THICK.**
- To scale, Saturn's rings are 10,000 times thinner than a standard piece of paper.
- The outmost rings sit 175,000 miles away from the planet.
- Every 15 years or so, the rings disappear from view. Since the planet tilts on its axis throughout its orbit, its rings occasionally align edge on relative to Earth—a phenomenon that stumped Galileo.
- When Galileo first spotted Saturn's rings through his telescope, their angle relative to Earth gave the illusion that they were two smaller orbs sandwiching the planet, like dual handles on a cosmic teacup.
- When Galileo viewed the rings again, the angle shifted and Saturn's two companions disappeared. The befuddled Galileo mused, in keeping with the ancient Roman myth, "Has Saturn swallowed his children?"
- Unbeknownst to Galileo, Saturn really is gobbling up its own rings. In about 100 million years, the rings will disintegrate entirely.
- Saturn's immense gravity and strong magnetic field constantly pull ring debris into the planet's surface.
- Its rings shine so brilliantly because they are mostly made of highly reflective water ice.
- They help shape Saturn's weather, as they rain icy particles into its upper atmosphere.
- Saturn's rings might have once been comets and asteroids, captured and shredded by the planet's gravity.
- Saturn's core is a soupy mix of rock and metals plus liquefied hydrogen and helium compressed under pressure.
- **THE PRESSURE UNDER SATURN'S CLOUDS IS SO INTENSE, IT COMPRESSES HELIUM GAS INTO A METALLIC LIQUID.**
- At Saturn's north pole, a massive jet stream traces a hexagon.
- A mega-hurricane swirls in the center of the planet's hexagonal jet stream.
- Saturn's northern jet stream may reach hundreds of miles down into the planet.
- **THE PLANET SPINS SO FAST THAT THE CENTRIFUGAL FORCE FLINGS THE CENTER MASS OUTWARD LIKE A BALL OF TOSSED PIZZA DOUGH.**
- Saturn isn't a perfect sphere: It bulges out noticeably farther at its equator than at its poles.

"NO OTHER PLANET LOOKS AS UNWORLDLY OR SURREAL AS SATURN."

—PLANETARY SCIENTIST CAROLYN PORCO

The Cassini spacecraft
checks out Saturn's night side,
including the B and C rings.

The moon Mimas, as photographed by the Cassini spacecraft in 2010

✦ MOON COLLECTOR

- Saturn has the most moons of any planet: at least 274 confirmed as of this writing, and still counting.

- In March 2025 alone, astronomers added 128 more confirmed moons to Saturn's total, further solidifying it as the most moon-full planet.

- **ONE OF SATURN'S MOONS, MIMAS, IS A SPHERE WITH A HUGE INDENTATION, SORT OF LIKE THE DEATH STAR IN *STAR WARS:* EPISODE IV *A NEW HOPE.***

- The impact that formed Mimas's giant crater probably almost shattered it and may have cracked its other side.

✦ ALL ABOUT TITAN

- Titan is Saturn's largest moon.

- Like Jupiter's moon Europa, Titan may have started as a minuscule grain of hail trapped by the planet's gravity.

- A dense atmosphere shrouds Titan, the only moon in our solar system with one.

- On Titan, your voice would sound slightly deeper, due to its denser atmosphere.

- What does it smell like on Titan? Eau de gasoline mixed with hints of bitter almonds and decomposing fish.

- Apart from Earth, Titan is the only known world to have liquid rivers, lakes, and seas. But they're filled not with water, but with liquid hydrocarbons (methane, for instance).

- It's so cold on Titan, any water is frozen into a thick crust of ice.
- Titan's icy crust may cover a subsurface ocean of water and ammonia.
- If it were to rain on Titan, the drops would fall in slow motion: about six times more slowly than on Earth.
- Titan has deserts, made not of the quartz sand found on Earth, but of water ice coated with hydrocarbons.
- It's dark on Titan. Very little visible light can shine through the atmosphere.
- The Huygens probe, designed exclusively to crash-land on Titan, parachuted from the Cassini spacecraft to Titan's surface in 2005.
- It sent back images and data during its 2.5-hour fall and became the first probe to ever land on a surface in the outer solar system.
- Huygens wasn't designed to survive the landing, but it transmitted images of its surroundings for three hours, revealing a vast, dry lake bed coated in ice-sand and speckled with smooth ice boulders.

✦ ALL ABOUT ENCELADUS

- Life elsewhere in the solar system? Saturn's moon Enceladus is a top candidate.
- **GIGANTIC GEYSERS OF SLUSHY WATER SPEW FROM CRACKS ON ENCELADUS.**
- The Cassini spacecraft once tasted the spray from Enceladus's geysers and discovered organic compounds.
- The icy crust of Enceladus makes it the shiniest body in our solar system.
- Enceladus's bright white surface is regularly refreshed by snow driven to the surface by its geysers.
- **ENCELADUS'S ERUPTING GEYSERS FORM ONE OF SATURN'S RINGS.**
- Gravitational tugs from Saturn and another nearby moon heat Enceladus's subsurface and drive its geysers.

✦ IAPETUS, DAPHNIS, AND MORE

- Like Earth's Moon, one side of Saturn's moons Iapetus and Dione always faces the planet and the other always faces away.
- The Saturn-facing side of Iapetus is stained reddish brown, likely deposited by collisions with other space rocks.
- **SATURN'S MOON DAPHNIS ORBITS WITHIN THE PLANET'S RINGS, LEAVING A WAKE BEHIND IN THE DEBRIS.**
- Saturn's moon Dione plows through debris, but its leading side is curiously smoother than its heavily cratered backside.
- Scientists think a powerful impact may have spun the moon around.
- Saturn's moon Pan is shaped just like a ravioli, with a circular, pillowy-looking center surrounded by frilled edges.
- Another of Saturn's moons—Atlas—is shaped like a cartoon flying saucer.
- Saturn has several so-called shepherd moons: Their gravity pushes the debris they encounter, tidying up the rings and clearing the gaps in between.
- Prometheus and Pandora shepherd Saturn's outermost ring, keeping it super skinny.
- Janus and Epimetheus, two more moons, share one orbital path, crisscrossing and swapping positions every four Earth years.
- Saturn's moon Hyperion is so cratered it looks like a sponge. It acts like one, too: Meteorite impacts squish its surface.
- **HYPERION TUMBLES SO CHAOTICALLY AS IT ORBITS SATURN THAT SCIENTISTS CAN'T PREDICT ITS FUTURE PATH.**
- Hyperion is so light that it could float in water.
- Hyperion may be a pile of rubble so small its gravity cannot fully compact it.

✦ CASSINI

- The Cassini spacecraft orbited Saturn from 2004 to 2017, darting between its many moons, rings, and icy satellites.
- Cassini revealed Saturn's rings to be much younger than any previous estimates—around 300 million years old, while Saturn itself is 4.5 billion.
- After more than a decade of studying Saturn, Cassini expended its final dregs of fuel in a dramatic final farewell.
- It dove through Saturn's atmosphere, sending back images and data that could only be obtained in a final daring death plunge.
- **IN THE *STAR TREK* UNIVERSE, COSMIC EXPLORERS UNDER STARFLEET COMMAND MUST OBEY THE PRIME DIRECTIVE, THEIR GUIDING PRINCIPLE: DO NOT INTERFERE WITH ALIEN CIVILIZATIONS, AT WHATEVER COST. CASSINI OBEYED THAT DIRECTIVE AS WELL.**
- Cassini's death dive wasn't just a flamboyant farewell or an attempt to gather otherwise impossible images, but also a protective measure for Saturn's moons—worlds unto themselves and some likely contenders to harbor alien life.
- Cassini snapped more than 450,000 images before its end.

CURRENT FACT COUNT 3217

SATURN: BY THE NUMBERS

7.5

Earth years: the length of each season on Titan

10

the number of times faster Saturn's jet streams travel than top winds on Earth

1.5

the number of Earth raindrops it would take to equal the size of Titan's raindrops, made of hydrocarbons

60

the percentage of Saturn's diameter made up of its core

800

miles an hour: the velocity of Enceladus's gushing geysers

10

the percentage by which Saturn is flattened, pole to pole relative to its equator, by the speed of its rotation

250

miles: the diameter of Mimas, making this little moon too small for its gravity to maintain a spherical shape

300

miles an hour: the wind speed of Saturn's hexagonal north pole jet stream

175,000

miles: the diameter of Saturn's rings

30

feet: the average thickness of Saturn's rings

80

the number of months NASA's Cassini spacecraft required to reach Saturn and enter into orbit around the planet

274

the number of known moons orbiting Saturn, by far the most of any planet in our solar system

764:1

the ratio of Earth's volume to that of Saturn

0.7

g/cm^3: the average density of Saturn relative to water, which means it would float in a big enough bathtub.

96

the percent of Saturn's volume occupied by hydrogen

3

the percent of Saturn's volume occupied by helium

10.7

hours: the length of a day on Saturn

5

the number of children eaten by the god Saturn (Cronus) out of fear one of them would ultimately overthrow him

CURRENT FACT COUNT 3223

ENCELADUS AND THE E RING

Scientific observations strongly suggest that Enceladus, like Europa, may have liquid water beneath its surface. Unlike other moons, though, Enceladus showcases its ocean by spraying so much salty water into space that it forms Saturn's diffuse and beautiful E ring.

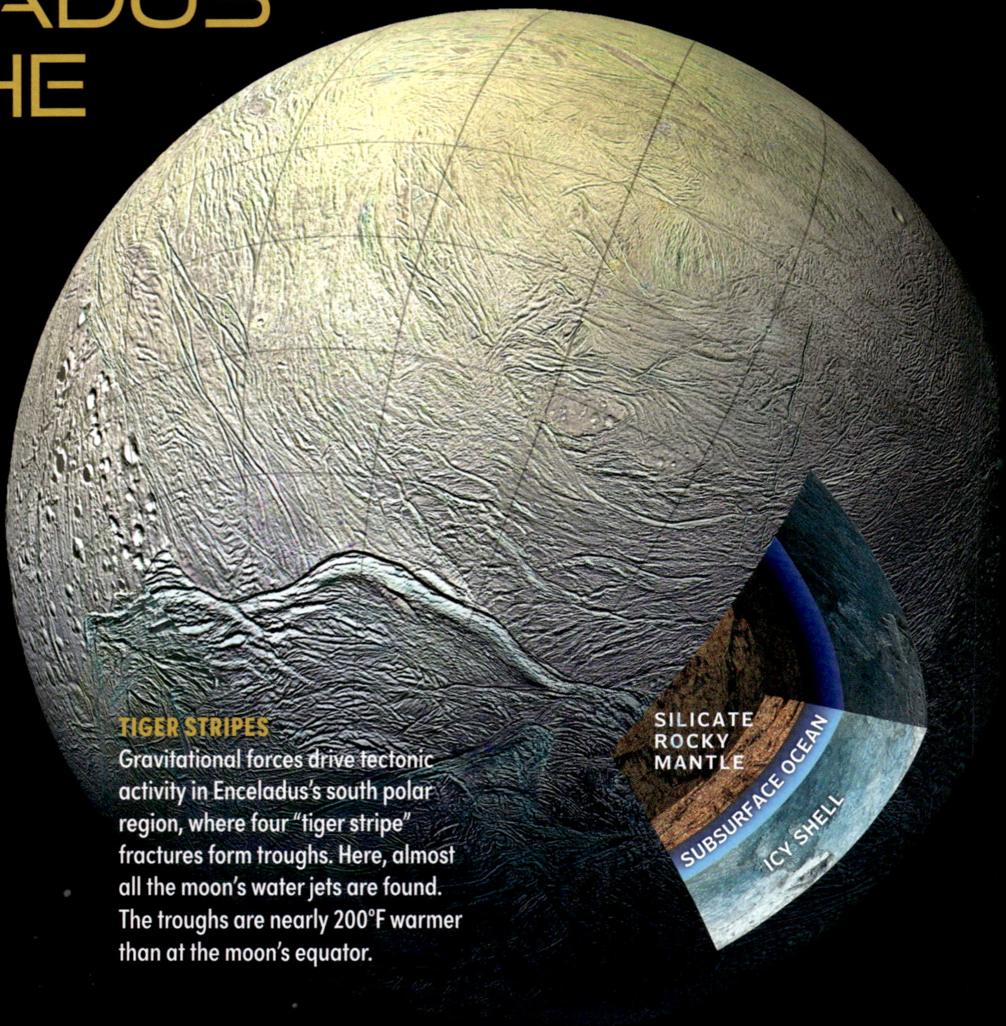

TIGER STRIPES
Gravitational forces drive tectonic activity in Enceladus's south polar region, where four "tiger stripe" fractures form troughs. Here, almost all the moon's water jets are found. The troughs are nearly 200°F warmer than at the moon's equator.

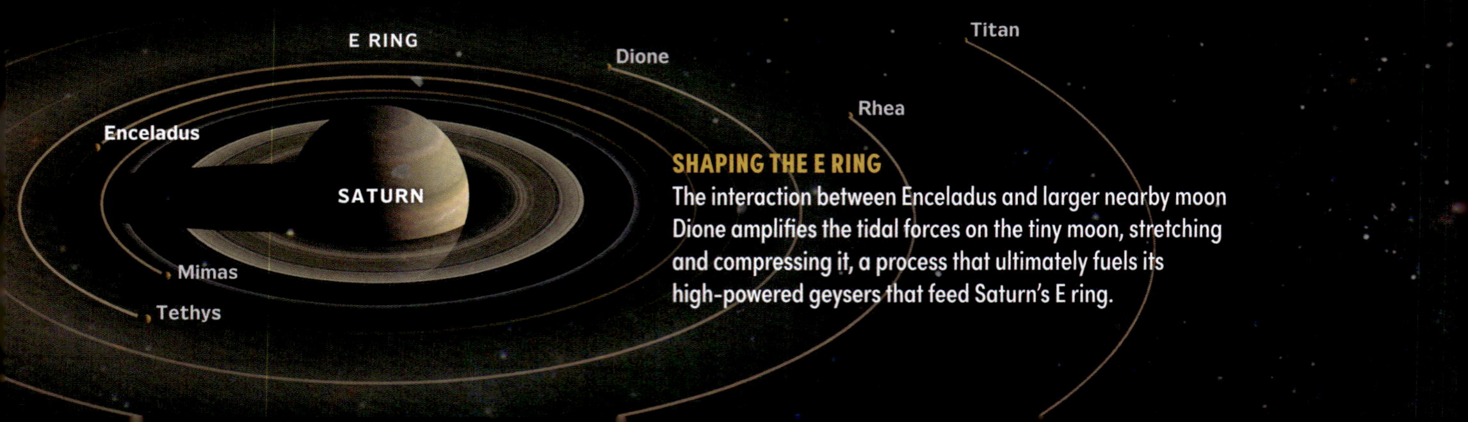

SHAPING THE E RING
The interaction between Enceladus and larger nearby moon Dione amplifies the tidal forces on the tiny moon, stretching and compressing it, a process that ultimately fuels its high-powered geysers that feed Saturn's E ring.

OCEAN CIRCULATION
Upwelling water warmed by Enceladus's mantle is thought to thin the ice sheet over its southern pole. Large-scale ocean circulation may create thicker ice along the equator.
FROZEN PLUME
Most of the gas and water particles that the moon's geysers erupt create an enormous plume of fine ice dust.
ORGANICS IN ICE
The density of organic material Cassini found in the plume during several flybys was about 20 times what scientists expected. Additionally, analysis of Saturn's E ring revealed silica nanograins, which can only form where rock and water interact at temperatures above 200°F, a sign of the moon's internal heat.

CURRENT FACT COUNT 3 2 3 1

TITAN AND THE OCTOCOPTER

Like Europa and Enceladus, Saturn's moon Titan is believed to have an interior saltwater ocean. But where Earth has a water cycle, Titan's methane cycle rains complex molecules onto its surface, creating methane lakes, rivers, and surface dunes made of granular hydrocarbons. NASA's thousand-pound Dragonfly octocopter will analyze the scientists' first ever samples of Titan's exotic terrain.

DRAGONFLY

Titan's atmosphere is 60 percent denser than Earth's at its surface, making the moon a perfect candidate for drone exploration. Powered by a 94-pound radioactive fuel source, the Dragonfly octocopter will return valuable data from in situ analysis of surface samples.

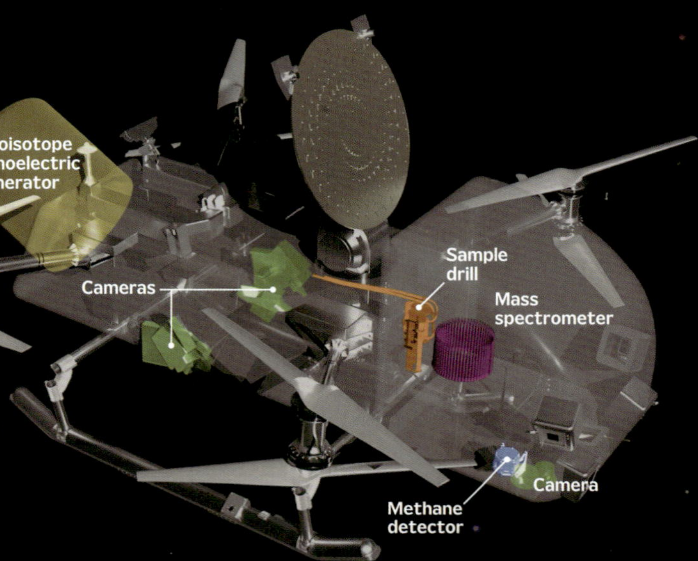

ONBOARD LAB

Thick insulation and a durable power source will warm Dragonfly's instruments to between minus 40°F and minus 22°F so they can operate on Titan's minus 290°F surface. A mass spectrometer will analyze surface samples for signs of life.

HIGH AND FAST

Dragonfly's eight-rotor design will allow it to fly at up to 22 miles an hour and reach altitudes about 13,000 feet above Titan's surface, enabling it to scout sites for surface analysis.

ONBOARD SENSORS

These will scan nearby surface minerals, search for subsurface liquid water, and monitor "Titanquakes" for signs of tectonic activity.

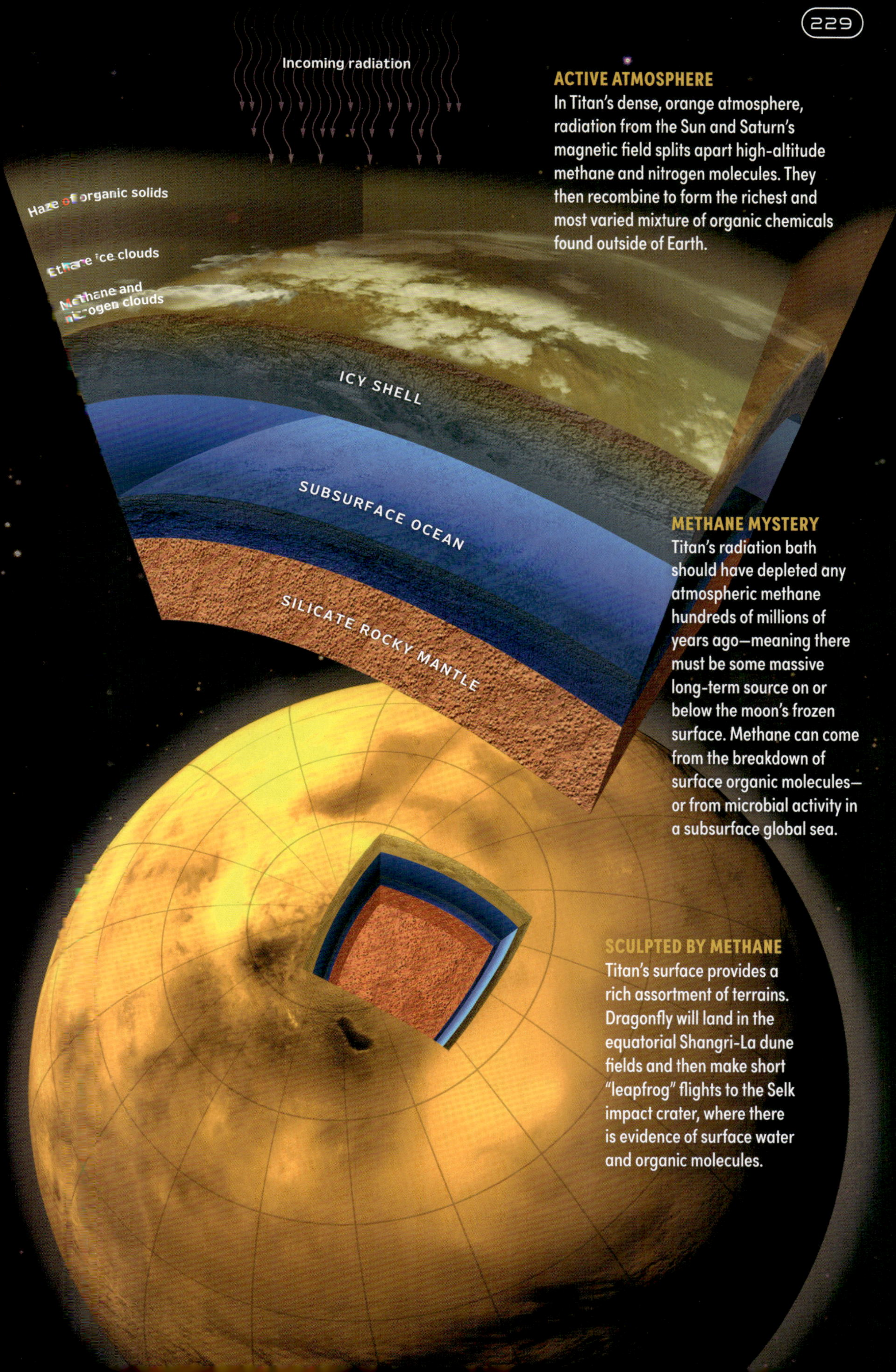

ACTIVE ATMOSPHERE

In Titan's dense, orange atmosphere, radiation from the Sun and Saturn's magnetic field splits apart high-altitude methane and nitrogen molecules. They then recombine to form the richest and most varied mixture of organic chemicals found outside of Earth.

METHANE MYSTERY

Titan's radiation bath should have depleted any atmospheric methane hundreds of millions of years ago—meaning there must be some massive long-term source on or below the moon's frozen surface. Methane can come from the breakdown of surface organic molecules—or from microbial activity in a subsurface global sea.

SCULPTED BY METHANE

Titan's surface provides a rich assortment of terrains. Dragonfly will land in the equatorial Shangri-La dune fields and then make short "leapfrog" flights to the Selk impact crater, where there is evidence of surface water and organic molecules.

CURRENT FACT COUNT 3237

OCEANS BEYOND EARTH

Over the next two decades, spacecraft will explore the watery moons of our solar system—some with perhaps even more water than we have here on Earth.

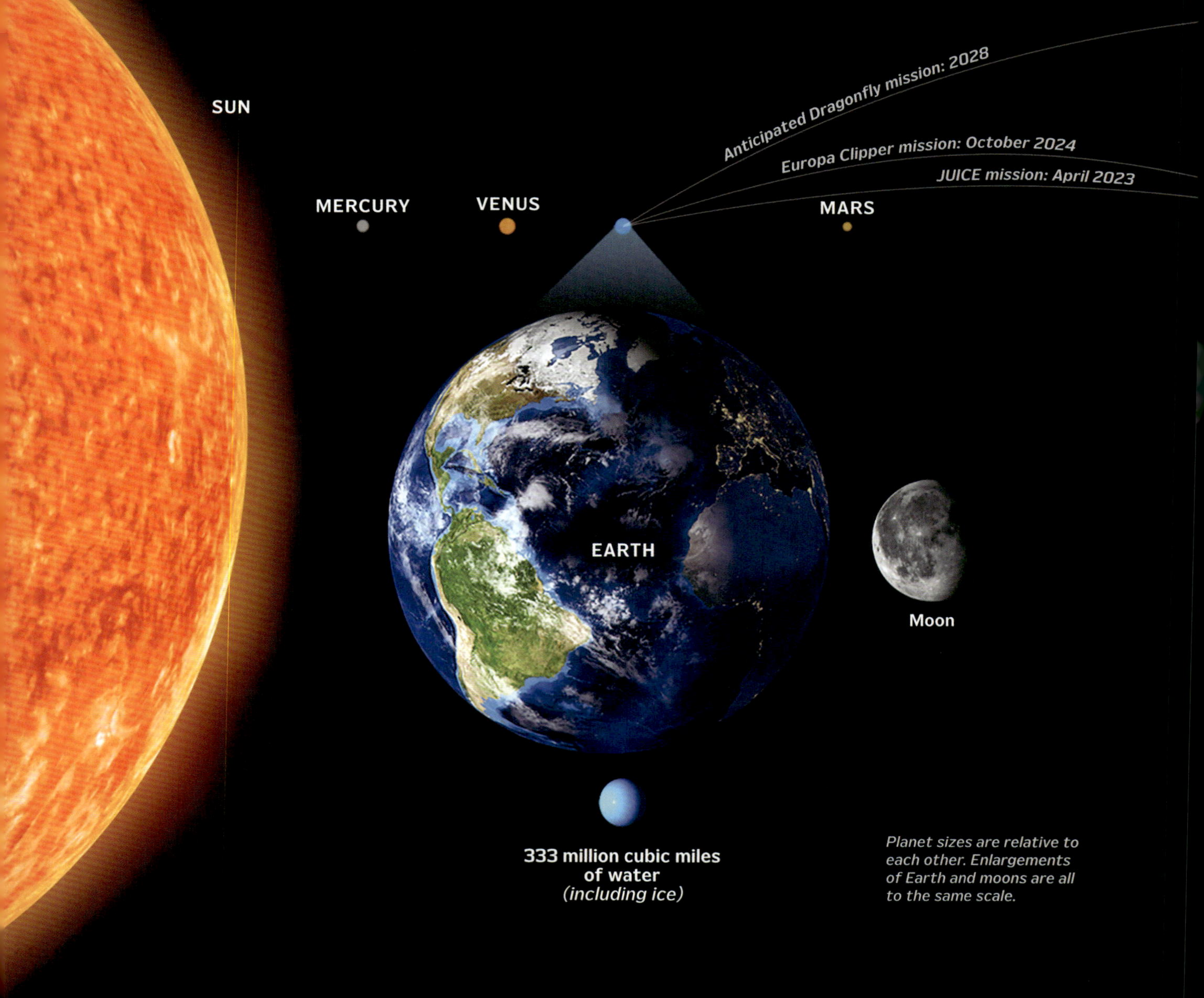

Planet sizes are relative to each other. Enlargements of Earth and moons are all to the same scale.

JUPITER

Jupiter has 95 moons, with most less than 100 miles in diameter. The largest are the planet's four innermost moons, first discovered by the astronomer Galileo in 1610, of which three are known to have subsurface oceans.

SATURN

Saturn's conspicuous rings, made of ice particles, are shaped and fed by some of its 274 moons, including water-spewing Enceladus. Scientists think that giant Titan and tiny Enceladus are promising candidates in the search for life in Saturn's system.

CURRENT FACT COUNT 3293

URANUS

An unconventional misfit of a world

✦ ICE GIANT

- Uranus was the first planet humanity discovered. Every other known planet was easily visible to the naked eye.
- **WILLIAM HERSCHEL DISCOVERED URANUS WITH A TELESCOPE IN 1781.**
- Caroline Herschel worked side by side with her brother and received a salary of 50 pounds a year, the first woman to be paid for scientific work.
- She was also the first woman to publish work in the *Philosophical Transactions of the Royal Society* and the first woman to be awarded the Gold Medal of the Royal Astronomical Society.
- William, upon discovering Uranus, was in denial that it could be a planet and presumed it a comet instead.
- His 1781 discovery paper of Uranus was titled "Account of a Comet."
- He initially named it Georgium Sidus after King George III of England. But the name didn't stick.
- For better or worse, we now call the planet Uranus, the latinized version of Ouranos, the ancient Greek god of the sky.
- Uranus is the only planet named for a Greek god instead of the Roman god equivalent. If we followed the traditional naming scheme, Uranus should really be called Caelus.
- **URANUS'S MOONS SIMILARLY BREAK NAMING TRADITIONS BY TAKING THE NAMES OF SHAKESPEAREAN AND OTHER LITERARY CHARACTERS.**
- Uranus is four times the size of Earth but about 15 times its mass.
- If Earth were the size of a nickel, Uranus would be the size of a softball.

An 18th-century engraving of William Herschel discovering Uranus

Uraninite, a uranium-rich mineral and ore

- Uranus is composed of a dense fluid of water, methane, and ammonia surrounding a small rocky core.
- **AT LEAST 80 PERCENT OF URANUS'S MASS IS DUE TO THE FLUID ON ITS SURFACE.**
- All the other giant planets (Neptune, Jupiter, and Saturn) emit twice as much heat as they absorb. Why Uranus doesn't is a mystery.
- A spacecraft approaching Uranus's vast ocean would be crushed by atmospheric pressure and shattered by its frigid temperatures before arrival.

✦ SPIN ME RIGHT ROUND

- While the other planets in our solar system rotate on vertical axes like spinning tops, Uranus rotates on its side, as if it's a ball rolling along its orbit.
- **URANUS IS THE ONLY PLANET THAT IS TIPPED ON ITS AXIS MORE THAN 90 DEGREES.**
- An ancient impact with one or more celestial bodies likely knocked Uranus and its rings off-kilter by 98 degrees.

- The impact that knocked Uranus sideways (if an impact was responsible) may also have destroyed its internal reservoir of heat: It's the coldest planet in the solar system.
- One of this book's authors—Neil deGrasse Tyson—published a paper that concerned the tilt of Uranus titled "On the Possibility of a Major Impact on Uranus in the Past Century," in which this possibility was ruled out.
- Another hypothesis is that the tilt is due to the gravitational influence of a large, now missing moon that gradually whirled Uranus askew.
- Uranus's dramatic tilt creates bizarre and dramatic seasons, each lasting a quarter of its year.
- So, a human resident on Uranus would spend a quarter of their life in one Uranian season.
- Uranus's moons orbit the planet's tilted equator like the cars of a Ferris wheel.
- **AT NIGHT, IF YOU KNOW EXACTLY WHERE TO LOOK, THE PLANET IS JUST BARELY VISIBLE TO THE UNAIDED EYE.**
- Methane in Uranus's atmosphere is responsible for its beautiful pale blue color.
- With its sideways spin, Uranus heats unevenly and kicks up stormy weather.
- The planet has a fluffy atmosphere of hydrogen, helium, and methane.
- Uranus and Neptune are known as ice giants because their atmospheres have plenty of ice-forming molecules, including water.
- **URANUS HAS ITS OWN SPECIAL KIND OF AURORAS THAT ONLY GLOW IN THE ULTRAVIOLET AND INFRARED LIGHT SPECTRUM.**

The Hubble Space Telescope shot this enhanced image of Uranus with its rings and moons in 2003.

- Uranium, discovered in 1789, was named for the planet Uranus.
- A cyclone swirls at Uranus's poles.
- Occasional bright, stormy spots speckle the planet.
- Uranus's rings are the remnants of destroyed moons.
- Uranus's core may have lost most of its primordial heat after a devastating impact, but not all scientists agree on that.
- Voyager 2 is the only spacecraft ever to visit Uranus.
- The Voyager mission visited all four outer planets, capitalizing on an alignment that only happens every 175 years.

The James Webb Space Telescope captured Uranus and six of its moons.

✦ LITERATE MOONS

- Uranus has 29 known moons.
- **ITS FIVE MAJOR MOONS ARE CALLED THE LITERARY MOONS, BECAUSE THEY ARE NAMED FOR CHARACTERS IN ENGLISH LITERARY CLASSICS.**
- Neil deGrasse Tyson's daughter is named Miranda, after one of the moons of Uranus, discovered in 1948 by Dutch American planetary scientist Gerard Kuiper.
- Miranda was also the heroine in Shakespeare's *The Tempest.*
- Oberon and Titania are the largest of the moons and the first to be discovered, in 1787.
- Their discoverer, William Herschel, named them for the king and queen of the forest in Shakespeare's *A Midsummer Night's Dream.*
- Uranus's moon Ariel is named for a sprite in Shakespeare's *The Tempest.*
- Umbriel is a sprite in Alexander Pope's *The Rape of the Lock.*
- In 2024, scientists announced the discovery of a new moon, the first in more than 20 years. Yet another was discovered in 2025.
- Aside from the two new discoveries and three other moons named from Alexander Pope's poem, all the other moons of Uranus have Shakespearean names.
- Many of Uranus's moons are hard to see because they're darker than asphalt.
- Miranda's landscape includes canyons up to 12 times as deep as the Grand Canyon.
- **GRAVITY ON MIRANDA IS SO WEAK THAT IF YOU TOSSED A ROCK OFF THE EDGE OF ITS HIGHEST CANYON, IT WOULD TAKE 10 MINUTES TO LAND.**
- Eruptions of an ammonia-water mix blast onto Miranda's surface, likely driven by internal melting from space rock impacts.
- Some of Uranus's moons may be ocean worlds.
- Its inner moons all seem to be made of half water ice and half rock.
- Uranus's outer moons all seem to be captured asteroids.
- Cordelia and Ophelia are likely shepherd moons, their gravitational force tidying up one of the planet's rings.

"NO OTHER PLANET HAS AS MANY SMALL INNER MOONS AS URANUS, AND THEIR COMPLEX INTER-RELATIONSHIPS WITH THE RINGS HINT AT A CHAOTIC HISTORY THAT BLURS THE BOUNDARY BETWEEN A RING SYSTEM AND A SYSTEM OF MOONS."

—PLANETARY SCIENTIST MATTHEW TISCARENO

CURRENT FACT COUNT 3311

URANUS: BY THE NUMBERS

21

the number of Earth years that sunny summers last on Uranus

84

the number of Earth years in a single Uranus year

21

the number of Earth years that pitch-dark winters last on Uranus

10

the number of minutes it would take for a rock, dropped off Miranda's highest cliff, to fall to the surface, thanks to the moon's low gravity

160

minutes: how long it takes sunlight to reach Uranus

9,000°F

the estimated temperature of Uranus's core

-320°F

the mean temperature on the cold and windy planet

130

pounds: your weight on Uranus, if on Earth you weigh 150 pounds

-371.56°F

the lowest temperature ever measured on Uranus, the most frigid planet in the solar system

19

astronomical units: Uranus's distance from the Sun

17

hours: the length of one day on Uranus

7TH

Uranus's planetary position from the Sun

1977

the year Uranus's first rings were discovered

1.8

billion miles: the average distance of Uranus from the Sun

3

the number of times Uranus has orbited the Sun since its discovery in 1781

13

the number of faint rings encircling the planet

560

miles an hour: wind speed near Uranus's equator

1,785,600,000

miles: the distance from the Sun to Uranus

CURRENT FACT COUNT 3367

NEPTUNE

A mysterious, underexplored world and the most isolated planet, Neptune lies a billion miles beyond its closest neighbor (Uranus) and nearly two billion miles from the dwarf planet Pluto.

✦ THE DISTANT WORLD

- Neptune was the first planet discovered using math—not with telescopes, not with the naked eye, but with cold, hard calculations.
- In 1846, after the discovery of Uranus, two astronomers independently tracked irregularities in Uranus's orbit that defied Newton's laws of gravity.
- To account for Uranus's odd orbital path, they concluded that either Newton's laws failed at those huge distances from the Sun or there was a yet-to-be-discovered planet tugging on Uranus.
- The astronomers' names were Urbain Le Verrier and John Couch Adams.
- When they communicated a likely position in the sky to Johann Gottfried Galle at the Berlin Observatory, he observed planet Neptune that very night: September 23, 1846.
- **ONCE NEPTUNE WAS IDENTIFIED, SCIENTISTS TRACED PAST OBSERVATIONS OF IT GOING BACK TO GALILEO.**
- Neptune is named for the Roman god of the sea.

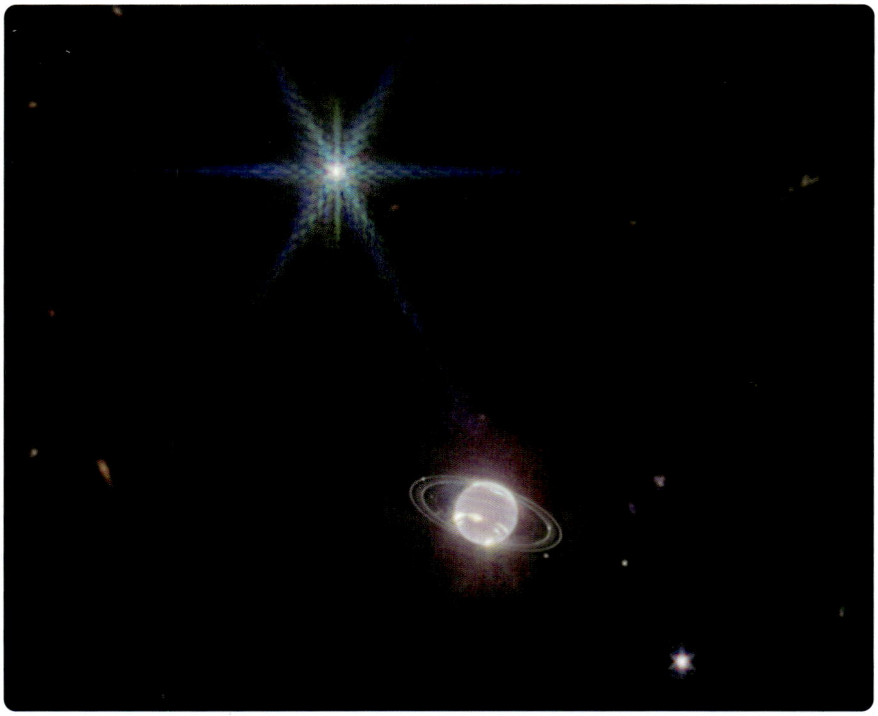

The James Webb Space Telescope's first image of Neptune (with its moon Triton above) shows the planet's bright rings and dust bands.

- Neptune's moons followed suit and, of those named, have been given monikers for mythological water gods and creatures.
- **ONLY ONE SPACECRAFT HAS EVER VISITED NEPTUNE: VOYAGER 2 FLEW PAST THE PLANET IN 1989.**
- Voyager 2 captured images of a raging storm on Neptune, dubbed the Great Dark Spot.
- Later, Hubble images suggested that the storm has abated.
- Neptune's Great Dark Spot was large enough to swallow Earth.
- Since Neptune is so far from the Sun, solar heat likely can't fuel storms, leaving their cause a mystery.
- In 2011, Neptune completed its first orbit around the Sun since its 1846 discovery.
- If Earth were the size of a nickel, Neptune would be the size of a baseball.
- Neptune is the solar system's only planet that can't be seen with the naked eye.
- Its rings are mysteriously clumpy.
- At 30,775 miles across, Neptune is about four times wider than Earth.
- While most pictures depict Neptune as deeper blue than Uranus, they may actually be the same pale aqua color. Recalibrations of old Voyager 2 data and new Hubble data revealed the more accurate hue.
- **IT MIGHT RAIN DIAMONDS ON NEPTUNE AND URANUS, THANKS TO PRESSURES AND TEMPERATURES STRONG ENOUGH TO SQUASH CARBON ATOMS INTO GEMSTONES.**
- On Neptune, wind speeds are supersonic—faster than the speed of sound on Earth. They can also be up to three times faster than wind speeds on Jupiter.

Neptune, depicted here via images shot through the ultraviolet, violet, and green filters of Voyager 2's wide-angle camera

- Sunlight takes four hours to travel from the Sun to Neptune.
- The planet's temperature is close to Uranus's, even though it orbits farther from the Sun.
- Its core is about the same mass as Earth's.
- Take a dip in Neptune's ocean? You'll be boiled alive and crushed under pressure.
- On Neptune, the pressure under the atmosphere is so strong that liquids can't evaporate.
- You can't count on your compass on these outer planets: While Earth has two magnetic poles (north and south), Neptune and Uranus have many, sending the compass needle whirling.
- **NEPTUNE ROTATES ONCE EVERY 16 HOURS: THE LENGTH OF A NEPTUNIAN DAY.**
- The planet revolves around the Sun every 165 Earth years. That's how much longer a Neptunian year is than ours.

✦ THE FIVE RINGS

- There are five known rings circling Neptune: Adams, Arago, Galle, Lassell, and Le Verrier.
- These rings contain clumps, called arcs, of dusty material that appear to defy the laws of motion and remain bundled together, possibly due to the gravitational effects of the nearby moon Galatea.
- The arcs in the outermost ring, Adams, are called Liberté (Liberty), Egalité (Equality), Fraternité (Fraternity), and Courage.

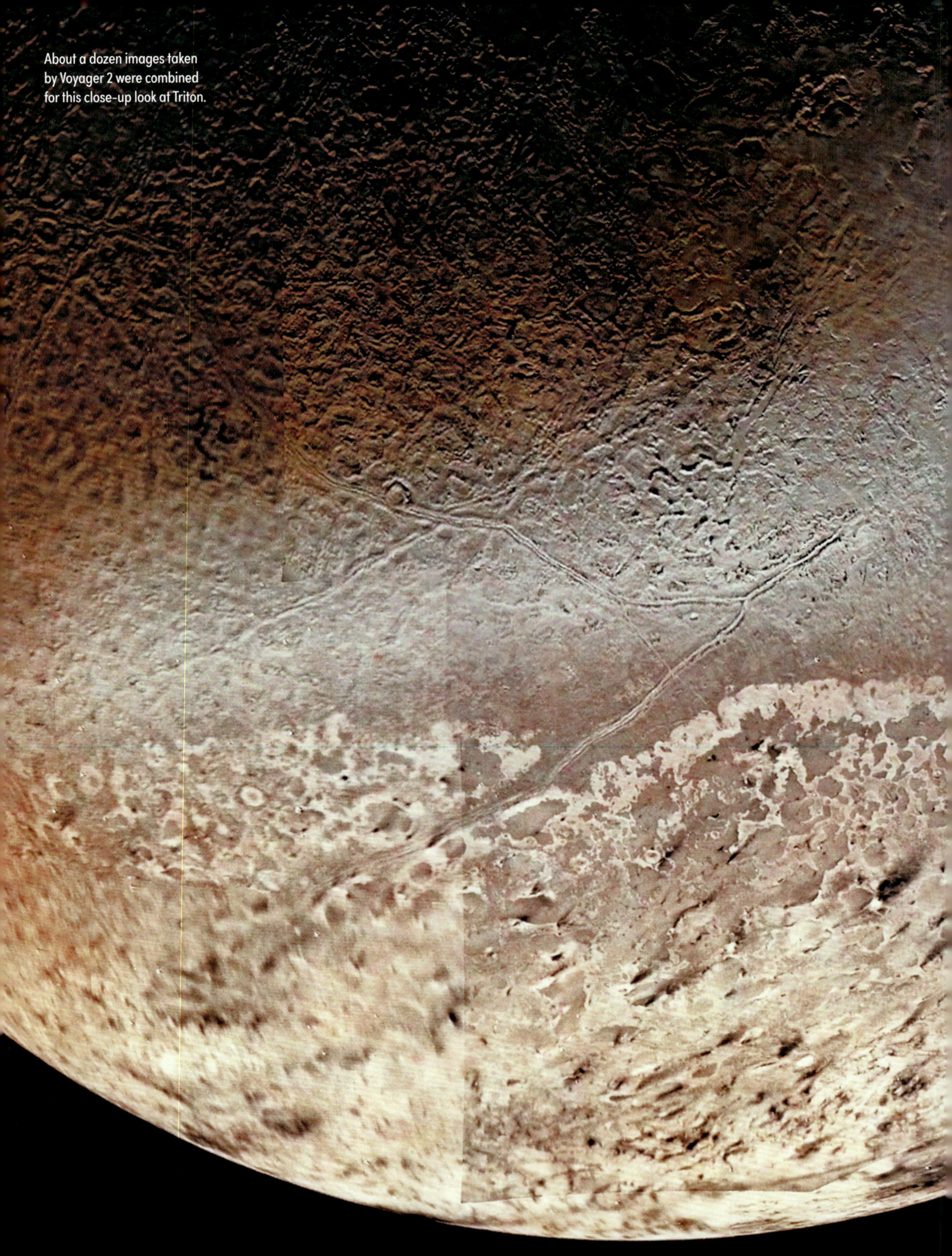

About a dozen images taken by Voyager 2 were combined for this close-up look at Triton.

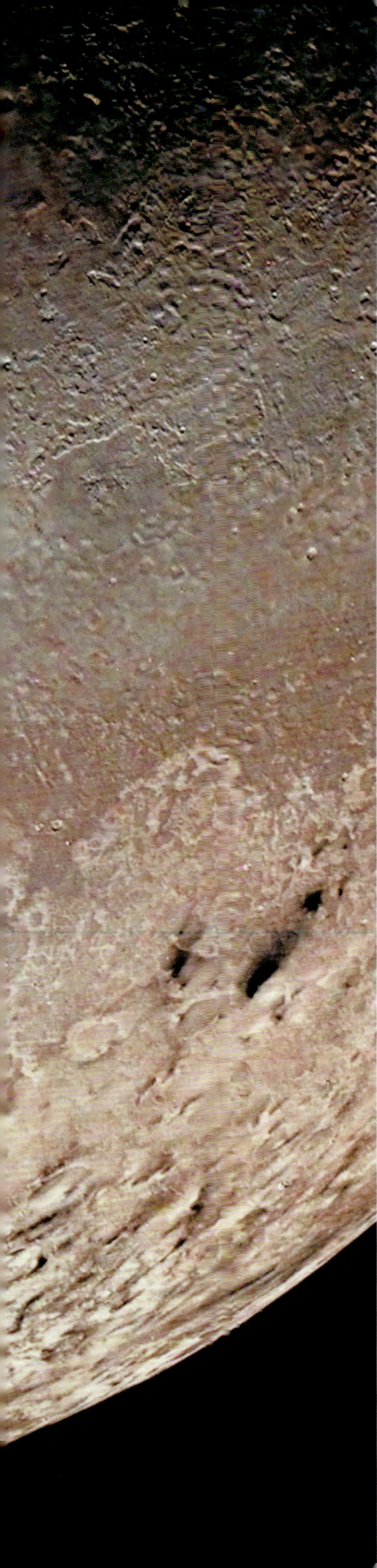

✦ NEPTUNE'S MOONS

- Neptune has 16 moons that we know of, but only 14 have been named.

- All of the planet's moons are named after lesser sea gods and water nymphs from Greek mythology.

- **THEIR NAMES, FROM MOST MASSIVE TO LEAST MASSIVE, ARE:**
  - Triton
  - Proteus
  - Nereid
  - Larissa
  - Galatea
  - Despina
  - Thalassa
  - Naiad
  - Halimede
  - Neso
  - Sao
  - Laomedeia
  - Hippocamp
  - Psamathe

- Neso takes 27 Earth years to orbit Neptune, the longest orbit of any known moon in the solar system.

- The moon Proteus, discovered in 1989 by Voyager 2, is one of the darkest objects in our solar system.

- Triton was actually discovered 10 days after its home planet by English astronomer William Lassell.

A depiction of the power of the geysers on Triton

- Triton orbits in the opposite direction of Neptune's rotation, called a retrograde orbit.

- **VOLCANOES ON TRITON SPEW NOT LAVA BUT ICE.**

- The cryovolcanoes can belch liquid nitrogen and a slurry of frozen particles more than five miles into the air.

- It's so cold on Triton that most of its nitrogen has condensed into shiny frost.

- The icy surface reflects 70 percent of the sunlight that reaches it.

"SIR, THE PLANET WHOSE POSITION YOU MARKED OUT ACTUALLY EXISTS. ON THE DAY YOUR LETTER REACHED ME, I FOUND . . . THE PLANET OF WHICH WE WERE IN QUEST."

—JOHANN GOTTFRIED GALLE, IN A LETTER TO URBAIN LE VERRIER

CURRENT FACT COUNT 3387

NEPTUNE: BY THE NUMBERS

8

hours: how long it takes for Neptune's tiny moon Despina to complete one full orbit

12

the number of Earth years a modern rocket requires to reach Neptune

1

the number of times Galileo spotted Neptune in the sky but didn't know it was a planet

1,200

miles an hour: the speed of winds on Neptune, the windiest planet in our solar system

237

years: the time that elapsed after Galileo unwittingly saw Neptune before it was "officially" discovered to be a planet

-391°F

the surface temperature of Triton, one of Neptune's moons

165

Earth years: the orbital period of Neptune

1949

the year Neptune's second moon, Nereid, was discovered

1

the number of complete orbits made by Neptune since it was discovered in 1846

16

the number of known moons orbiting Neptune

5

the number of spacecraft that have passed Neptune and exited the solar system, all NASA: Pioneer 10 and 11; Voyager 1 and 2; and New Horizons

9.3

miles: the radial width of Neptune's ring Galle, the closest ring to the planet

9

years: how long it took for the New Horizons mission to pass the orbit of Neptune en route to Pluto (the fastest ever)

205

pounds: how much you'd weigh on Neptune if you weigh 150 pounds on Earth

500

minutes: the time it takes a beam of light to cross the diameter of Neptune's orbit

16

hours: the length of a day on Neptune

57

the number of times larger, in volume, Neptune is compared with Earth

80

percent: Neptune's volume that's occupied by hydrogen

12,120

miles an hour: the speed of Neptune in orbit around the Sun

17

the number of times greater Neptune's mass is compared with Earth's

CURRENT FACT COUNT 3441

KUIPER BELT & OORT CLOUD

Like the asteroid belt, the Kuiper belt ferries ancient vestiges and ruins of would-be planets. But don't confuse the Kuiper belt with the Oort cloud, which is 1,000 times more distant—so distant, it is yet to be observed at all.

- The Kuiper belt and Oort cloud are two distinct, massive arrays of objects wrapping around our solar system.
- The Kuiper belt is a doughnut-shaped belt of icy objects at the outer edge of our solar system.
- The Oort cloud is a vast, distant spherical shell of billions of comets encasing our entire solar system and extending trillions of miles beyond the Kuiper belt.
- **BOTH VOYAGER 1 AND 2 HAVE EXITED THE KUIPER BELT BUT WON'T REACH THE HYPOTHETICAL OORT CLOUD FOR ANOTHER 300 YEARS—THEY'LL NEED ANOTHER 30,000 YEARS TO EXIT IT.**
- The Kuiper belt is slowly disappearing, as objects are blown away by solar winds or collide to create fragments that may zip away as comets.
- The objects of the Kuiper belt may still contain ancient ice, little changed since the dawn of our solar system.
- The Kuiper belt begins at Neptune's orbit (which is about 30 times the distance from the Sun to Earth) and stretches to more than 100 times that distance.
- It's the largest mapped structure in the solar system.

✦ THE KUIPER BELT

- Most dwarf planets—including Pluto—lie within the Kuiper belt.
- Neptune's orbit defines the inner edge of the Kuiper belt.
- Most bodies of the Kuiper belt are icy—except for 2004 EW95, a carbon-covered asteroid likely flung from the inner solar system.
- Comets in the Kuiper belt take less than 200 years to orbit the Sun.

✦ KBOs

- Like the asteroid belt, Kuiper belt objects (KBOs) are probably leftovers from our solar system's formation.
- There may be hundreds of thousands of KBOs larger than 60 miles across and millions even smaller.
- The mass of all KBOs taken together is likely no more than 10 percent of Earth's.
- **COLLIDING KBOs MIGHT BREAK INTO CHUNKS THAT WHIZ THROUGH THE INNER SOLAR SYSTEM, PROPELLED BY NEPTUNE AND JUPITER'S GRAVITY.**
- If it weren't for Neptune's gravitational meddling, KBOs may have eventually combined to form another planet.
- In 1992, David Jewitt and Jane Luu discovered the first KBO. They wanted to name it Smiley, but instead, it's called 1992 QB1.
- Many KBOs have moons and even the occasional ring.
- NASA's New Horizons visited a KBO up close for the first time in 2015: Arrokoth.
- In 2019, New Horizons sent the first pictures of a KBO other than Pluto: Arrokoth, looking like a lumpy snowman.
- Arrokoth is 22 miles long.
- Estimates of Arrokoth's age put it at 4.5 billion years.
- It orbits four billion miles from the Sun.
- Its two lobes probably didn't collide but swirled closer and closer to one another until they gently smooshed in a cosmic kiss.
- **THE FARTHEST ASTEROID EVER OBSERVED IN THE SOLAR SYSTEM IS A KBO APTLY NAMED FARFAROUT.**
- Farfarout—officially known as 2018 AG37—is four times farther from the Sun than Pluto.
- Farfarout is even farther away than Farout, the previous record holder for the solar system's farthest observed object.
- Farfarout takes a millennium to complete a single orbit of the Sun.

✦ PLANET X: A BRIEF HISTORY

- Some astronomers are searching for a hypothesized "Planet X" to explain orbital anomalies of objects in the Kuiper belt. This hypothesized planet would be about six times the mass of Earth and orbit 15 times farther away from the Sun than Neptune.

So why can't we find Planet X? If it is indeed 15 times farther away than Neptune and about the same size as Neptune, that means it's about 50,000 times fainter than Neptune.

1846: Based on the odd orbital behavior of Uranus, a more distant planet invisible to the naked eye was hypothesized to exist. Then, Le Verrier discovered the predicted planet Neptune.

1906: Following the same logic that led to Neptune's discovery, astronomer Percival Lowell documented anomalies in Uranus and Neptune's orbits, leading him to believe that an even more distant Neptune-size planet must exist. It was nicknamed "Planet X."

1930: PLUTO WAS DISCOVERED BY CLYDE TOMBAUGH. IT WAS HERALDED AS THE LONG-LOST PLANET X.

1978: Pluto's moon Charon was discovered, allowing astronomers to accurately estimate Pluto's mass for the first time.

It turns out, Pluto's mass is a mere fraction of one percent of Earth's. It's also thousands of times too small to account for any orbital anomalies observed in its neighbors. The search for Planet X continues.

1993: E. Miles Standish, Jr. used new Voyager data to correct previously erroneous measurements of Neptune's and Uranus's masses, proving that their orbits are exactly as they should be and killing any hopes for Planet X.

2006: The International Astronomical Union defined, for the first time, what the word "planet" means.

WITH THAT DEFINITION PLUTO WAS DEMOTED TO A DWARF PLANET.

- 2016: Hope for a Planet X was revived by some of the same astronomers responsible for Pluto's demotion. They proposed that orbital anomalies in the distant Kuiper belt could be attributed to a planet larger than Earth. The search goes on ...

✦ THE OORT CLOUD

- The Oort cloud was first predicted in the 1950s by Jan Oort.
- It is purely theoretical. It's located so far away that no one has ever directly observed it, so its existence has only been inferred by models.
- **MORE THAN A TRILLION ICY OBJECTS MIGHT POPULATE THE OORT CLOUD.**
- The cloud starts between 2,000 to 5,000 times the distance between Earth and the Sun.
- And the cloud stretches outward to 100,000 times or more than the distance between Earth and the Sun.
- The Oort cloud's icy objects are more likely interstellar visitors than solar system residents.
- The objects in the cloud barely feel the Sun's gravity and are easily disturbed by things like passing stars.
- Oort cloud objects can have unusual orbits that don't all follow the same plane as the planets.
- Unlike the disk shape of our solar system, the Oort cloud objects form a giant bubble around us.
- **COMETS IN THE OORT CLOUD CAN TAKE FROM 200 TO TENS OF MILLIONS OF YEARS TO ORBIT THE SUN.**
- In 2013, C/2013 A1 Siding Spring—a likely former Oort cloud resident—whizzed close by Mars, racing along at 35 miles a second.
- After comet Siding Spring's close encounter with the planet, Mars was bathed in cometary gas and dust for hours.
- It may be another 740,000 years before the Siding Spring comet zips through the solar system again.
- Sunlight takes almost a month to reach the inner edge of the Oort cloud—and up to a year and a half to cross its outer edge.

"JUST LIKE THE ASTEROIDS AND THE EARTH ARE DIFFERENT, THE COMETS AND THE SMALL PLANETS OF THE KUIPER BELT ARE VERY DIFFERENT. BUT, THEY'RE ALL TEACHING US ABOUT THE ORIGIN OF THE SOLAR SYSTEM."

—PLANETARY SCIENTIST ALAN STERN

CURRENT FACT COUNT 3555

DWARF PLANETS

Astronomers were forced to rewrite the membership rules of planethood in response to the existence of these small yet dynamic worlds.

✦ DEFINING PLANETS

- Before 2006, the word "planet" had no official definition.
- So until 2006, Pluto was considered the ninth planet in the solar system.
- **BUT THE 2005 DISCOVERY OF ERIS, A MORE MASSIVE OBJECT FARTHER FROM THE SUN THAN PLUTO, INSTIGATED DELIBERATIONS.**
- On August 24, 2006, the International Astronomical Union redefined Pluto as a dwarf planet.
- Now, a planet in our solar system must meet three criteria.
- **FIRST, IT MUST BE ROUND.**
    - To be a sphere in space is a feat only very large objects can accomplish once the pull of their own mass overcomes the pull of external forces.
    - That's why asteroids and small moons (like Mars's two potato-shaped moons) take on odd shapes.
- **NEXT, IT MUST PRIMARILY ORBIT THE SUN AND NOT ANOTHER OBJECT.**
    - Seems obvious, but a moon is a moon and not a planet for this fact alone.
    - Even Ganymede, larger than Mercury or Pluto, pledges its orbital allegiance to Jupiter, not the Sun.
    - Sometimes, moons or asteroids have their own moon. In the case of the former, they're sometimes called "moonmoons."
- **AND LASTLY, IT MUST CLEAR ITS ORBIT.**
    - In other words, the object's gravity must be so powerful that no other objects within its orbital space can compete.
    - It's not just a large sphere, but the ringleader.
    - When a planet encounters debris, asteroids, or other small bodies of any kind along its orbital path, the planet will either accrete the smaller body or kick the smaller body out into a new orbit.
- If a body meets criteria one and two, but not the third, that body is what's now known as a dwarf planet.
- **PLUTO FAILED CRITERION NUMBER THREE.**
- The Kuiper belt, where Pluto resides, contains objects similar in size to Pluto.
- The dwarf planet Eris is more massive than Pluto.
- We have officially discovered and named five dwarf planets.
- But hundreds more likely lurk undiscovered in our outer solar system.
- **A NONEXHAUSTIVE LIST OF CELESTIAL OBJECTS RECLASSIFIED FROM "PLANET" STATUS:**
    - The Sun: promoted to a "star" in 1543
    - The Moon: demoted to ... well, a "moon" in 1543
    - Vulcan: demoted to "an utterly nonexistent figment of imagination" in 1916
    - Vesta: demoted to "asteroid" in 2006
    - Pallas: demoted to "asteroid" in 2006
    - Juno: demoted to "asteroid" in the 1850s
    - Ceres: demoted to "asteroid" in the 1850s
    - Ceres again, this time promoted to "dwarf planet" in 2006

Ceres's 57-mile-wide Occator crater is centered in this projection by the Dawn spacecraft.

NASA's Dawn spacecraft took this orthographic projection of Ceres's Occator crater.

American astronomer Clyde Tombaugh

✦ PLUTO: EVERYBODY'S FAVORITE DWARF PLANET

- Pluto is roughly three billion miles from Earth.
- Despite decades of debate, it remains a dwarf planet.
- If the word "planet" were redefined to include Pluto, hundreds of other objects would become planets.
- Some argue if Earth orbited as far away as Pluto, it would lose its planetary status, too, because its gravity couldn't clear its path.
- The American Dialect Society chose "Plutoed"—meaning "to be demoted"—as the 2006 Word of the Year.
- **A SHORT TIMELINE OF THE DISCOVERY AND DEMOTION OF PLUTO:**
  - 1905: Observations of Neptune and Uranus's orbits hint at a ninth planet.
  - 1930: Astronomer Clyde Tombaugh discovers the elusive Planet X.
  - 2005: The discovery of Eris, initially touted as a 10th planet, sparks debate about what defines a planet.
  - 2006: Pluto, along with Eris and Ceres, are reclassified as dwarf planets.
- **SOME MONUMENTAL WORKS OF ART STILL STAND THE TEST OF TIME, HAVING BEEN CREATED AFTER THE DISCOVERY OF NEPTUNE BUT BEFORE THE DISCOVERY OF PLUTO:**
  - Bronze plaques showing eight planets at Chicago's Adler Planetarium, created in 1930
  - Gustav Holst's orchestral suite *The Planets,* composed from 1914 to 1916
- Early 20th-century images of the sky included Pluto at least 16 times before it was officially discovered by Tombaugh in 1930.
- The earliest image that included Pluto was taken by the Yerkes Observatory in Wisconsin in 1909.
- From Pluto's surface, the Sun would be dim, but still about 250 times brighter than Earth's full moon.
- Pluto was first thought to be hundreds of times more massive than it is now believed to be.
- In 1930, the Lowell Observatory held a naming contest for the newly discovered celestial body that would become known as Pluto.
- They received more than 1,000 ideas.
- Eleven-year-old Venetia Burney suggested the planet be named after the god of the underworld.
- Burney was given five pounds as a reward for the winning name.
- First detected in 1941, the element plutonium was named after Pluto.
- Some say Disney's Pluto the dog was named after the newly discovered planet.
- **ASTRONOMERS INITIALLY THOUGHT PLUTO WAS LARGER THAN MERCURY.**
- We now know Pluto is about two-thirds the diameter of Earth's Moon.
- Pluto's highly elongated orbit crosses inside Neptune's, bringing Pluto closer to the Sun than Neptune for 20 years every 248 years.
- 1979–1999: the last time Pluto was closer to the Sun than Neptune
- 2226–2246: the next time Pluto will be closer to the Sun than Neptune

Fans of the demoted planet sometimes used Hubble images to offer their support, like in this bumper sticker.

- For every three times Neptune orbits the Sun, Pluto orbits twice, keeping the worlds far from collision.
- Pluto's orbit is tilted by 17 degrees relative to the other planets, which also helps prevent collisions with Neptune.
- When Pluto gets its closest to the Sun, its surface ice becomes a gas, forming a thin atmosphere.
- **WHEN PLUTO IS FARTHEST FROM THE SUN, MOISTURE IN THE ATMOSPHERE LIKELY FREEZES AND FALLS AS SNOW.**
- Pluto's mountains are giant blocks of water ice, occasionally frosted with frozen gases such as methane.
- Pluto has giant featureless plains of frozen nitrogen, likely resurfaced by slow circulation of the gas.
- Pluto has a vast heart-shaped plain, frozen and craterless, informally named Tombaugh Regio.
- The region drives atmospheric circulation on Pluto, sweeping in the opposite direction of the dwarf planet's rotation.
- Icy mountains near the Tombaugh Regio have been informally named after the first two people to summit Mount Everest, Tenzing Norgay and Sir Edmund Hillary.
- An area within the Tombaugh Regio has been informally named Sputnik Planum, after the first satellite launched into space.
- The ice filling the heart-shaped Tombaugh Regio weighs so much it may have tipped the dwarf planet.
- **IN 2178, PLUTO WILL COMPLETE ITS FIRST ORBIT OF THE SUN SINCE ITS DISCOVERY.**
- If Pluto's orbit changed to bring it close to the Sun, it would grow a tail and look like a jumbo comet.

A close-up look at 260 miles of Pluto's surface

- Similar to Earth's Moon, Pluto's moons likely formed from a collision early in its history.
- In keeping with its name, all Pluto's moons are named for mythological characters of the underworld. The dwarf planet's largest moon, Charon, is named for the ferryman who took souls to the underworld.
- Charon orbits Pluto every 6.4 Earth days.
- Like Earth and the Moon, Pluto and Charon are tidally locked: The same side of Charon faces Pluto through its entire orbit.
- Pluto's moons spin surprisingly fast, making multiple rotations each time they revolve around the dwarf planet.
- Hydra is Pluto's fastest-spinning moon, rotating 89 times per trip around the dwarf planet.

✦ OTHER DWARF PLANETS

- The International Astronomical Union (IAU) officially recognizes four dwarf planets besides Pluto: Haumea, Ceres, Eris, Makemake.
- There are many other candidates for dwarf planet status. But so far, the IAU has studied, named, and declared these four.
- All these dwarf planets, save Ceres, lie in the Kuiper belt.
- Possible dwarf planet Sedna is among the farthest-located bodies in our solar system to be discovered.

✦ HAUMEA

- One of the fastest-spinning objects known in our solar system, Haumea makes a full rotation every four hours.
- A collision caused the icy dwarf planet Haumea to spin so fast it squashed itself into the shape of an American football.
- **HAUMEA TAKES 285 EARTH YEARS TO ORBIT THE SUN, SO ONE HAUMEA YEAR HAS 104,025 DAYS.**
- Haumea is the first Kuiper belt object discovered that has rings.
- Sunlight takes six hours to reach Haumea's icy surface.
- Who discovered Haumea? Two teams claim credit: one from the U.S.'s California Institute of Technology and one from Spain's Instituto de Astrofísica de Andalucía.
- **HAUMEA, MADE OF ICE-COVERED ROCK, IS NAMED FOR THE HAWAIIAN GODDESS OF FERTILITY, WHO ALSO RULES THE ELEMENT OF STONE.**
- Haumea has two satellite moons named for the goddess's children, Hi'iaka and Namaka.

✦ CERES

- Discovered in 1801, Ceres was first considered a planet.
- Companions to Ceres were soon discovered, and by 1845, they were all deemed asteroids, a new class of celestial objects.
- So, the 2006 definition of "dwarf planet" was a downgrade for Pluto but an upgrade for Ceres.
- Cerium, an element discovered in 1803, was named after the dwarf planet.
- A day on Ceres lasts nine hours four minutes.
- On Ceres, dozens of volcanoes erupt a slushy combination of frosty mud and briny water.
- Ceres's ice volcanoes may be powered by internal heat caused by the decay of radioactive elements.
- In 2015, NASA's Dawn spacecraft spotted bright splotches on Ceres that may be salt deposits from its icy volcanic eruptions.
- Ceres may have layers like its rocky planet neighbors, but the dwarf planet's mantle is likely made of ice, not rock.
- As much as a quarter of Ceres's mass could be water. If so, it holds more water than Earth.
- **CERES LIKELY FORMED NEAR THE OUTER EDGES OF THE SOLAR SYSTEM BUT WAS LATER CAPTURED IN THE ASTEROID BELT.**
- Ceres was the first dwarf planet to be orbited by a spacecraft, when Dawn visited in 2015.

From left: Earth (for scale) followed by Pluto, Eris, Makemake, Ceres, and Haumea

✦ ERIS

- Dwarf planet Eris was initially nicknamed Xena, after the popular heroine in the television show *Xena: Warrior Princess*.
- Eris's moon was dubbed Gabrielle, after Xena's sidekick.
- **AFTER DELIBERATION IN THE INTERNATIONAL ASTRONOMICAL UNION, THE DWARF PLANET EARNED THE OFFICIAL NAME ERIS, AFTER THE GODDESS OF STRIFE AND TURMOIL.**

Actresses Renee O'Connor (at left) as Gabrielle and Lucy Lawless as Xena in cult-favorite TV show *Xena: Warrior Princess*

- Eris is aptly named, since its discovery triggered the heated 2006 debate over the definition of a planet.
- **ERIS'S MOON IS OFFICIALLY NAMED DYSNOMIA, ERIS'S DAUGHTER AND THE DEMON GODDESS OF LAWLESSNESS.**
- Sunlight takes nine hours to travel to the far-flung surface of Eris.
- Eris orbits outside the plane of the solar system's planets.
- Eris takes 557 Earth years to orbit the Sun.
- Eris and Pluto are the largest Kuiper belt objects yet discovered, with diameters around 1,400 miles.

✦ MAKEMAKE

- Makemake was first called Easter Bunny, since it was discovered just after Easter in 2005.
- Easter Bunny, indeed: Pellets speckle Makemake's surface. They aren't droppings, but frosty bits of methane.
- A Makemake day is similar in length to an Earth day.
- **MAKEMAKE TAKES 305 EARTH YEARS TO ORBIT THE SUN, BUT IT ONLY TAKES 6 HOURS 20 MINUTES FOR SUNLIGHT TO REACH MAKEMAKE.**
- One moon has been found orbiting Makemake, and it's only about 100 miles across.
- Visible only by telescope, Makemake looks reddish pink.

CURRENT FACT COUNT 3640

EXOPLANETS

Our universe is teeming with worlds. There are more planets in the universe than stars. Not billions, not trillions, but billions of trillions of exoplanets may exist beyond the solar system. But until the mid-1990s, exoplanets were just science fiction.

✦ MAKING HISTORY

- An exoplanet is any planet that orbits a star that is not our Sun.
- **BEFORE 1992, WE HAD NO SOLID EVIDENCE THAT THE EIGHT PLANETS OF OUR SOLAR SYSTEM WEREN'T THE ONLY ONES IN THE UNIVERSE.**

A poster from the NASA Jet Propulsion Lab's travel series "Visions of the Future"

- That's right, we've only known about exoplanets since 1992—the same year the first text messages were sent.
- We've since discovered thousands—and expect to continue uncovering thousands more every single year.
- Most exoplanets discovered so far are dozens or hundreds of light-years from Earth.
- In April 1984, the first planetary disk outside the solar system was discovered via an image of dust and gas around the star Beta Pictoris.
- In 1992, Aleksander Wolszczan and Dale Frail announced the discovery of two planets in the constellation Virgo.
- The planets were discovered orbiting a pulsar (a dead star), so they named them Poltergeist (a word for a "mischievous ghost") and Phobetor, meaning "frightening."
- A third, smaller planet, was found orbiting the same pulsar. They named it Draugr, after a zombie in Norse mythology.
- There's no unified system for naming exoplanets.
- In 1995, the first exoplanet orbiting a sunlike host star was found: 51 Pegasi b.
- 51 Pegasi b sits about 50 light-years away in the constellation Pegasus.
- Swiss astronomers Michel Mayor and Didier Queloz won the 2019 Nobel Prize in Physics for their discovery of 51 Pegasi b.
- In 1999, astronomers observed five planets orbiting a single host star: the first multi-exoplanet system around a star like our Sun.
- **IN 2001, ASTRONOMERS FOUND HD 28185 B, THE FIRST PLANET SEEN ORBITING ITS SUN WITHIN A HABITABLE ZONE.**
- In 2001, thanks to the Hubble Space Telescope, scientists first analyzed an exoplanet's atmosphere.
- Where might life be found? On an exoplanet in the Goldilocks zone: not too close, not too far from its host star, but just the right position to retain liquid water on its surface.

"IF WE WANT TO FIND OUT IF WE'RE ALONE IN THE UNIVERSE, ALL OF THIS KNOWLEDGE IS ESSENTIAL."

—DAWN GELINO, HEAD OF NASA'S EXOPLANET EXPLORATION PROGRAM

The first image of an exoplanet, 2M1207 b, taken by the Very Large Telescope in 2004

✦ KEPLER'S SEARCH FOR HABITABLE ZONES

- NASA's Kepler Space Telescope, launched on March 6, 2009, was designed to study a patch of sky rich with stars within the constellation Cygnus.
- In fact, Kepler turned out to be the first telescope to spot exoplanets in the Milky Way.
- Ultimately, it revealed that there are billions of planets in our universe.
- **KEPLER'S FINDINGS SUGGEST THERE ARE MORE PLANETS THAN STARS IN THE MILKY WAY.**
- During its nine-year mission (2009–2018), Kepler cataloged more than 2,600 planets outside our solar system.
- Key to these discoveries, Kepler observed transits: dips in star brightness when a planet crosses in front of it.
- Kepler identified five exoplanets in its first six weeks of operation.
- In 2013, Kepler's positioning mechanism failed. So engineers developed the K2 mission, using the Sun's energy to direct the telescope.
- In October 2018, as Kepler ran out of fuel, NASA ended its mission and left the telescope in its orbit.

✦ COOL KEPLER FINDS

- Kepler-9: the first confirmed planetary system with more than one planet around a single star
- Kepler-11: a yellow dwarf star with six planets orbiting it
- Kepler-22 b: the first Kepler-confirmed planet in the habitable zone of its host star, about 2.4 times the size of Earth
- Kepler-16 b: a planet orbiting binary stars, meaning you could see a *Star Wars*–like double sunset from the planet, surface conditions permitting
- Kepler-10 b: the first confirmed rocky planet, with a nearly one-day orbit, discovered in 2011
- Kepler-10 b sits 20 times closer to its host star than Mercury does to ours, so it must have scorching surface temperatures.
- It's as dense as an iron dumbbell. That's about 4.5 times more dense than Earth.
- Kepler-452 b: a "super-Earth" similar to ours
- Kepler-452 b is about 1.6 times the size of Earth and orbits its star every 385 days.
- **ULTIMATELY, KEPLER'S FINDINGS SUGGEST THAT AS MANY AS 50 PERCENT OF THE STARS IN THE MILKY WAY HOST ROCKY PLANETS IN THEIR HABITABLE ZONES.**
- While the Kepler mission has come to an end, scientists will continue to pore over its vast trove of data for decades to come.

✦ SUPER-EARTHS

- Many extra-stellar systems have super-Earths: planets of gas and rock that are larger than Earth but smaller than Neptune.
- In 2014, the exoplanet Kepler-186 f was discovered. It's the first Earth-size planet found within a habitable zone.
- Kepler-186 f is about 500 light-years from Earth and about 10 percent larger.
- **IF OUR SOLAR SYSTEM HAD A SUPER-EARTH, THEN DEPENDING ON ITS SIZE AND LOCATION, IT COULD:**
    - Drive Uranus and Neptune out of our solar system altogether.
    - Alter the orbits of Saturn and Jupiter enough to eject Mercury, Venus, Earth, and Mars from the solar system.
    - Influence the shape of Earth's orbit, altering the climate and making it less friendly to life.

✦ SURVEYING EXOPLANETS

- In 2018, NASA and MIT jointly launched the Transiting Exoplanet Survey Satellite (TESS) to hunt for exoplanet transits across the sky.

An artist's illustration of the surface of the exoplanet LHS 475 b, which orbits a red dwarf star in the constellation Octans

- Then the James Webb Space Telescope (JWST) ushered in a new era of exoplanet observation.
- In 2023, scientists announced JWST had observed its first exoplanet in 2022.
- That exoplanet, LHS 475 b, is a rocky planet 99 percent the size of Earth.
- JWST determined that LHS 475 b is much warmer than Earth—and faster. It orbits its star in two days.
- Another exoplanet explored thanks to JWST is HIP 65426 b, a gas giant.
- HIP 65426 b is about 15 to 20 million years old—significantly younger than Earth's 4.5 billion years.
- It has a mass nine times that of Jupiter. (NASA expresses the mass of exoplanets in terms of Earths and Jupiters. Exoplanet 14 Herculis b = 8,053 Jupiters; 47 Ursae Majoris c = just over half a Jupiter.)
- HIP 65426 b was discovered in 2017. JWST provided the first direct images of it.
- NASA's Nancy Grace Roman Space Telescope (RST) was specifically designed for exoplanet exploration in mind.
- Like JWST, RST is optimized for infrared light.
- It is designed to explore not just new exoplanets and their atmospheres, but also their host stars and even dark energy.
- RST's design allows it to view exoplanets by blocking out the host star's light using a coronagraph.
- The coronagraph, a design already used to observe the Sun, uses prisms, masks, and detectors to suppress the glare of an exoplanet's host star.

An illustration of Kepler-69 c, 2,700 light-years from Earth in the constellation Cygnus

✦ IN GALAXIES FAR, FAR AWAY

- The first exoplanet discovered, 51 Pegasi b, orbits so close to its star, it's called a roaster.

- It turns out that among exoplanets, roasters are pretty common.

- It takes just over four days for 51 Pegasi b to circle its host star.

- 51 Pegasi b is bigger than Jupiter but has half the mass.

- **SCIENTISTS ESTIMATE THE TEMPERATURE ON 51 PEGASI B IS 1800°F.**

- The nearest exoplanet, Proxima b, orbits the nearest star, Proxima Centauri, about four light-years away.

- One of the farthest known exoplanets in the Milky Way is Kepler-443 b.

- At more than 2,500 light-years from Earth, Kepler-443 b would require 28 billion years of travel at 60 miles an hour to reach.

- **THE HUBBLE SPACE TELESCOPE HAS DETECTED HELIUM IN THE ATMOSPHERE OF GAS GIANT EXOPLANET WASP-107 B, ABOUT 200 LIGHT-YEARS AWAY.**

- Eight planets have been discovered in the Kepler-90 system, just like in our solar system.

- But in this case, all eight are closer to their star than Earth is to the Sun.

- Still in planning, the Habitable Worlds Observatory would be a large infrared/optical/ultraviolet space telescope seeking Earthlike rocky exoplanets with atmospheres.

✦ EXOPLANET HUNTING

- Scientists use five main techniques to spot exoplanets.

- **DIRECT IMAGING**
  - Only a few exoplanets have been directly imaged. The relative brightness of a star compared with the far less reflective planetary body makes discovery in this way prohibitively difficult except for in very special cases.
  - Our first direct image of an exoplanet, in 2004, was of 2M1207 b, 230 light-years from Earth.

- **TRANSIT METHOD**
  - When a planet passes in front of its host star from our perspective, we can measure the slight drop in light detected from that star. The larger the shift, the larger the planet.

- **DOPPLER METHOD**
  - Like the transit method, the Doppler method measures fluctuations in the host star caused by orbiting planets. Instead of measuring luminosity changes, however, the Doppler method calculates shifts in a star's radiation frequencies as an orbiting planet's gravity distorts it.

- **GRAVITATIONAL MICROLENSING**
  - This method detects starlight bent by the gravity exerted by a planet.

- The first exoplanets were discovered with the Doppler method, and it has accounted for about one-fifth of all known exoplanets discovered.

- Most of the rest—about three-quarters of all known exoplanets—were discovered via the transit method.

- **LESS THAN 5 PERCENT OF KNOWN EXOPLANETS WERE DISCOVERED USING GRAVITATIONAL MICROLENSING.**

- Astrometry, by which minute shifts in star position suggest exoplanets, has discovered only a handful.

✦ WEIRD EXOPLANETS

- TrES-2 b: The darkest known exoplanet. It is less reflective than coal.

- WASP-12 b: An egg-shaped world in peril. As you read this, the planet is being ripped apart by its host star, stretching the exoplanet into an oblong shape while the star slowly consumes it piece by piece.

- 55 Cancri e: A tidally locked world covered in molten lava. Its dark side is full of glittering silicate clouds—evidence that beauty can be deceiving.

- TIC 241249530 b: Unlike almost every other planet yet discovered, this exoplanet (another with an egg-shaped orbit) cruises around its star in the opposite direction of the star's rotation.

- HD 189773 b: The blue exoplanet's extremely hot temperature causes it to rain glass.

- **TOI-3757 B: WITH THE DENSITY OF A MARSHMALLOW, THIS CUSHY EXOPLANET IS THE LEAST DENSE PLANET EVER OBSERVED.**

NASA Jet Propulsion Lab travel poster

4

One hundred million years after the Big Bang, some pockets of the universe held just the right conditions for the first stars to form. These early, ancient giants were likely hundreds of times more massive than our Sun and millions of times more luminous. Their explosive deaths would ripple outward, shaping the evolution of the early cosmos.

More than 1,000 newborn stars are birthed 1,500 light-years away in the cosmic cloud of the Orion Nebula.

SUPER

STARS

CURRENT FACT COUNT 3695

THE STUFF OF STARS

When Carl Sagan famously said "we are made of star-stuff," he wasn't just conveying a poetic sentiment. He meant it literally. You are the result of an ancient, stellar cycle of births and deaths that formed the atoms that make you, you.

- Within every star, throughout its entire volume, lies a battle of opposing forces: the outward pressure caused by the heat of thermonuclear fusion in the core and gravity's inward pull.
- This war of push and pull wages on throughout a star's life until it exhausts all fuel in its core, precipitating a catastrophic gravitational collapse.
- **HYDROGEN FUEL POWERS STARS IN THEIR PRIME.**
- Hydrogen is the simplest, most abundant element in the universe.

- The fiery furnaces of stars create all other natural elements, up through uranium.
- Carbon, nitrogen, and oxygen, foundational elements in organic chemistry, are all forged in violent supernova explosions: the death throes of massive stars.
- When two pulsars collide, they create abundant rare earth elements as well as stupendous quantities of gold and platinum, vastly exceeding all that has ever been mined on Earth.
- In 1766, British chemist Henry Cavendish first identified hydrogen, which he called "inflammable air."
- In 1797, he would also make the first measurement of Newton's gravitational constant.
- The word "hydrogen" comes from the Greek words for "water-forming."
- In the presence of oxygen, hydrogen is highly flammable.
- Together, hydrogen and helium make up around 98 percent of all ordinary matter in the universe: around 73 percent hydrogen atoms, 25 percent helium.

A 1924 illustration of balloons inflated with helium and strapped to people's backs

Heavy element: King Tut's solid gold coffin

- Helium is second to hydrogen by a long shot, but it's still 12 times more abundant than all other elements combined.
- The fusion of hydrogen nuclei into helium in stars releases energy as heat and light.
- In 1925, Cecilia Payne wrote what has since been revered by many as the most brilliant PhD thesis in astronomy of all time. She proposed that the stars are composed mostly of hydrogen and helium.
- For the next 20 years, astronomers hypothesized, but could not prove, the cosmic origins of the elements.
- **IN 1948, A SCIENTIFIC THEORY SHOWED THAT ALL EXISTING HYDROGEN, SOME HELIUM, AND A SMIDGEN OF LITHIUM LIKELY ORIGINATED IN THE BIG BANG.**
- In a seminal paper from 1957, Geoff Burbidge, Margaret Burbidge, William Fowler, and Fred Hoyle demonstrated that stellar nucleosynthesis can manufacture all the heavier elements in our natural world and disperse them through the universe via supernova explosions.

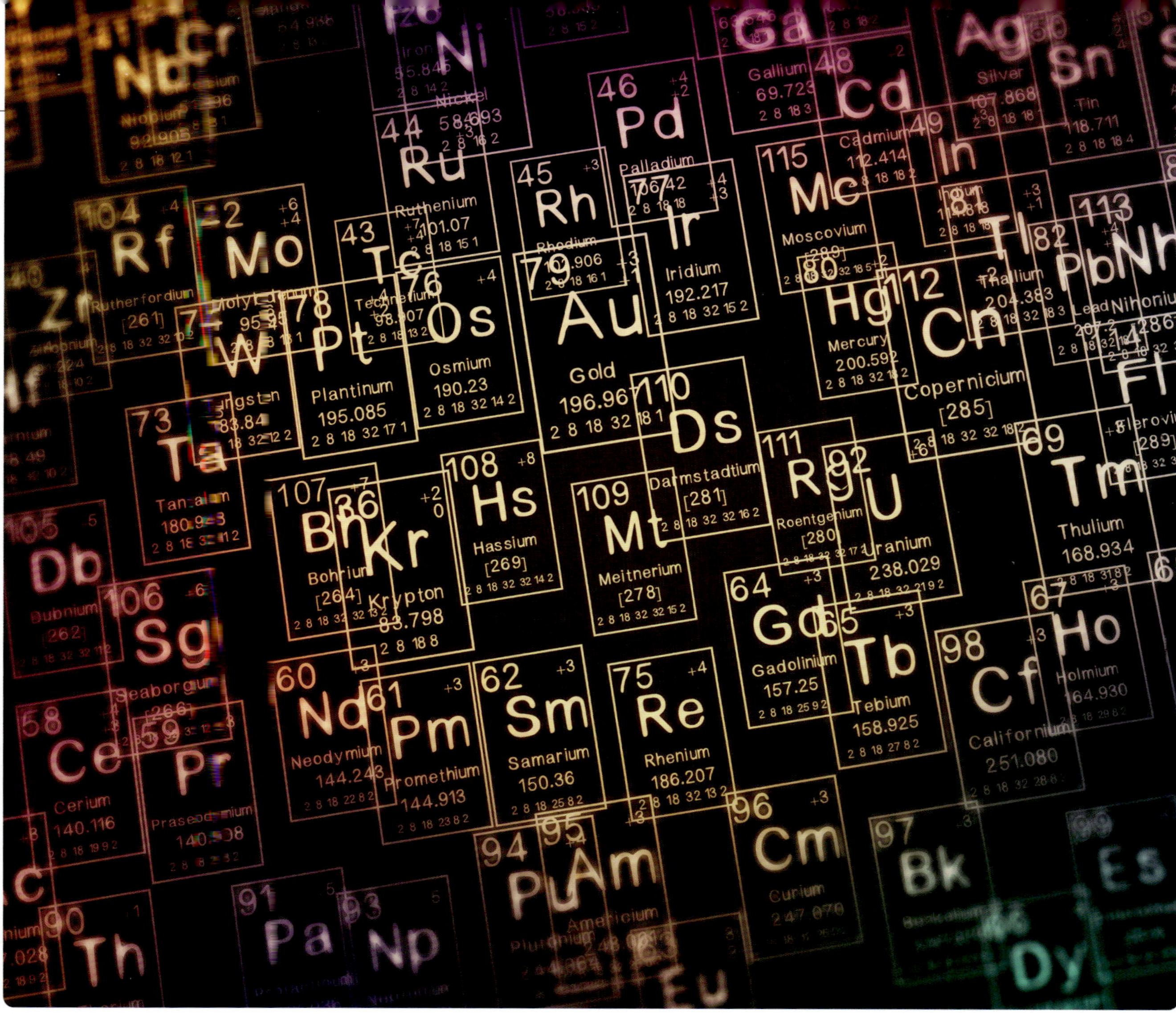

A 3D rendering of the periodic table of the elements

- Helium was first detected in 1868, when two scientists separately noticed a feature in the Sun's spectrum with no counterpart to laboratory measurements on Earth.
- One helium discoverer was in India, observing a total solar eclipse; the other was in London, observing through dense fog.
- **JOSEPH LOCKYER, THE ASTRONOMER IN LONDON, NAMED IT HELIUM, FOR HELIOS, THE GREEK SUN GOD.**
- It was the first element discovered in the cosmos before it was found here on Earth.
- Because it's chemically inert and rare on Earth, scientists were not able to isolate helium until 30 years later.

✦ YOU ARE STARDUST

- Proof that we're made of stardust: Every atom of our body can be traced to the Big Bang and the starry furnaces that expelled atomic creations during their death throes.
- The calcium in our bones, carbon in our DNA, and iron in our blood were all forged in the fiery furnaces of massive stars.
- The most common, chemically reactive elements in the cosmos—hydrogen, oxygen, carbon, and nitrogen—are also the four most common elements of life on Earth.
- Our bodies are 60 percent water (mostly hydrogen and oxygen), an elemental combination originating with the Big Bang and created inside very hot stars.

- **BOTH CARBON AND OXYGEN ARE SOCIABLE ELEMENTS: THEY LOVE TO BOND AND MAKE MOLECULES WITH OTHER ATOMS.**
- On Earth, plants and algae create an abundance of free oxygen as a by-product of photosynthesis.
- The greatest episode of climate change Earth has ever known was caused not by humans but by photosynthesizing cyanobacteria in the oceans, three billion years ago, changing an atmosphere of carbon dioxide to one with oxygen, thereby enabling oxygen-metabolizing animals to evolve and thrive.

A Hubble image of Cepheid variable star RS Puppis

✦ CEPHEID VARIABLES

- Cepheid variables are stars with predictable changes to their brightness and temperature. These predictable pulses are used to help calculate vast distances in the universe.

- Before Cepheid variables were discovered, astronomers struggled to accurately measure the distances to faraway stars.

- In the early 20th century, Henrietta Swan Leavitt established the foundation of what is now called Leavitt's law: the period-luminosity relationship of stars.

- The period-luminosity relationship of Cepheid variables unlocked a new method to measure cosmic distances.

- How does it work? First, by understanding that we observe only a star's apparent brightness, or the total photons that hit our telescopes. A star's intrinsic luminosity is exponentially greater but diminishes exponentially over distance.

- Knowing the length of a Cepheid star's pulsing period (from bright to dull to bright again) allows astronomers to establish its actual intrinsic luminosity. The longer the pulsing period, the more luminous the star.

- Using a simple formula, we compare the star's brightness in the sky with its intrinsic luminosity to calculate the star's distance from Earth.

- All stars with fluctuating brightness are called variable stars.

- **CEPHEID VARIABLES ARE VARIABLE STARS WITH REGULAR PULSES.**

- Cepheids were key to Edwin Hubble's observation that the Andromeda Nebula (as it was known then) lay well outside the Milky Way.

- He realized three stars in his photographic plates of October 5, 1923, were Cepheids and used them to calculate that Andromeda was 900,000 light-years away—three times farther than the largest estimated size of the Milky Way at the time.

- We know now that Andromeda lies 2.5 million light-years away and 25 times farther away than the known size of the Milky Way.

✦ THE PARALLAX METHOD

- Have you ever winked at your outstretched thumb with one eye and then the other and noticed how it appears to move against the background? The closer your thumb is to your eye, the greater the apparent shift. Astronomers used this phenomenon to measure the distance of nearby stars.

- Like your thumb, stars appear to change position relative to their backgrounds, depending on our own angle of view.

- Ancient Greeks used this method, called parallax, too. We now know it's useful mainly for closer stars.

- **FOR STARS FARTHER AWAY, THE DIFFERENCE IN PERSPECTIVE IS TOO SMALL FOR ACCURATE MEASUREMENTS.**

- Aristotle justified the idea of a stationary Earth by the absence of parallax shift in stars. He didn't consider that the stars were simply too far away.

✦ WALK THIS WAY

- A photon executes a "random walk" from the Sun's core to its surface, batted to and fro by charged particles on the way.

- It takes tens to hundreds of thousands of years for photons created in the Sun's core to reach its surface—and a mere 500 seconds to reach Earth at the speed of light.

American astronomer Henrietta Swan Leavitt discovered the period-luminosity relationship of Cepheids.

- In the 19th century, before we confirmed stars fused hydrogen into helium and that they're made of mostly hydrogen, astronomers presumed stars were burning lumps of coal.

- **BASED ON ITS LUMINOSITY, IF THE SUN WERE A BURNING LUMP OF COAL, IT COULD ONLY BURN FOR ABOUT 5,000 YEARS OR SO.**

- This hypothesis was discarded once geologists established the age of Earth was at least many millions, if not billions, of years old. And of course, Earth could not be older than the Sun.

- In the 2015 film *Star Wars:* Episode VII *The Force Awakens*, the First Order occupies Starkiller Base, which is capable of simultaneously destroying an entire star system of planets.

- This new base obtains its energy by nestling up to and sucking all the energy from other stars.

- This is space fantasy, of course, but in reality, a star holds enough energy to obliterate thousands of planets, not just a few.

CURRENT FACT COUNT 3760

THE LIFE CYCLE OF A STAR

Most stars do not end in a dazzling supernova explosion. A star's birth mass determines how it lives and how it dies.

- Some stars are more than 10 billion years old; others are born at this very moment.
- Similar stars follow similar patterns of birth, development, and death.
- Modern telescopes allow snapshots of all kinds of stars living out their lives.

✦ BIRTH

- The birth of stars begins in enormous clouds of interstellar gas and dust called nebulae.
- Within these clouds, the dust and gas particles begin to glom together, generating heat as molecules collide.
- The gravity of these higher-density clumps attracts more material, generating more density, more gravity, more heat, and more pressure.
- **THE CYCLE CONTINUES AND THE TEMPERATURES BUILD UNTIL THERMONUCLEAR FUSION IGNITES. A STAR IS BORN.**
- Because star birth happens deep inside dense clouds of dust and gas impenetrable by visible light, infant stars are mostly shrouded from our view.
- They can remain hidden for the first million years of their young lives, until stellar winds and radiation blow away enough of their dusty veils to reveal the newborn star.

Stellar nursery Messier 78, with ionized atomic hydrogen shown in blue and the absorption and scattering of light in red

✦ NEWBORNS

- Newborn stars come swaddled in blankets of gas and dust: the remains of the cosmic cloud that gave rise to them.
- Stellar nurseries are the subjects of some of the most extraordinary images captured since we started observing the cosmos, such as the Pillars of Creation, the Orion Nebula, and the Cosmic Cliffs.
- **MODERN INFRARED SPACE TELESCOPES, LIKE THE JAMES WEBB SPACE TELESCOPE, ARE NEEDED TO SPY THE NEWBORN STARS DEEP INSIDE THEIR COLORFUL VEILS AND WISPS OF NEBULAE.**
- Stars are born in batches and gradually drift out of their nurseries, shedding their birth sacs of gas and dust over millions of years.
- As they emerge, these star batches can remain gravitationally bound to one another for as long as a billion years, like celestial siblings.
- At around 100 million years old, the stars in the Pleiades are at this stage; they'll drift apart over another few hundred million years.
- **MOST OF THE STARS MAKING UP THE BIG DIPPER—EACH ABOUT A HALF BILLION YEARS OLD—WERE ONCE NURSERY COMPANIONS BUT HAVE SEPARATED FROM THEIR STELLAR BIRTH CLUSTER.**
- Some of our Sun's siblings, born in the same stellar nursery some 4.6 billion years ago, may have drifted across the Milky Way by now.
- For its first anniversary, the James Webb Space Telescope provided a close-up view in July 2023 of the birth of Sunlike stars in the Rho Ophiuchi cloud complex, the stellar nursery closest to Earth.
- The small stellar nursery sits about 390 light-years from Earth.

The iconic Pillars of Creation image shot by the Hubble Space Telescope

✦ PILLARS OF CREATION

- The Pillars of Creation is a luminous, active birthing region set within the Eagle Nebula, 6,500 light-years away.
- The Hubble Space Telescope's iconic images of the region, from 1995 onward, display three towering pillars of gas and dust stretching four to five light-years high.
- The Pillars are just a small part of the Eagle Nebula, which stretches roughly 55 by 70 light-years.
- The James Webb Space Telescope's near-infrared image of the Pillars, released in 2022, reveals thousands of newborn stars, some as young as a few hundred thousand years old.

✦ THE ORION NEBULA

- The Orion Nebula offers one of our best views of star births in the Milky Way galaxy.
- Hubble's images capture more than 3,000 stars in various stages of formation.
- **THE ORION NEBULA IS CONSIDERED CLOSE TO EARTH, AT APPROXIMATELY 1,500 LIGHT-YEARS AWAY.**
- You can see it as a fuzzy patch in the Orion constellation, just under the three bright stars that form the hunter's distinctive belt.

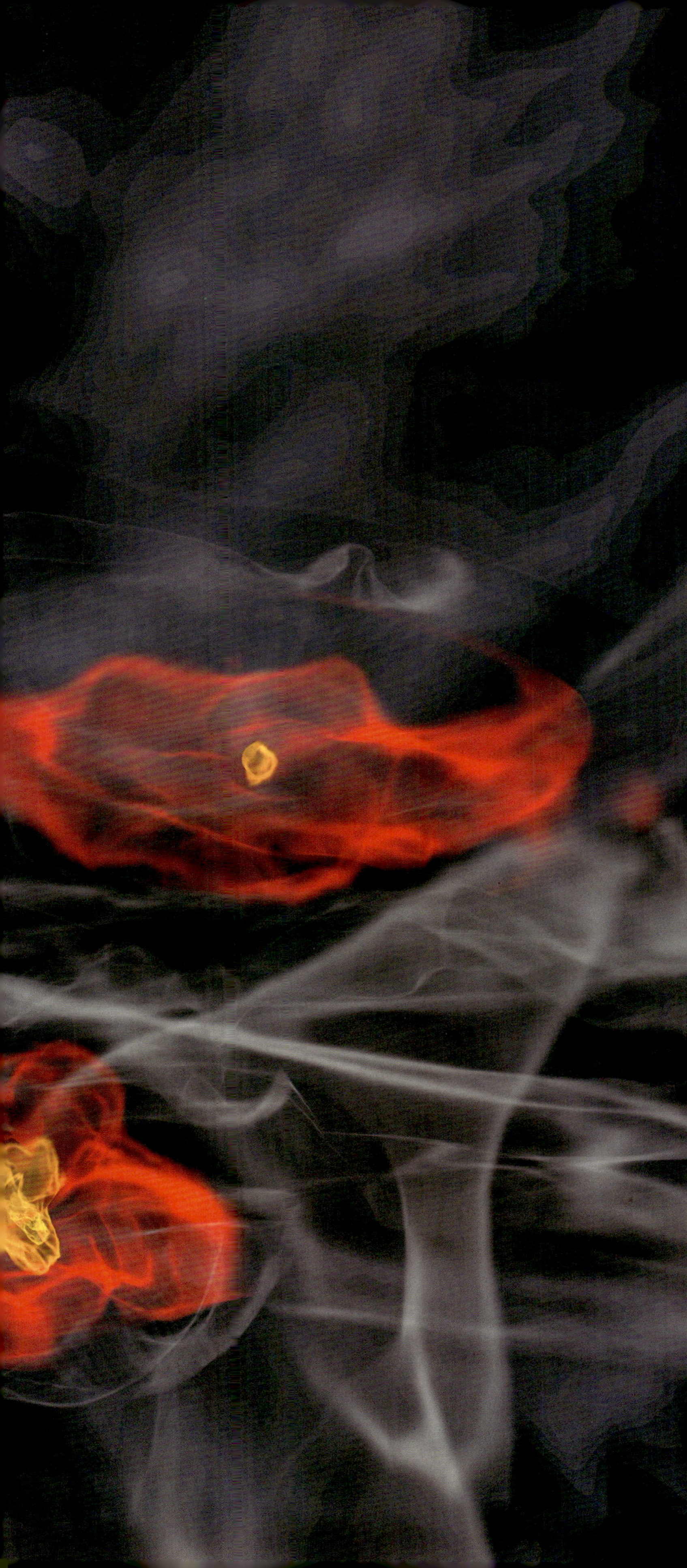

✦ FIRST STARS

This computer simulation depicts three primordial clouds just 100 million years after the Big Bang.

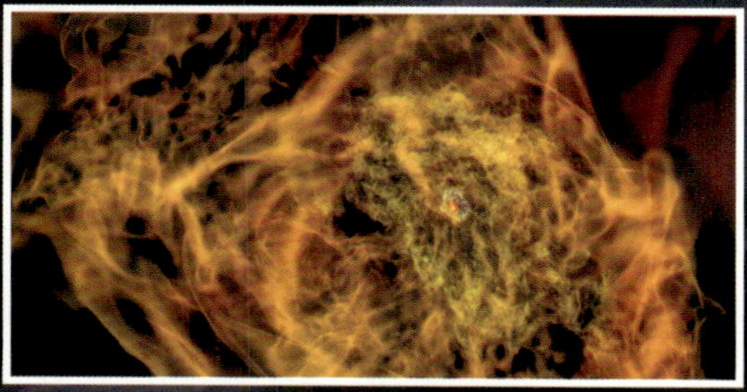

A close-up (above) of the far-left cloud reveals billowing layers of hydrogen and helium gas being driven inward by surrounding halos of unseen dark matter. The gas accumulates in higher and higher densities (shown in red and yellow).

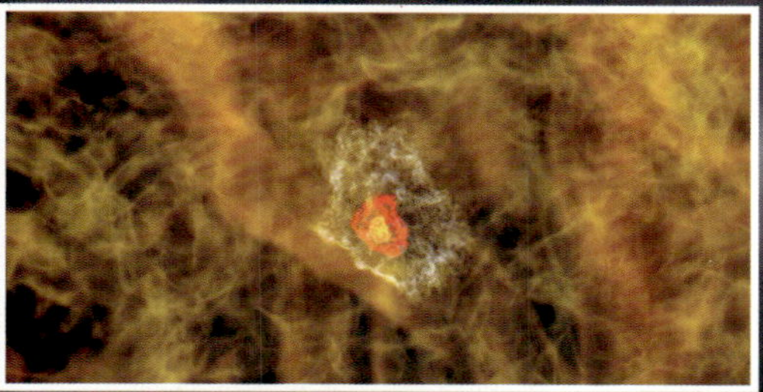

Deeper in the cloud's structure, hydrogen molecules cool the gas, allowing it to collapse. Within this cold cloud, massive stars will form.

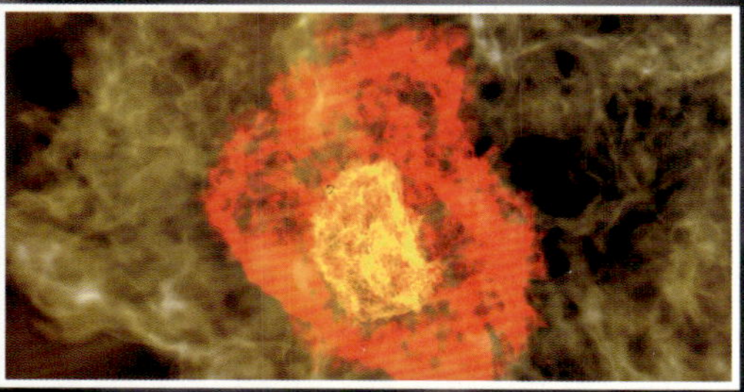

Hundreds of solar masses compress in the yellow area. Most of them will end up in the first stars.

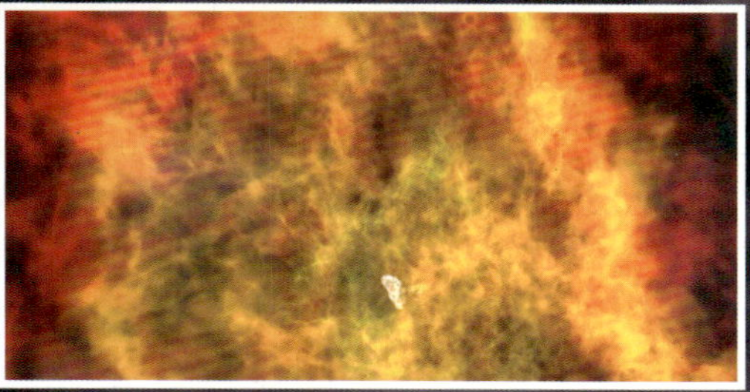

The small, bright region in the center of this cosmic pileup is roughly the size of our solar system. The hydrogen eventually fuses into helium, birthing the first stars in the universe.

Open star cluster NGC 330, around 180,000 light-years away inside the Small Magellanic Cloud

✦ LIFE: THE MAIN SEQUENCE

- The interior heat and pressure in a newborn star will eventually emit enough outward pressure to support the star against the urge to collapse with its own gravity.
- **ASTROPHYSICISTS CALL THIS AGE OF STABILITY THE STAR'S "MAIN SEQUENCE."**
- The lifetimes of stars span millions to billions of years.
- Stars spend about 90 percent of their lives fusing hydrogen into helium in their cores.
- Most stars in the universe are main-sequence stars.

✦ DEATH

- How a star dies depends on its starting mass.
- **THE MOST MASSIVE STARS HAVE THE SHORTEST LIFE SPANS, FROM ONE MILLION TO A FEW MILLION YEARS.**
- The least massive ones live for billions of years.

- When a star burns the last of its hydrogen fuel, an extraordinary transformation begins. Successive waves of stellar fusion transform heavier elements: helium into carbon, carbon into oxygen, and up the periodic table.

- Large stars, with masses some eight to 20 times that of our Sun, forge heavier elements all the way up to iron, the 26th element.

- Stars are in the business of making energy. Without hydrogen fusion a star collapses under its own weight, which increases the temperature of its core until it's hot enough to fuse the next heavier element that has accumulated in the core

- The process of stars fusing elements into new ones is called stellar nucleosynthesis.

- The fusion of iron, unlike the fusion of all other elements leading up to it, absorbs rather than emits energy.

- **THE FUSION OF SILICON INTO IRON IS A MASSIVE STAR'S DEATH RATTLE.**

- Fusing iron from a pair of silicon atoms requires temperatures of billions of degrees.

- Once a star's core contains only iron, the star's inner regions collapse catastrophically, leading to a rapid heating and explosive rebound known as a supernova.

- A supernova can temporarily (for several days) outshine the collective light of all other stars in its home galaxy.

- Supernovae are rare, occurring at the end of the life of rare, high-mass stars.

- **WE EXPECT GALAXIES SUCH AS THE MILKY WAY TO HOST A SUPERNOVA ONCE OR TWICE PER CENTURY.**

- The last known supernova in the Milky Way occurred in 1604, as documented and studied by the German mathematician and astronomer Johannes Kepler.

- The Milky Way is overdue for a supernova.

- A star does not die in vain. Its final battle forges the material necessary for the births of future stars and planets.

✦ OUR SUN'S DEATH

- Yellow dwarf stars like our Sun do not end in supernovae, but rather balloon in size and become red giants.

- When hydrogen molecules fuse at the Sun's core, the resulting helium molecule is slightly less massive. That extra mass is converted to energy and expelled as light—exactly as prescribed by Einstein's most famous equation, $E = mc^2$.

- The Sun is slowly losing mass through this fusion and, as is the fate of every star, will run dry of thermonuclear fuel.

- The inner core will then collapse, causing a surge of heat that ends not in implosion, but expansion.

- The outer layers will balloon to several hundred times the star's original size. From that moment on, our Sun will be a red giant.

- The red giant Sun will shine about 1,000 times brighter than it does now.

- **MERCURY, VENUS, AND EARTH WILL BE ENGULFED IN THIS NEW BLAZE.**

- Frigid Pluto, half composed of water ice, will begin to thaw toward balmy temperatures more conducive to life as we know it.

- As the outer layers of the Sun drift away, it will leave behind the compressed, ultra-dense core.

- Now the size of Earth, the Sun will endure as a white dwarf.

- If our Sun *could* go supernova at the end of its life, all objects in its orbit would either incinerate or fly off at a tangent to be lost forever in interstellar space.

- Luckily for us, only stars at least eight times the size of our Sun can go supernova.

Hubble captured this star's demise and the resulting planetary nebula, known as NGC 6565.

"THE COSMOS IS ALSO WITHIN US. WE ARE MADE OF STAR-STUFF. WE ARE A WAY FOR THE COSMOS TO KNOW ITSELF."

—CARL SAGAN, *COSMOS: A PERSONAL VOYAGE*

The Carina Nebula's Cosmic Cliffs, captured by the James Webb Space Telescope

CURRENT FACT COUNT 3792

CLASSIFYING THE STARS

All stars are defined by their ability to churn out energy through nuclear fusion in their cores. But they vary significantly in color, size, temperature, mass, and luminosity.

- In the 19th century, Joseph von Fraunhofer was the first to analyze and identify patterns in the solar spectrum. To this day, these features are still called Fraunhofer lines.
- High-mass stars fuse their hydrogen more efficiently and more swiftly than low-mass stars, giving them a relatively short life span of a few million years rather than a few billion or even trillion years.
- Stars can be classified by their surface temperature, their spectra, their mass, or their stage of life—all of which are interrelated.
- One modern system combines two approaches: the Harvard classification, based on a star's temperature, and the Morgan-Keenan system, based on its luminosity.
- The 19th-century Jesuit priest and Italian astronomer Angelo Secchi pioneered grouping stars according to their spectra.
- The Harvard classification system expanded Secchi's four categories and designated stellar types with letters indicating temperature: O, B, A, F, G, K, M.
- O is the hottest; M, the coolest.
- **OUR SUN IS A G-TYPE STAR.**
- More letters have since been added for supernovae and other types of stars.
- A traditional mnemonic for remembering the seven main types of stars is "Oh Be A Fine Girl/Guy, Kiss Me" (OBAFGKM).
- The seemingly random ordering of letters is actually a holdover from an even earlier classification system that sorted stars alphabetically based on total hydrogen composition.
- Total hydrogen does not correlate with generated temperature, so the letters now appear out of order.

✦ SUPERGIANTS

- Stars come in many sizes, from a bit larger than the largest planets to massive giants that dwarf our Sun.
- The largest stars can be hundreds of times more massive than the smallest ones.
- Stars with more than eight times the mass of our Sun are considered, by definition, massive.
- Massive stars are blindingly bright—as much as a million times brighter than our Sun.
- **THESE STARS FORM ULTRADENSE NEUTRON STARS AND BLACK HOLES WHEN THEY DIE.**
- Massive stars live fast, shine bright, and explode in dazzling supernovae at the end.

✦ BIG, BIGGER, BIGGEST

- The red giant Betelgeuse is not only one of the brightest stars, it's also one of the biggest.
- A supergiant star, it's visible as the left shoulder of the constellation Orion.
- Betelgeuse is only 10 million years old.
- It is much younger than our Sun, but 15 times more massive—700 times as large.
- Since it's so massive, Betelgeuse will burn through its elemental components much faster than our Sun.
- If you replaced our Sun with Betelgeuse, it would engulf the entire orbits of Mercury, Venus, Earth, and Mars.
- Observers in 2019 spotted a dramatic dimming of Betelgeuse, called the Great Dimming.
- **A PATTERN OF PERIODIC DIMMING AND BRIGHTENING OF BETELGEUSE THROUGH LONGER PERIODS OF TIME HAS BEEN CALLED THE STAR'S HEARTBEAT.**
- Now scientists hypothesize that another star orbits Betelgeuse, perhaps a cause of the dimming.
- The biggest known star in the universe is called UY Scuti.
- It's a rare example of a hypergiant star.
- UY Scuti lies near the center of the Milky Way, 9,500 light-years from Earth.
- It's so big, you could put five billion Suns inside it.
- If you replaced our Sun with UY Scuti, it would engulf Mercury, Venus, Earth, Mars, the asteroid belt, Jupiter, and just graze Saturn's orbit.

The starscape over
Zion National Park, Utah

CURRENT FACT COUNT 3816

RUNNING THEIR COURSE

A star's fate is determined by its mass. Their lives might end as a white dwarf like our Sun's—or as a black hole, supernova, or neutron star.

- In about five billion years or so, our Sun will become a red giant.
- As it does, it will balloon to more than 100 times its original size.
- While red giants increase in size, they decrease in temperature as they develop, signaling the first step of the star's death.
- Red giants gradually shed their outer layers, which form an expanding cloud of gas and dust called a planetary nebula.
- Despite the name, planetary nebulae have nothing to do with planets.
- When 18th-century astronomer William Herschel first observed them, he found their blue-green radiance reminiscent of the planet Uranus, which he'd discovered five years prior.
- As a red giant exhausts all its fuel, its core collapses inward and it becomes a white dwarf star.

Planetary nebula PK 329-02.2 (also called Menzel 2), located in the constellation Norma

✦ NEUTRON STARS

- The explosion of a supernova leaves behind a nucleus that collapses into an ultradense neutron star. (If the star is massive enough, it collapses into a black hole.)
- **NEUTRON STARS ARE STELLAR REMNANTS SO DENSE, THEY'RE LIKE PACKING 60 MILLION ELEPHANTS INTO A LIPSTICK TUBE.**
- If you could collect one teaspoon of a neutron star and return it to Earth, it would weigh about a billion tons, or two trillion pounds, about the weight of Mount Everest.
- Neutron stars are cocooned in a dense, solid, outer crust 10 billion times stronger than steel.
- A magnetar is a type of neutron star with powerful magnetic fields—the strongest in the universe.
- Magnetars can undergo starquakes, whereby sudden changes in their magnetic field will cause stress along their dense crusts, causing rips and cracks.
- When a neutron star cracks, it throws gamma and x-rays into space.
- The neutron star at the heart of the Crab Nebula is like an enormous atomic nucleus the size of Manhattan, spinning rapidly.

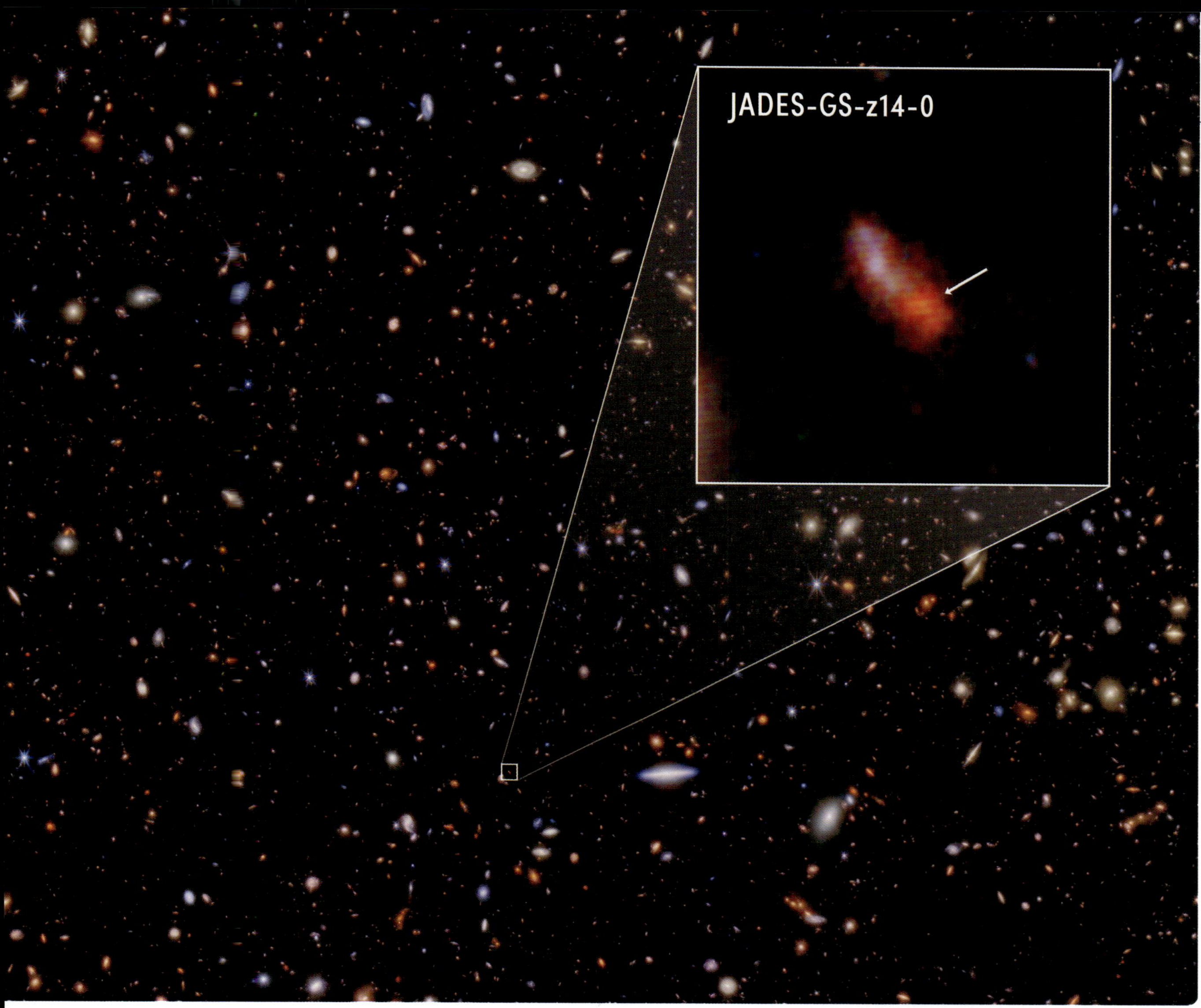

JWST took this image of JADES-GS-z14-0, one of the most distant galaxies currently known.

✦ PULSARS

- Electrons caught in the neutron star's powerful magnetic field beam radiation like a cosmic lighthouse. We call this a pulsar.
- The flash of rotating light we see from pulsars is remarkably steady, like a cosmic clock keeping time.

✦ DARK STARS

- Dark stars are hypothetical.
- **THEY'RE SO NAMED FOR THEIR POWER SOURCE: DARK MATTER.**
- They wouldn't appear dark at all, but would shine billions of times brighter than the Sun, with millions of times more mass.
- JADES-GS-z14-0 is the rather boring name of a fascinating and mysterious object.
- It's one of the farthest objects in the universe we've ever seen, a possible dark star more than 33 billion light-years away.
- It hails from when the universe was a mere 300 million years old, a time called the "dark ages," believed to be too early to form galaxies.
- JADES-GS-z14-0 is being carried away from us by the expanding universe at 98 percent the speed of light.

"FOR THE FIRST TIME WE FOUND A POTENTIAL SMOKING GUN SIGNATURE OF A DARK STAR. WHICH, IN ITSELF, IS REMARKABLE."

—SCIENTIST COSMIN ILIE

"THE REMOTER STARS SEEM A NEBULA OF UNITED LIGHT, YET THERE IS NO GROUP WHICH A TELESCOPE WILL NOT RESOLVE."

—RALPH WALDO EMERSON, *SOCIETY AND SOLITUDE*

Most of the energy in the BFS 29 nebula is probably from the star CE-Camelopardalis.

CURRENT FACT COUNT 3845

OUT WITH A BANG

Only the most massive stars in the universe—those at least eight times more massive than our Sun—die in a cataclysm. When they do, the cosmos quakes.

- A dying star often explodes as a supernova, among the most spectacular of celestial phenomena.
- The explosion can hurl gas and dust into space at nearly 10 percent the speed of light.
- A supernova explosion can outshine a billion suns for days on end.
- Hypothetically, the brightest supernova could outshine trillions of suns, even our entire Milky Way galaxy.
- Hydrogen and helium molecules make up 98 percent of all atoms in the universe, but the remaining 2 percent were forged by supernovae.
- Just how energetic is a supernova explosion? Typically, about the same as the Sun's total output over 10 billion years multiplied by 100.

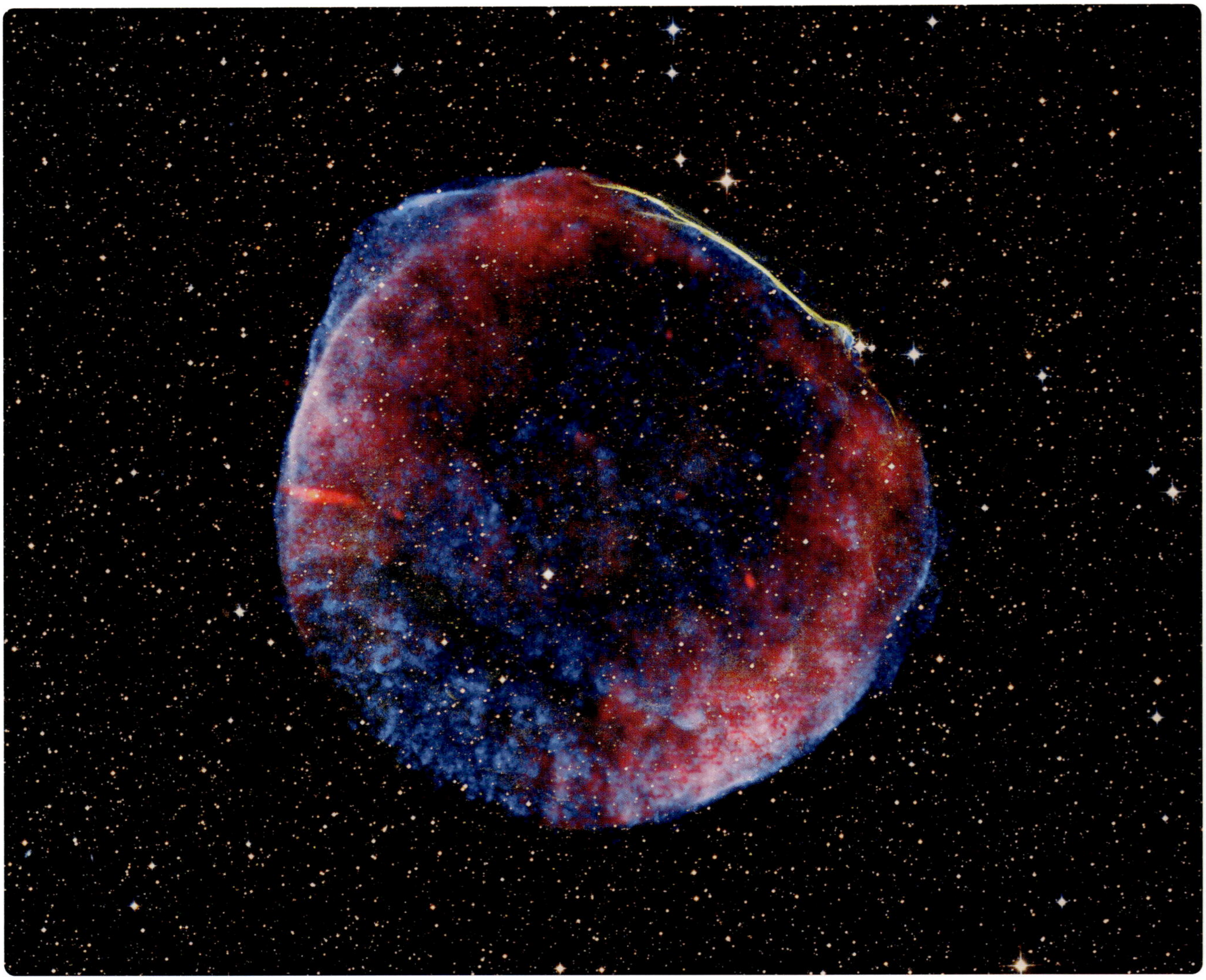

A composite image of the SN 1006 supernova remnant, likely the brightest supernova yet recorded

A ninth-century "Thor's hammer" amulet

- A supernova explosion jettisons most of the star's matter into space, including enormous amounts of heavier elements like carbon, silicon, and iron that enrich the raw material for stars and planets to come.

✦ THE BRIGHTEST STELLAR EVENTS IN HISTORY

- Ancient records from Europe, Asia, and Arabia describe bright "temporary" or "guest" stars. We now know them as supernovae.
- **SUPERNOVAE ARE RELATIVELY RARE IN OUR COSMIC TIME FRAME.**
- The last one observed by humans in the Milky Way exploded about 400 years ago, before Galileo looked through a telescope.
- The brightest stellar event perceived with the naked eye was SN 1006—SN for "supernova" and 1006 for the year recorded (A.D. 1006).
- Various records say it was so bright it outshone Venus and even the quarter moon.
- Two and a half million years ago, a supernova may have erupted near enough to Earth to have caused a massive extinction event.
- The gust of radioactive particles would have flayed our ozone layer and lambasted the DNA of vulnerable living organisms below—particularly ocean dwellers.
- Indeed, the megalodon may have succumbed to this supernova-fueled extinction.
- **THE SUPERNOVA WITNESSED ON EARTH IN 1054 WAS INITIALLY BRIGHT ENOUGH TO BE SEEN IN DAYLIGHT AND REMAINED VISIBLE FOR MANY WEEKS.**
  - It's now known as the Crab Nebula.
- At its peak luminosity, the Crab Nebula was visible during the daytime for almost three weeks and outshone every object in the night sky except the Moon.
- Although ancient Chinese and Islamic sky-gazers noted its appearance in 1054, no mention appears in any European records.
- Sometimes, stars explode more than once.
- Called recurrent novae, these periodic outbursts occur only in multiple star systems when a white dwarf accretes enough material from its companion star (or stars). Like a supernova, the star eventually undergoes collapse and explodes.
- Unlike a full-blown supernova, however, the blast releases pressure without destroying the star.
- **AS MANY AS A QUARTER OF ALL NOVAE MAY BE RECURRENT, THOUGH WE'VE WITNESSED ONLY 10 WITHIN OUR OWN MILKY WAY.**
- Most recurrent novae cycles will last 10 to 100 years.
- Located in the Corona Borealis (Northern Crown) constellation, the recurrent nova nicknamed the Blaze Star bursts every 80 years or so, when it briefly becomes about as bright as the North Star for a few days before fading away.

✦ THE MIGHTY MJÖLNIR

- In the Marvel Cinematic Universe, Thor's hammer, Mjölnir (from the Old Norse word for "lightning"), is forged in the heart of a dying star.
- If we interpret that as a neutron star, Thor's hammer should weigh as much as 300 billion elephants on Earth.
- Marvel Comics released a trading card that revealed that the hammer was made of a fictional Asgardian element called uru and that the hammer weighed a whopping 42 pounds—but they didn't specify *where* it weighs 42 pounds.
- Something that weighs 42 pounds on Mars's moon, Deimos, for example, will weigh 140,000 pounds on Earth.

Thor chucks his infamous hammer on this 1991 trading card.

CURRENT FACT COUNT 3882

DWARF STARS

In astrophysics, we call it as we see it. Thus this single category of all the smallest stars in the universe. But when sorted by color, what seems similar becomes very different.

- As their name suggests, dwarf stars are the smallest among the stellar population.
- **THERE ARE FIVE TYPES OF DWARF STARS: WHITE, RED, YELLOW, ORANGE, AND BROWN.**
- Technically, brown is the outlier, because brown dwarfs are not, by definition, stars.

✦ WHITE DWARF

- A white dwarf is the stellar core that remains after a red giant has shed all its layers.
- It is the final stage of a star's death.
- A white dwarf can linger for billions of years as it cools down.
- In theory, a white dwarf will continue to cool until it becomes a "black dwarf"—a hypothetical object that emits so little energy it's nearly nonexistent.
- The universe is not yet old enough for black dwarfs to exist.
- The nearest white dwarf to Earth orbits a far fainter star that has the same mass as our Sun but is a million times smaller and, therefore, a million times denser.
- **TYPICALLY, AN EARTH-SIZE WHITE DWARF IS SO DENSE THAT A TEASPOON OF MATERIAL FROM IT WOULD WEIGH MORE THAN A PICKUP TRUCK.**
- The smallest white dwarf ever reported has a mass 1.35 times greater than our Sun's squeezed into a ball only slightly larger than the Moon.
- The extreme magnetic field of the star registers as a billion times stronger than our Sun's magnetic field.
- Scientists believe that this white dwarf was actually created when two white dwarf stars merged.

A faux travel poster for Promixa b

- If one star of a binary system is a white dwarf, then cataclysm can ensue.
- A white dwarf on its own won't go supernova, but with a companion at its side, transferring more and more material to it, the dwarf star may just explode.

✦ RED DWARF

- Red dwarfs are much smaller and cooler than our Sun.
- They burn steadily through their hydrogen for billions of years with little internal change.
- **SOME RED DWARFS ARE PREDICTED TO LAST MORE THAN A TRILLION YEARS.**
- The red dwarf is probably the most common type of star in the universe.
- One example is Proxima Centauri, our nearest stellar neighbor at four light-years away.

✦ YELLOW AND ORANGE DWARF

- Our Sun is currently a yellow dwarf star.
- The term "yellow dwarf" is misleading. Plenty of yellow dwarfs are white, including our Sun.
- Orange dwarf stars burn hotter than red dwarfs but cooler than yellow dwarfs.
- Despite a yellow dwarf star hosting the only known planet to contain life (Earth), we expect orange dwarf stars to provide the most hospitable Goldilocks zones.
- With their slightly hotter temperatures, an orange dwarf's Goldilocks zone lies farther from the star than a yellow dwarf's, making

An artist's conception of a brown dwarf

any resident planets less vulnerable to crippling solar weather events.

- They have the longest, most stable life spans, up to seven times longer than yellow dwarfs'.

- **ORANGE DWARFS ARE ALSO ABOUT THREE TIMES MORE PLENTIFUL IN THE UNIVERSE THAN YELLOW DWARFS.**

- Despite their promising Goldilocks zones, young orange dwarfs tend to emit higher amounts of UV light for much longer periods of time than yellow or red stars.

- So, if we do find life around an orange dwarf, it's likely to be hosted by a star old enough for UV emission to have dwindled—far older than our Sun—or on a planet with a strong magnetic field to minimize damaging radiation.

✦ BROWN DWARF

- Brown dwarf is a category of celestial object all its own.

- **THEY ARE MORE MASSIVE THAN PLANETS, YET ARE NOT QUITE STARS.**

- To become a star, a gaseous mass must be held together by gravity and hot enough for nuclear fusion to ignite—requirements not met by brown dwarfs.

- While brown dwarfs form like a star, they are not massive enough to ignite the nuclear fusion that defines stardom.

- They have been referred to as "failed stars," but you could instead call them "over-performing planets."

- A brown dwarf actually appears very red.

- They glow so feebly, they are difficult to find.

- Despite the name and their lack of nuclear fusion, brown dwarfs can become up to 70 times more massive than Jupiter.

"BECAUSE THEY ARE THE FOSSILS OF ALL PRIOR GENERATIONS OF STARS, WHITE DWARFS ARE KEY TO SORTING OUT THE HISTORY AND EVOLUTION OF GALAXIES SUCH AS OUR OWN."

—ASTRONOMER TERRY OSWALT, *SCIENCE*

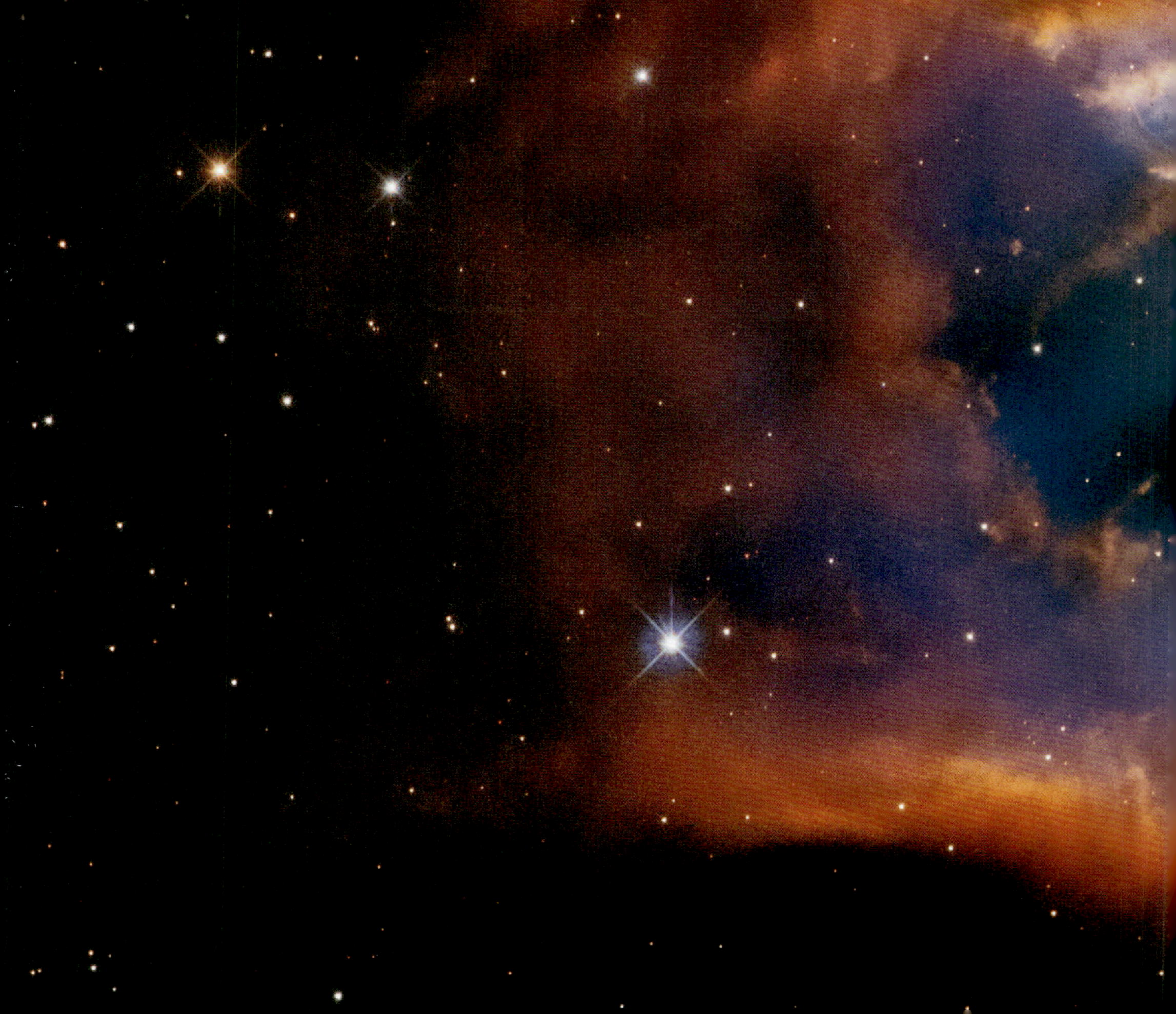

A Hubble image of the moth-like planetary nebula NGC 2899 with a white dwarf at its center

CURRENT FACT COUNT 3920

STAR SYSTEMS

Our solar system is a single star system. But we are an exception: Most stars in the universe likely have at least one, if not more, stellar companions.

- The universe holds as many as a septillion stars distributed among all its estimated one trillion galaxies.
- **ONE SEPTILLION LOOKS LIKE A 1 FOLLOWED BY 24 ZEROS: 1,000,000,000,000,000,000,000,000.**
- At least half of all stars are believed to be part of a binary or multiple star system: a system containing two (binary) or more (multiple) gravitationally bound stars.
- Polaris—also called the North Star—is actually a triple star system.
- Astronomers have observed a couple of seven-star systems, but there may be even larger ones out there.
- The chaotic and unpredictable orbits of three or more celestial bodies is known as the three-body problem.
- It's also the title of a popular science fiction story and television series by Cixin Liu that creatively explores the phenomenon.
- Although our solar system appears to have settled into a relatively stable configuration, there's a tiny chance that three to four billion years from now, Jupiter's gravity will drive orbital chaos, causing Venus or Mars to smash into Earth.
- Our nearest stellar neighbor, Proxima Centauri, is part of a three-star system called Alpha Centauri.
- The two stars, Alpha Centauri A and Alpha Centauri B, orbit each other to form a separate binary system within the larger three-star system.
- Their combined luminosity is considered the third brightest star in our night sky.
- The Proxima Centauri system hosts multiple planets.
- **THE MOST NOTABLE PLANET IN THE SYSTEM IS PROXIMA B, AN EXOPLANET JUST FOUR LIGHT-YEARS AWAY.**
- The nearest known exoplanet, it shares many similarities with our home planet, including its size and distance from the star it orbits.
- Proxima b is probably tidally locked to its host star, much like the Moon is to Earth. If so, that means half of the planet constantly boils under unrelenting starlight while the other half remains cloaked in cold and darkness.
- As such, both sides are as unforgiving and inhospitable a hellscape as the other, with no temperate regions.

"THESE STARS ARE AMONG THE HEAVIEST WE'VE EVER MEASURED."

—PHIL MASSEY, SPEAKING OF THE MILKY WAY'S BINARY STAR SYSTEM NGC 3603-A1

✦ THE CHALLENGE OF MULTISTAR SYSTEMS

- Just because a system is chaotic does not mean it is random—at least not in the world of mathematics.
- Chaos theory is an entire branch of scientific research that explores how minuscule changes within a system can lead to wildly different outcomes.
- **A PLANET ENTANGLED IN A MULTISTAR SYSTEM'S ORBITAL EMBRACE WILL MANEUVER IN ONE OF THREE POSSIBLE CONFIGURATIONS:**
    - It can retain allegiance to one star exclusively, while that star continues its dance around its partner star or stars.
    - It can orbit the entire multistar system at a distance sufficiently far enough away to feel the combined gravity of all the stars as one force rather than separate varied forces.
    - It can escape the gravitational pull of all the stars entirely and eject into interstellar space, where it will roam in darkness as a rogue planet.
- Not one of these relationships to the multistar system would be conducive to life as we know it.

The original concept art of Luke Skywalker looking across the desert toward Mos Eisley in ***Star Wars:*** **Episode IV *A New Hope***

✦ BINARY STAR SYSTEMS

- Binary star Sirius is the brightest star in our sky.
- Some binary star systems orbit one another so closely that they look like a giant peanut.
- Stars in a close binary system feed off one another's mass and energy like cosmic vampires.

✦ A REAL-LIFE TATOOINE

- Tatooine (Anakin and Luke Skywalker's home planet in a galaxy far, far away, aka *Star Wars*) is a circumbinary planet: It orbits two stars instead of one.
- Since Tatooine orbits two stars, there are often two suns in the sky.
- **CIRCUMBINARY PLANETS LIKE TATOOINE WERE FOUND ONLY IN SCIENCE FICTION UNTIL 2005, WHEN A JUPITER-SIZE PLANET WAS FOUND IN ORBIT AROUND A SUNLIKE STAR AND A BROWN DWARF.**
- We know Tatooine-like worlds can exist, but they remain rarities among discovered exoplanets.
- Planets orbiting binary or multiple-star systems are thought to be highly unlikely to host extraterrestrial life, as the conditions would be too extreme and erratic.
- **IF OUR SOLAR SYSTEM WERE A DOUBLE-STAR SYSTEM, AND IF BOTH STARS WERE SIMILAR TO OUR SUN, THEN:**
  - We would need to wear double the SPF protection.
  - Days would be twice as bright.
  - We would receive double the radiation.
  - The planet might undergo long seasons of mostly day or mostly night.
  - We would experience extreme temperature fluctuations ranging from inhospitably hot to freezing cold, depending on our orbital positions.
  - We would see double eclipses, with one star blocking the other from our view.
  - Our orbit would become so erratic that we might fly off into interstellar space or become engulfed by one of our twin-star hosts.

5

Two hundred million years after the Big Bang, dense, star-forming pockets of the universe, seeded by the gravitational influence of dark matter, congealed into the first galaxies. The meldings and mergings of the earliest galactic fragments culminated in a stupendous web of diverse galaxies that continue to surprise us.

A cornucopia of stellar magic: Two spiral galaxies, three merging galaxies, galaxy groups, Milky Way stars, and more are visible in this photo of the Virgo cluster.

GALA

CTICA

CURRENT FACT COUNT 3984

OUR HOME GALAXY

The Milky Way: our star's home among at least 100 billion gravitationally bound stellar companions

✦ GALAXY BASICS

- Galaxies are some of the most ancient structures in the universe.
- **A GALAXY IS MADE OF STARS, PLANETS, GAS, AND DUST HELD TOGETHER BY GRAVITY.**
- The distance between galaxies grows ever greater because the universe is expanding.
- To glimpse these distant celestial objects is to peer into the universe's distant past.
- For instance, the large barred spiral galaxy NGC 1015 is 120 million light-years away.
- This means our observation of NGC 1015 shows the galaxy as it was nearly 120 million years ago, when dinosaurs roamed Earth and Australia was still connected to Antarctica.
- At 65 million light-years away, the galaxies of the Virgo supercluster appear as they were around the time that the dinosaurs perished.
- **ONE OF THE FAINTEST OBJECTS EVER OBSERVED, GALAXY A2744_YD4, EXISTED WHEN THE UNIVERSE WAS JUST 600 MILLION YEARS OLD.**
- The farthest known galaxies, thriving in an era of the early universe called the "dark ages," defy conventional explanation, requiring astrophysicists to revisit their models of galaxy formation.

✦ STELLAR BUMPER CARS

- The matter composing the Milky Way is caught in a game of tug-of-war among neighboring galaxies.
- Galaxies routinely collide. Before colliding galaxies could be accurately simulated by computers, they were cataloged as peculiar.
- **UNTIL THE EARLY 20TH CENTURY, MOST ASTRONOMERS CONSIDERED THE MILKY WAY THE ENTIRE UNIVERSE.**
- We were driving cars before we had evidence that galaxies existed beyond our own.
- Current estimated number of galaxies in the universe: several hundred billion (give or take a few hundred billion)
- There are as many galaxies in the universe as there are stars within the Milky Way.
- A large galaxy contains trillions of stars.
- A dwarf galaxy contains as few as a million or even less.
- There are at least 10 times more dwarf galaxies than large galaxies in the universe.
- Dwarf galaxies were a research specialty of Neil deGrasse Tyson's in the 1990s. Among his published papers was "On the Possibility of Gas-Rich Dwarf Galaxies in the Lyman-alpha Forest," which explored whether dwarf galaxies vastly outnumbered all other objects in the universe. (Yes, in fact, they do.)

✦ WHAT'S IN A NAME?

- The word "galaxy" comes from *galaxias*—Greek for "milky."
- Visible from Earth, the stars of the Milky Way spill across the sky in a white haze, like milk, inspiring the galaxy's name.
- The Western term "Milky Way" was a translation from the Roman Latin name Via Lactea (Milky Way).
- **A CHEROKEE FOLKTALE RECOUNTS THE STORY OF A DOG WHO STEALS CORNMEAL, SCATTERING IT AS HE RUNS AWAY. THE MILKY WAY IS THUS CALLED "WHERE THE DOG RAN."**
- Many ancient cultures used words like "road," "river," "path," or "way" to name the band that is our galaxy.
- In China, where milk is a less popular beverage than in Europe, our home galaxy is called the Silver River.
- The ancient Inca called the Milky Way the Celestial River.

✦ MILKY WAY STATS

- Age of the Milky Way: 13.6 billion years
- The Milky Way is a barred spiral galaxy with an enormous central bar of stars and gas.
- The central bar of the galaxy appears as a band of light stretching across the night sky.
- The galaxy's spiral structure wasn't recognized until the mid-20th century.
- **LENGTH OF THE MILKY WAY: ABOUT 100,000 LIGHT-YEARS ACROSS**
- The Milky Way spreads 100 times its thickness at the edges and 10 times its thickness at the center. This makes it flatter than a pancake—more like a crepe or a tortilla.

The Milky Way pours across the sky of New Mexico's White Sands National Park.

- Our galaxy probably started out as a rotating sphere, flattening out as it spun, like pizza dough.

- Our cosmic carousel makes a full rotation once every 220 million years.

- **WHEN DINOSAURS ROAMED THE EARTH, THE SOLAR SYSTEM WAS LOCATED ON THE OPPOSITE SIDE OF THE MILKY WAY.**

- Living on a plane of the Milky Way, as the solar system does, our views of the distant universe all point in directions away from the Milky Way's band of light.

- When you see the Milky Way during the summer months in the Northern Hemisphere, you're looking into the core of the galaxy. During winter, you see the outer edges.

- The solar system resides about 26,000 light-years from the galaxy's center.

- At any time of year, when you look on either side of the band, you are peering above and below the Milky Way.

- The Milky Way in the night sky has low surface brightness, leaving it forever lost amid the city lights of the world.

- Though we can see surrounding galaxies, we can't see the Milky Way in its entirety because we're embedded within it.

An artist's concept illustrating the spiral arms of the Milky Way galaxy

✦ THE GALACTIC CENTER

- At first, nobody knew which direction in the sky to find the Galactic Center. It's visible all around us. Like walking into a forest, you can't see the center because the trees are in the way.

- In 1918, American astronomer Harlow Shapley deduced that the Galactic Center was in the direction of the constellation Sagittarius.

- **HARLOW SHAPLEY PRESUMED THAT THE MILKY WAY'S TITANIC ASSEMBLAGES OF STARS, KNOWN AS GLOBULAR CLUSTERS, WOULD BE DISTRIBUTED UNIFORMLY AROUND THE CENTER. HE WAS RIGHT.**

- The Galactic Center sits beyond the famous "Teapot" asterism of the constellation Sagittarius, confirmed by gas-penetrating radio telescopes that can "see the forest for the trees."

- Our understanding of the Milky Way's structure continues to evolve as scientists gather more data from instruments such as the Gaia and Spitzer space telescopes.

- The European Space Agency's Gaia spacecraft provides keen insights on the positions and speeds of nearly two billion stars in our Milky Way.

- **USING MODERN TOOLS, WE CAN BUILD 3D MAPS OF STARS AND INTERSTELLAR GAS IN OUR LOCAL MILKY WAY NEIGHBORHOOD.**

- The solar system has orbited the Galactic Center 20 times in its lifetime, making it 20 galaxial years old.

✦ THE GALAXY'S ARMS

- The Milky Way has two major spiral arms, Perseus and Scutum-Centaurus, filled with young and old stars.

- Two minor arms, Norma and Sagittarius, consist primarily of gas clouds and star-birthing regions.

- We sit near a partial arm called the Orion Arm, or Orion Spur.

✦ THE RADCLIFFE WAVE

- The largest coherent structure observed in our galaxy was discovered in 2019.

- It is a gigantic chain of star-birthing clouds that forms an undulating pattern.

- **THE SCIENTISTS WHO DISCOVERED IT NAMED IT THE RADCLIFFE WAVE AFTER THEIR AFFILIATED COLLEGE, THE RADCLIFFE INSTITUTE FOR ADVANCED STUDY.**

- The Radcliffe Wave is 9,000 light-years long.

✦ OMEGA CENTAURI

- Globular clusters are densely packed, spherical globs of thousands to millions of stars that orbit within a galaxy.

- **THEY ARE AMONG THE OLDEST OBJECTS IN THE UNIVERSE.**

- Every large galaxy contains some of these globs, and the Milky Way hosts about 150.

- They contain some of the Milky Way's oldest stars, dating back to the time our galaxy was forming.

- Some globs are likely remnants of dwarf galaxies that were swallowed up by the Milky Way.

- One such cluster is Omega Centauri.

- It is the largest and brightest globular cluster in the Milky Way.

- It's far more massive—possibly 10 times more—than typical globular clusters.

Omega Centauri was first listed in Ptolemy's catalog nearly 2,000 years ago as a single star.

"PRAISE BE TO THE STARS THAT IMPLODE. A NEW FREEDOM OPENS UP WITHIN THEM: ANNULLED FROM SPACE, EXONERATED FROM TIME, EXISTING, AT LAST, FOR THEMSELVES ALONE."

—ITALO CALVINO, "IMPLOSION"

The Atacama Large Millimeter/submillimeter Array (ALMA) Observatory in Chile, at an elevation of about 16,500 feet

CURRENT FACT COUNT 4036

GALAXIES BEYOND OUR OWN

As recently as the early 20th century, we presumed ourselves to be the entirety of the universe, but the humble Milky Way is one of perhaps a trillion or more other galaxies.

- Even without telescopes, early sky-watchers observed and named our three nearest galaxies.
- **TWO OF THEM, NOW KNOWN AS THE MAGELLANIC CLOUDS, ARE OBSERVED ONLY IN THE SOUTHERN HEMISPHERE.**
- They are named for Portuguese explorer Ferdinand Magellan and his crew, who were among the first Europeans to observe and describe them in the early 1500s.
- They are called the Large Magellanic Cloud and the Small Magellanic Cloud.
- The third, the Andromeda Nebula (now called the Andromeda galaxy), is named for the constellation where it appears to reside from our view on Earth.
- The earliest mention of the Andromeda galaxy is in the Persian *Book of Fixed Stars* (A.D. 964) by Abd al-Rahman al-Sufi, who described it as a "small cloud."

The Large Magellanic Cloud, photographed at the Salar de Uyuni, Bolivia

- **THE ANDROMEDA GALAXY IS OUR NEAREST GALACTIC NEIGHBOR.**
- Besides our own Milky Way, the Andromeda and Small and Large Magellanic Cloud galaxies are the only ones visible from Earth with the naked eye. All others require a telescope to see.
- Astronomers call the Andromeda galaxy M31.
- The *M* is for "Messier object," after the 18th-century French astronomer Charles Messier, who cataloged many galaxies.
- Astronomers still use the list of objects started by Messier in his *Catalogue of Nebulae and Star Clusters.*
- Revised several times, the final edition of Messier's catalog was published in 1781.
- Messier was an avid comet observer.
- He started his catalog to help astronomers distinguish cloudy objects in the sky from comets.
- The object known today as the Crab Nebula is also called M1 (Messier 1), the first in the Messier catalog.
- **THERE ARE CURRENTLY 110 OBJECTS TOTAL IN THE MESSIER CATALOG.**
- The Messier catalog includes globular clusters and nebulae as well as galaxies.

"THE REAL PRIZE IS FINDING SOMETHING NEW OUT THERE."

—ASTRONOMER VERA RUBIN

Clouds of gas and dust near the Tarantula Nebula in the Large Magellanic Cloud

EDWIN HUBBLE'S CLASSIFICATION SCHEME

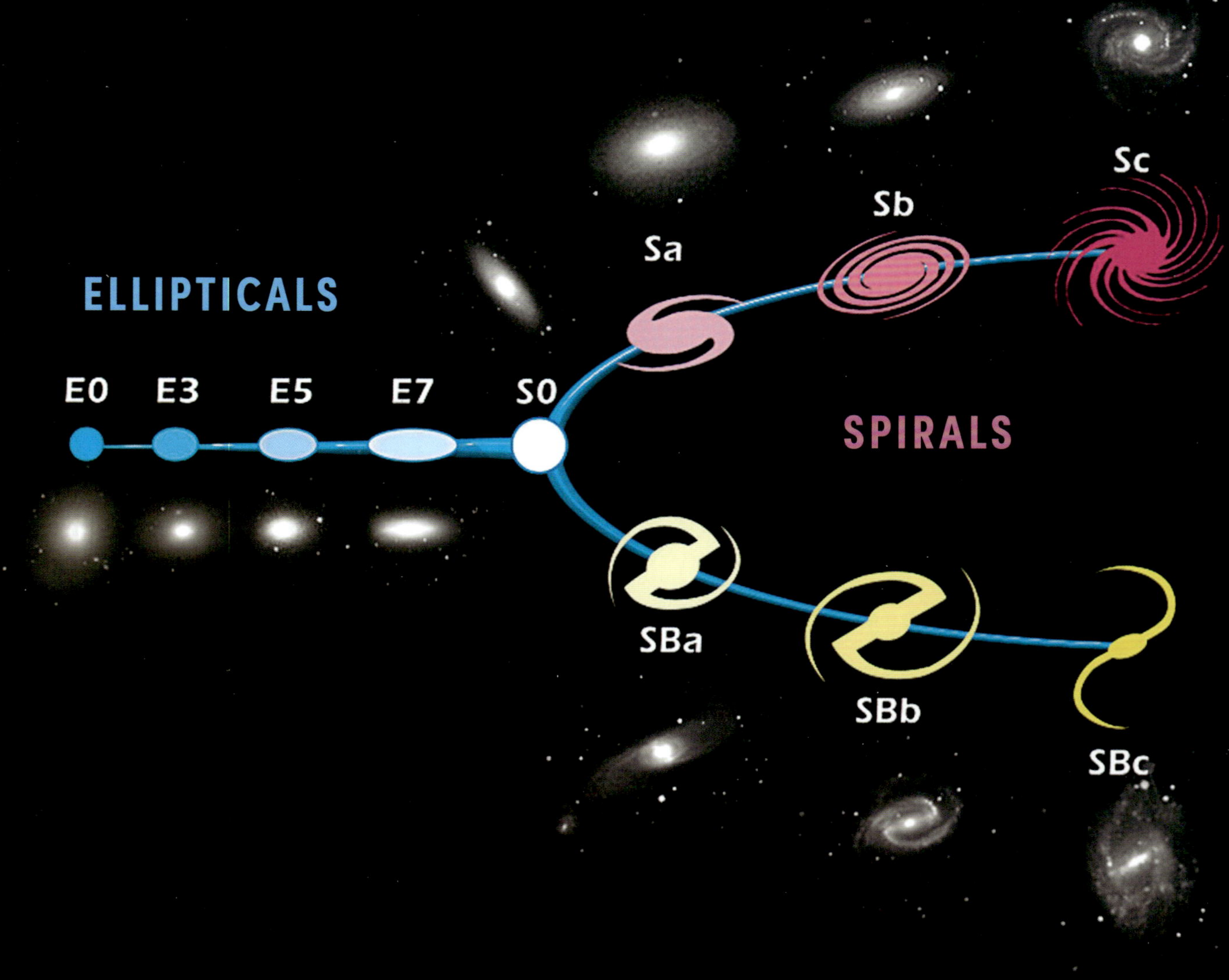

An artist's depiction of the Hubble tuning fork system of galaxy classification

✦ TYPES OF GALAXIES

- There are two ways to classify galaxies: by mass or by shape.
- Dwarf galaxies typically contain fewer than a billion stars, while supergiants contain a trillion or more.
- There are more dwarf galaxies in the universe than any other variety.
- Galaxies also come in three basic shapes: spiral, elliptical, and irregular.
- **ALTHOUGH THEY ARE CLASSIFIED INTO THREE TYPES, GALAXY SHAPES ARE ALL A BIT DIFFERENT. LIKE SNOW-FLAKES, NO TWO ARE IDENTICAL.**
- To classify their shapes, astronomers use the nearly century-old Hubble tuning fork system, named for its inventor, Edwin Hubble.
- The order of galaxies in Hubble's tuning fork is not meant to suggest galactic evolution.

✦ SPIRAL GALAXIES

- Spiral galaxies are shaped somewhat like pinwheels.
- **STATISTICALLY, WE ARE MORE LIKELY TO SEE SPIRAL GALAXIES EDGE ON RATHER THAN FACE ON.**
- To get a fuller picture of a spiral galaxy, we look at images in infrared and ultraviolet wavelengths together.
- Infrared light penetrates the cloud of dust, obscuring high-energy star formation activity, but we can sense this using ultraviolet wavelengths.
- Spiral galaxies are the most common shape, with bright regions of active star formation.
- They can have two, four, or more arms, tightly or loosely wound. Some are elegantly symmetrical; others distorted or messy in appearance.
- Spirals have two subgroups: normal and barred.
- Barred spiral galaxies have a pronounced bar of stars running through their central bulge.
- Spirals range in diameter from 15,000 light-years to a few hundred thousand light-years.
- Their centers, called the "central bulge," tend to contain a core of old stars.
- **SPIRAL GALAXIES ROTATE.**
- What we see can be deceiving: Our view of a galaxy depends on our orientation to it. We might have a clear, face-on view of one galaxy's spiral shape but see only a flattened disk and central bulge in a side view of another, as with the Sombrero galaxy (M104).
- How barred spiral galaxies form is a mystery, but we do know that barred spiral galaxies are much more common now than long ago.
- NGC 5584 is an example of a galaxy with loosely wound arms.
- Visible in the constellation Virgo, NGC 5584 is 70 million light-years away.
- Discovered in 1881 by American Edward Barnard, NGC 5584 played a key role in refining the expansion rate of the universe.
- Some spiral galaxies, such as NGC 2841, are called "flocculent galaxies." Their short arms, brimming with young, hot stars, appear fluffier than other classic grand spirals.
- Barred spiral galaxy NGC 6217 was the test subject for the Hubble Space Telescope after its camera was repaired.
- The spiral galaxy known as NGC 3972 lies 65 million light-years from Earth.
- In 2011, a supernova was observed in NGC 3972 that rivaled the brightness of the entire galaxy. That star's light is only just now reaching us, after a 65-million-year journey through space.
- **LIGHT FROM THE EXTINCTION OF THE DINOSAURS ON EARTH, 66 MILLION YEARS AGO, WOULD HAVE WASHED OVER EAVESDROPPING ALIENS IN NGC 3972 JUST A MILLION YEARS AGO.**

✦ ELLIPTICAL AND IRREGULAR GALAXIES

- Elliptical galaxies range in size from a few thousand to some 700,000 light-years across.
- As their name implies, elliptical galaxies resemble elongated spheres.
- Elliptical galaxies tend to contain older stars than spiral or irregular galaxies.
- Irregular galaxies is a catchall category for any that don't resemble spirals or elliptical galaxies.
- **ASTRONOMERS BELIEVE IRREGULAR GALAXIES WERE MORE ABUNDANT IN THE EARLY UNIVERSE UNTIL SPIRALS AND ELLIPTICAL GALAXIES DEVELOPED.**
- Irregular galaxies often lack any symmetry.
- Astronomers have begun to describe other shapes beyond the classics. Lenticular galaxies, for example, are believed to be a transition between elliptical and spiral.

CURRENT FACT COUNT 4081

GREAT GALAXIES

A diversity of galactic shapes and sizes has formed throughout the universe.

- Thanks to Hubble Deep Field (a census taken by the Hubble Space Telescope in 2016), astronomers now estimate there are two trillion galaxies in the observable universe.
- Scientists' estimate of galaxies in the universe increased by tenfold thanks to the Hubble Ultra Deep Field.

✦ MAGELLANIC CLOUDS

- The Large and Small Magellanic Clouds are not actually clouds, but irregularly shaped dwarf galaxies.
- The Large Magellanic Cloud is 160,000 light-years away.
- The Small Magellanic Cloud is 190,000 light-years from us.
- Visible from the Southern Hemisphere, the Magellanic Clouds look like a pair of fuzzy splotches.
- **POLYNESIAN SAILORS USED THE SMALL MAGELLANIC CLOUD FOR NAVIGATION.**
- Astronomers believe the Small Magellanic Cloud will ultimately be absorbed into the Milky Way galaxy.

One of two known photographs of the Leviathan of Parsonstown telescope (1880)

- The Persian astronomer Abd al-Rahman al-Sufi included the Large Magellanic Cloud in his *Book of Fixed Stars* (A.D. 964).
- **AL-SUFI DIDN'T SEE IT HIMSELF, BUT HEARD OF IT FROM MARINERS WHO HAD SEEN IT IN THE SOUTHERN HEMISPHERE.**
- He called the Large Magellanic Cloud al-Baqar al-Abyad (the "White Bull").
- The Large Magellanic Cloud is about 50 percent more massive than the Small Cloud.
- One supergiant star in the Large Magellanic Cloud, named Sanduleak -69° 202, exploded as a supernova in 1987.
- The explosion was so bright, it was visible from Earth without telescopes.
- Now named SN 1987A, it is the newest supernova known and the closest to us since the 1604 supernova observed by Kepler.
- At a distance of 160,000 light-years, supernova 1987A exploded about the time *Homo sapiens* began to migrate out of Africa.
- The Magellanic Clouds were described by a scribe aboard explorer Ferdinand Magellan's round-the-world voyage.
- Without a telescope, the scribe had no idea the "clouds" described were entire galaxies.

✦ WHIRLPOOL GALAXY

- A picturesque galaxy, Whirlpool (M51) has a classic spiral structure and was first identified by Charles Messier in 1773.
- The galaxy's elegant, spiraling arms harbor brilliant regions of new star formation.
- **THE WHIRLPOOL GALAXY LIES 25 TO 30 MILLION LIGHT-YEARS FROM EARTH.**
- In 1845, Irishman William Parsons and his wife, Mary, studied the Whirlpool galaxy.

- They observed it through the 72-inch telescope they built, known as the Leviathan of Parsonstown.
- Until the Hooker telescope opened in 1917, the Leviathan was the largest in the world.
- **THE PARSONS' 1845 PENCIL SKETCH OF THE GALAXY SHOWED ONE OF THE SPIRAL ARMS IS SLIGHTLY DISTORTED.**
- Since then, observations by the Hubble telescope confirmed Whirlpool's asymmetry is caused by smaller nearby galaxy NGC 5195.

✦ PINWHEEL GALAXY

- The Pinwheel galaxy (M101) is a giant spiral 170,000 light-years across.
- Pinwheel is about 1.5 times the size of the Milky Way in diameter.
- Astronomers estimate the Pinwheel galaxy contains at least a trillion stars.
- **IT'S A FAST-ROTATING GALAXY, PRESUMABLY DUE TO LARGE AMOUNTS OF DARK MATTER.**
- At 25 million light-years away, it's relatively close to us and visible in the constellation Ursa Major.
- Hubble's image of the Pinwheel galaxy was one of the largest it ever produced of a spiral galaxy. It comprised a mosaic of 51 exposures taken over a decade.

✦ ANDROMEDA GALAXY

- Andromeda (M31) is the closest spiral galaxy to us, at 2.5 million light-years away.
- The light we see from Andromeda started its journey to Earth when our evolutionary ancestors, *Homo habilis*, were fashioning the first stone tools.

A face-on view of the Whirlpool spiral galaxy (M51)

- With an estimated one trillion stars, Andromeda is as massive as the Milky Way.
- Before 1924, it was called the Andromeda Nebula, or the Great Nebula.
- The farthest thing we can see with the naked eye is Andromeda, as a fuzzy patch between the constellations Pleiades and Andromeda.
- Even at 2.5 million light-years apart, the Milky Way and Andromeda galaxies exert gravitational influence over each other.
- The two are on an inevitable collision course. Andromeda is hurtling toward us at a speed of approximately 250,000 miles an hour (about 70 miles a second). Plan for impact in about four billion years.
- Galaxy collisions are messy train wrecks, casting wayward stars hither and yon.
- After galaxies collide, their central regions ultimately merge, combining the masses of whatever black holes lurked in their centers.
- American astronomer Vera Rubin used observations of Andromeda as a basis for the first direct evidence of dark matter.

✦ NEW DISCOVERIES

- For centuries, other galaxies were considered part of the Milky Way, with no proof that any existed beyond our galaxy.
- During the 1920s, American astronomer Edwin Hubble's work proved that galaxies did exist beyond the Milky Way.
- Hubble's discovery depended on his observations of the Andromeda galaxy.

CURRENT FACT COUNT 4138

GALAXY GROUPS

The groups, clusters, and superclusters that galaxies congregate in make up a grand web of cosmic building blocks.

- Galaxies belong to groups. These groups congregate in clusters, and clusters form parts of superclusters.
- **THIS COSMIC WEB IS A TWISTING NETWORK OF GALAXIES, CLUSTERS, WALLS, AND FILAMENTS, WITH COSMIC VOIDS IN BETWEEN, ACTING AS THE SCAFFOLDING OF THE UNIVERSE.**
- Between galaxies, the intergalactic medium is full of gas and plasma.
- Intergalactic gas is extremely hot and thinly distributed.

"FAR AWAY . . . STARS WERE SILENTLY SHOOTING AND FALLING AND FINDING THEIR FATES, AMONG THESE BILLIONS AND BILLIONS."

—IRIS MURDOCH, *THE SEA, THE SEA*

✦ GROUPS

- A galaxy group generally contains 100 or fewer gravitationally bound galaxies with a brightness equal to or greater than that of the Milky Way.
- The Milky Way is part of the Local Group, a collection of about 80 galaxies, gravitationally bound to one another and spanning 10 million light-years.
- Members of the Local Group include the Milky Way, Andromeda, and Triangulum galaxies as well as the Large and Small Magellanic Clouds.
- All members of galaxy groups orbit their common center of mass—and sometimes collide.

✦ CLUSTERS

- A galaxy cluster contains hundreds or even thousands of galaxies and abounds in hot gases and dark matter.
- There's so much hot gas in the space between galaxies that it exceeds the mass of the member galaxies within the cluster.
- The center of a galaxy cluster often holds large elliptical giants, with smaller spiral galaxies along the edges.
- You'll also find stars and planetary nebulae in a cluster's intergalactic space.

✦ SUPERCLUSTERS

- Superclusters were first discovered in the late 1950s.
- **A GALAXY SUPERCLUSTER CONTAINS GROUPS, CLUSTERS, AND GALAXIES THAT ARE NOT NECESSARILY GRAVITATIONALLY BOUND TO ONE ANOTHER.**

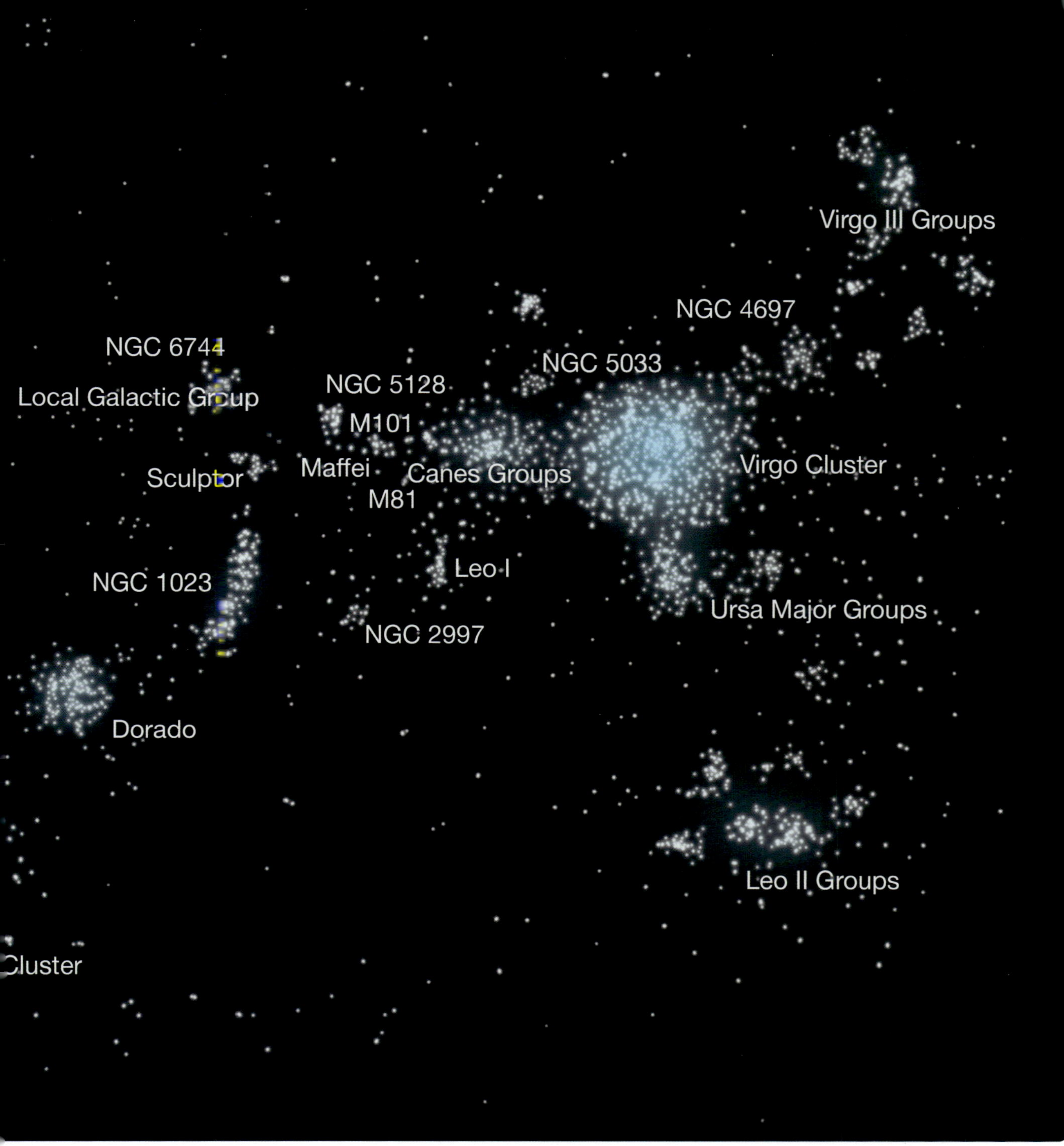

This illustration shows the location of our Local Group among the other galaxies forming the Virgo supercluster, 100 million light-years across.

- The Local Group resides within the Virgo supercluster, spanning 110 million light-years and composed of around 100 galaxy groups and smaller clusters.

- New observations suggest the Virgo supercluster is actually a lobe of the much larger supercluster called Laniakea.

- Laniakea means "immeasurable heaven" in Hawaiian. The supercluster consists of approximately 100,000 galaxies across a half billion light-years of space.

- This makes Earth's cosmic address: Planet Earth, Solar System, Milky Way Galaxy, Local Group, Virgo supercluster, Laniakea supercluster, Observable Universe.

- **THE MASS OF THE LANIAKEA SUPERCLUSTER IS EQUAL TO THAT OF A HUNDRED QUADRILLION (10^{17}) SUNS.**

- Other superclusters include Pisces-Cetus and Perseus-Pisces.

✦ EXPANDING GALAXIES

- More than 100 billion galaxies are gathered in clusters throughout the universe—all flying apart as it expands.
- The filaments and walls of superclusters are believed to vein the entire universe.

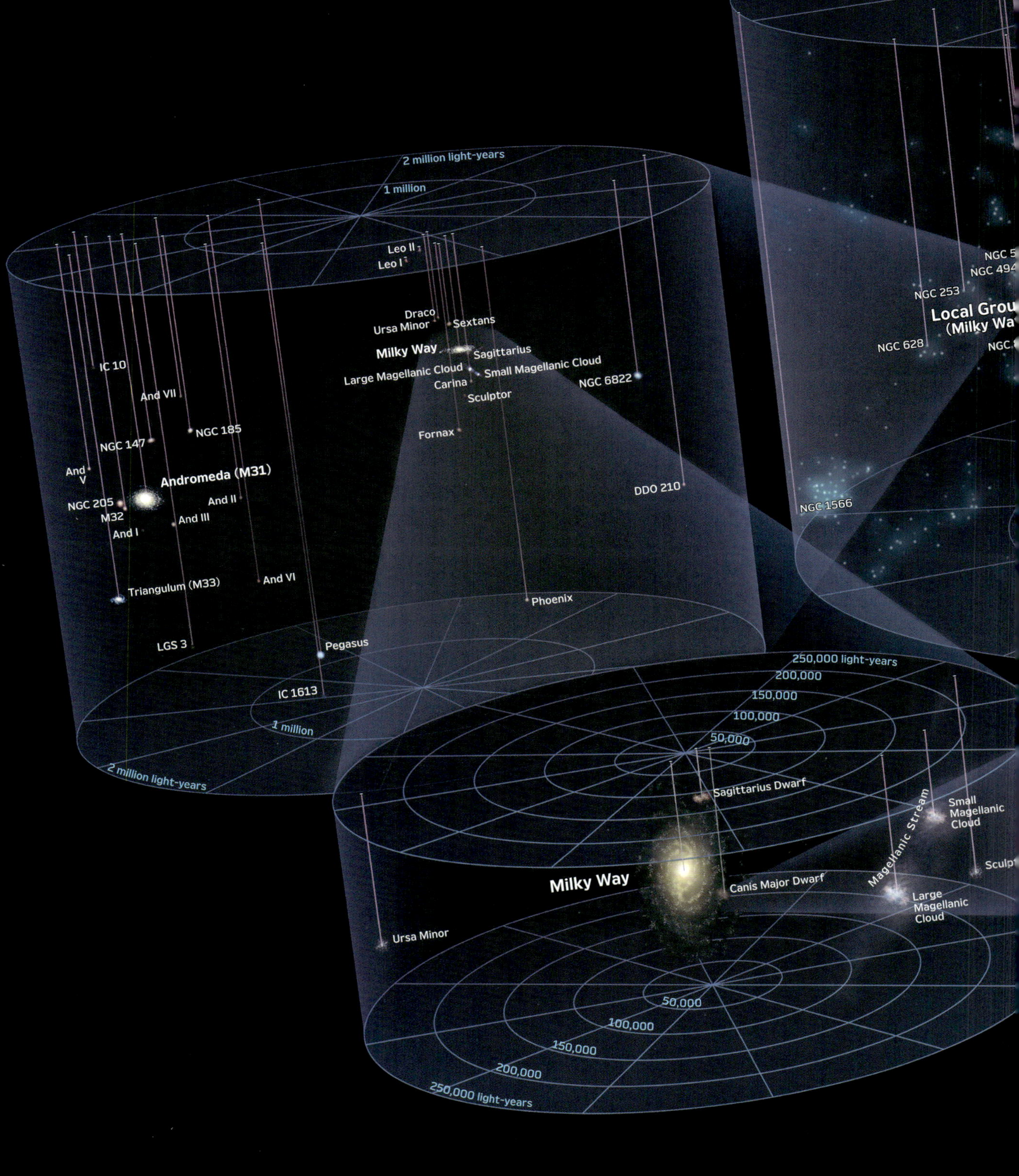

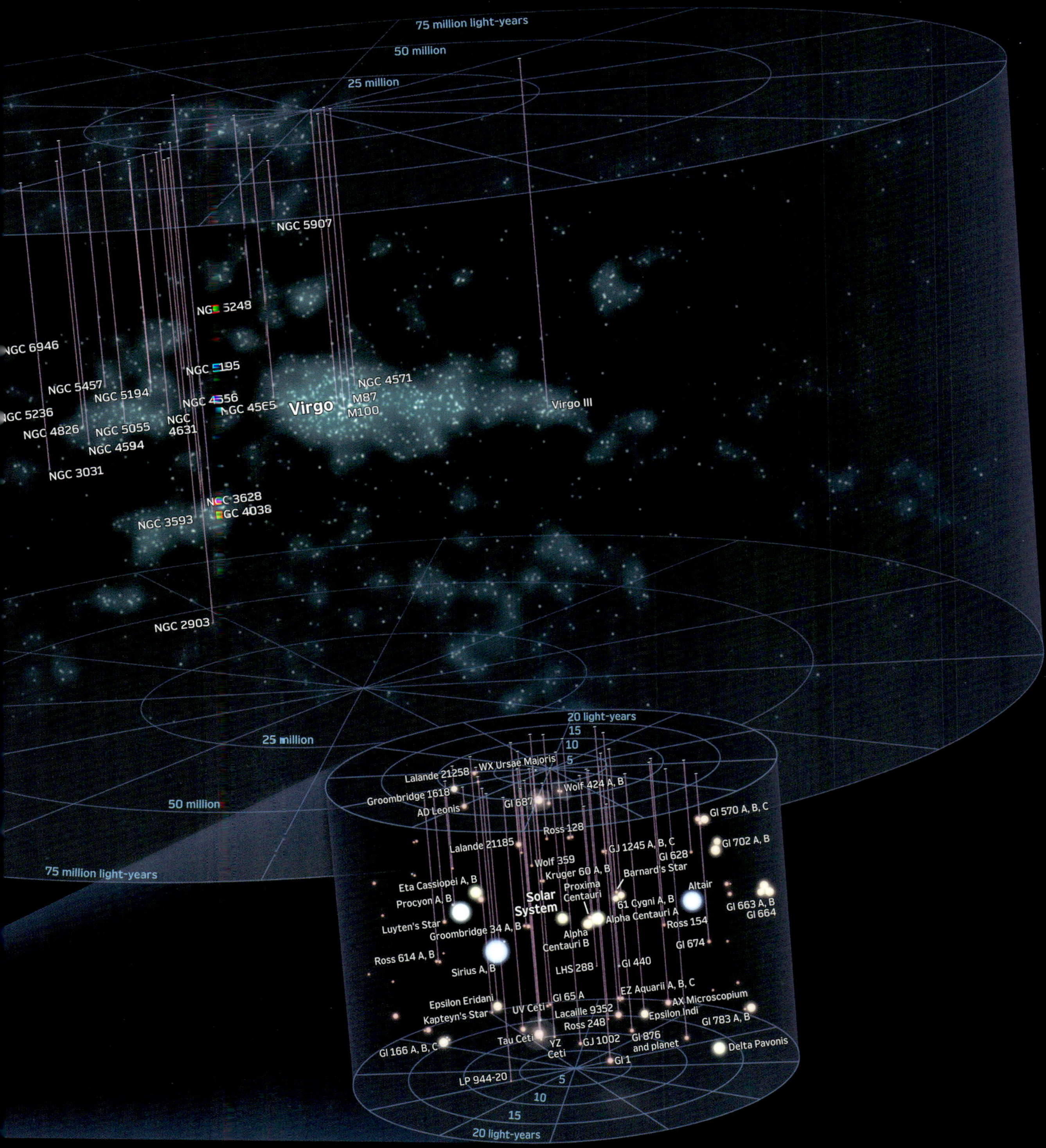

The Milky Way (bottom left) belongs to the Local Group (upper left). These belong to a galactic supercluster (upper right).

A new image of SN 1987A, taken by the Hubble Space Telescope in 2017 using its Wide Field Camera 3

✦ THE WALLS OF SUPERCLUSTERS

- The walls of superclusters form the largest known structures in the universe, some stretching for more than a billion light-years.
- Before the discovery of galactic walls, it had been presumed that galaxies peppered the universe at random.
- The first galactic wall identified was the Coma Wall, or CfA2, discovered in 1989.
- The specs of the Coma Wall: 500 million light-years across, 16 million light-years deep, 300 million light-years away
- The Sloan Great Wall, discovered in 2003, is 1.37 billion light-years across.
- The BOSS Great Wall is four superclusters comprising 830 galaxies connected by hot gas filaments. It's a twisting structure more than one billion light-years across.
- The Giant Arc, discovered in 2021, is a nearly symmetrical arc of galaxies stretching 3.3 billion light-years across.
- Discovered in 2024, the Big Ring of distant galaxies has a diameter of 1.3 billion light-years and circumference of four billion light-years.
- **THE HERCULES-CORONA BOREALIS GREAT WALL, ABOUT 10 BILLION LIGHT-YEARS ACROSS, IS THE CURRENT RECORD HOLDER FOR LARGEST STRUCTURE IN THE UNIVERSE.**
- The discovery of the Hercules-Corona Borealis Great Wall was announced in 2013.
- Its distance from Earth: about 10 billion light-years
- The South Pole Wall was discovered in 2020.
- It's 1.4 billion light-years across and 500 million light-years from Earth.

✦ COSMIC VOIDS

- These giant galactic walls exist alongside regions of relative emptiness that are called cosmic voids.
- Outer space is not truly empty, not between stars or galaxies, and not even within so-called voids.
- **THE SPACE BETWEEN STARS AND GALAXIES IS FILLED WITH SUPER-HEATED GAS THAT'S SPREAD VERY, VERY THIN.**
- Outside the solar system, interstellar space contains about a million particles per cubic meter.
- Compare that to around 27 quintillion—more than all the sand grains on an average beach—in a teaspoon of Earth air at sea level.
- In intergalactic space, you'll still find at least a few hydrogen particles per cubic meter.
- The first large void discovered was the Boötes void, a nearly spherical region peppered with very few galaxies.
- Discovered in 1981, the Boötes void spans more than 300 million light-years.
- This cosmic void contains only some 60 galaxies, a smattering, compared to the 2,000 or so typical of such an area.
- **BOÖTES IS SO EMPTY, IT'S CALLED THE "GREAT NOTHING."**
- The void closest to Earth is the Local Void, at least 75 million light-years away.
- The largest void found is the Eridanus supervoid at 1.8 billion light-years wide.
- The Eridanus supervoid contains 30 percent less matter than surrounding regions.

✦ THE LOCAL GROUP

- Most galaxies in the universe, the Milky Way included, are isolated systems or live in loose groups.
- Groups with dozens of members are the most common organization of galaxies in the universe.
- Typically, a galaxy group is dominated gravitationally by one or two large members.
- The Local Group of galaxies, with the Milky Way and Andromeda as its principal members, contains about 80 galaxies across 100 million light-years.
- The Local Group is, therefore, mostly empty space.
- Most galaxies in the Local Group are dwarf galaxies, each with fewer than a billion stars (hundreds of millions is more typical).
- **GALAXIES WITHIN GROUPS ARE NOT SEPARATED FAR ENOUGH FOR THE EXPANSION OF THE UNIVERSE TO OVERCOME THEIR HAPPY GRAVITATIONAL ATTRACTION TO ONE ANOTHER.**
- The Local Group contains four galaxies visible to the unaided eye from Earth. The rest require a telescope to see.
- The four galaxies visible without the need of a telescope that reside in the Local Group are the Milky Way, Andromeda, the Large Magellanic Cloud, and the Small Magellanic Cloud.

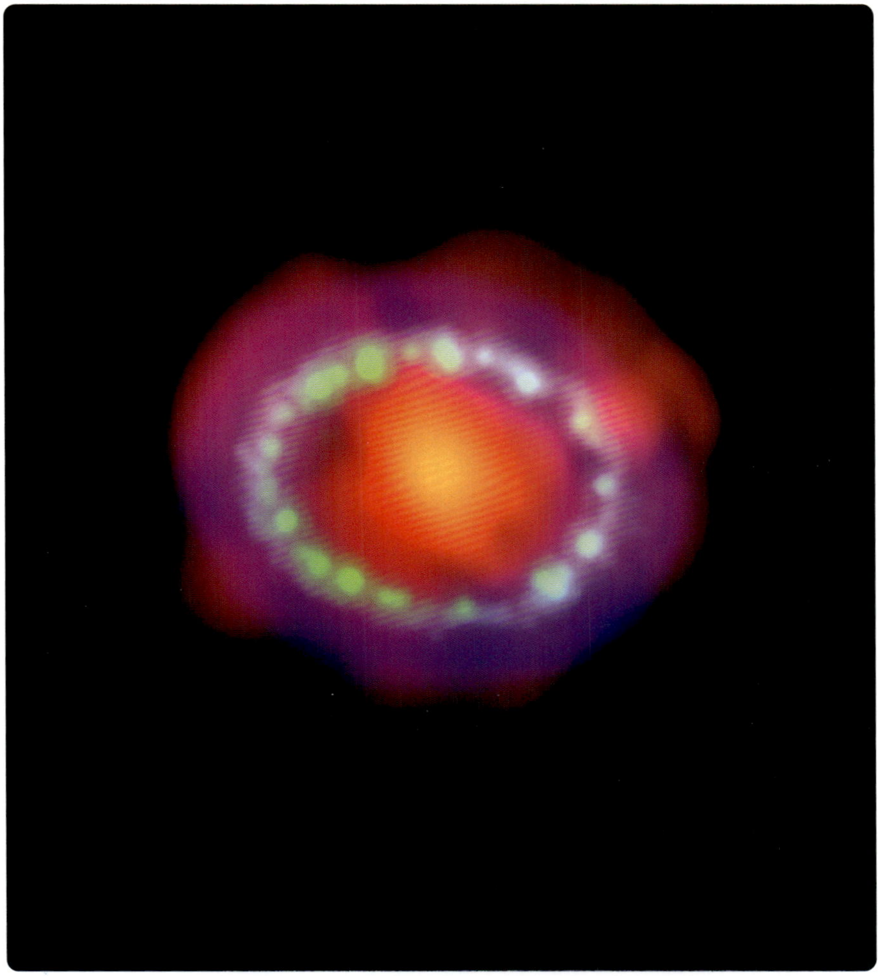

A composite image of the 1987 explosion of SN 1987A, created with images from the ALMA and Hubble telescopes and Chandra X-ray Observatory

CURRENT FACT COUNT 4175

WHEN GALAXIES COLLIDE

The formations and mergers of galaxies help shape the very fabric of spacetime itself.

- Today's galaxies emerged from collisions over billions of years.
- Galaxies interact by either colliding with or cannibalizing one another, slowly merging to form a new galaxy.
- These gravitational interactions reshape, distort, enlarge, or even shred galaxies as they enter each other's sphere of gravitational influence.
- Multiple collisions have already occurred in our galaxy's history.
- The Milky Way's next galaxy collision is not expected for four or so billion years from now, but its most recent significant collision occurred less than three billion years ago.
- **THIS COLLISION, ALONG WITH MORE ANCIENT ONES, LIKELY SHAPED OUR GALAXY INTO THE STRUCTURE WE OBSERVE TODAY.**
- Called the Virgo Radial Merger, this collision explains the erratic motions observed by distant stars, which were likely transferred to our galaxy in that cataclysm.
- Collisions leave ripples in their wake—ripples we can see and measure today.
- Galaxies can pass through each other, because stars are very far apart.
- **WHEN THEY PASS THROUGH EACH OTHER, EVERYTHING MINGLES: GRAVITATIONAL INFLUENCE, GAS, AND DUST.**
- This mingling creates variations in density.
- It also causes shock waves: ripe conditions for the birth of new stars.
- The gravity of interacting galaxies exerts influence much the same way the Moon's gravitational force creates tides on Earth.
- The effects of gravitational tidal forces depend on factors such as mass, speed, and orientation.
- The tidal effect of interacting galaxies can also deflect their orbits.
- The Whirlpool galaxy exemplifies the force of galactic tidal forces, its spiral arm stretched out by the gravitational pull of its companion galaxy, NGC 5195.
- The Cartwheel galaxy exemplifies a frontal collision that caused gas and dust to blow out into an enormous ring where new stars now form.
- Though common in the early universe, collisions and gravitational distortions between galaxies appear less frequent now because the universe is expanding.
- The Bullet cluster (in the southern constellation Carina) is an example of the aftermath of two galaxy clusters passing through each other after colliding for hundreds of millions of years.
- The spiral galaxy ESO 510-G13 looks like a warped record, suggesting interference from another galaxy's gravitational force.
- Arp 273 is a pair of asymmetrical spiral galaxies, mutually deformed as they pass each other.
- T UGC 10214, also called the Tadpole galaxy, has a long tail of gas and stars created by the tidal effect of a passing small galaxy.
- About 45 million light-years away, the Antennae galaxies form a shape like a reversed question mark.
- Now called NGC 4038 and NGC 4039, these conjoined galaxies were discovered as early as 1785 by William and Caroline Herschel.
- **THEIR COLLISION, HUNDREDS OF MILLIONS OF YEARS AGO, TRIGGERED A WAVE OF STAR BIRTHS IN EACH GALAXY.**
- Eventually the two will combine to form a giant elliptical galaxy.

"THEY'RE LOCKED IN A CLOSE INTERACTION, A . . . COSMIC DANCE DRIVEN BY THE GRAVITATIONAL FORCE."

—ASTRONOMER GIOVANNA GIARDINO, SPEAKING OF STEPHAN'S QUINTET

✦ STEPHAN'S QUINTET

- First discovered by Édouard Stephan in 1877, this configuration was first considered five interrelated galaxies.

This composite image of Stephan's Quintet was made from almost 1,000 different photographs taken by the James Webb Space Telescope.

- But the brightest of the five galaxies was just photobombing the four gravitationally bound galaxies behind it. The four are indeed allied, but the fifth just happens to be in front of them from our point of view.

- **THE STEPHAN'S QUINTET FOURSOME WAS THE FIRST COMPACT GALAXY GROUP DISCOVERED.**

- The galaxies are mutually deformed by their tidal forces—except for the top left blue galaxy, which lies only 40 million light-years away from Earth.

- The rest of Stephan's Quintet is located 300 million light-years from Earth.

- In the 1946 classic film *It's a Wonderful Life*, the angels who answer the prayers of friends and family on behalf of protagonist George Bailey are personifications of the galaxies of Stephan's Quintet.

- Stephan's Quintet was one of the first five images revealed to the public from the James Webb Space Telescope (JWST) in 2022.

✦ BACK TO THE BIG BANG

- In 2024, JWST spotted the galaxy named JADES-GS-z14-0.

- JADES-GS-z14-0 is one of the farthest objects in the universe ever seen.

- Its light represents a time when the universe was about 300 million years old—just 2 percent of its age today.

- Its name derives from the study that observed it: the JWST Advanced Deep Extragalactic Survey.

"SEEING OXYGEN IN A GALAXY THIS YOUNG IS LIKE IF YOU ARE AN ANTHROPOLOGIST AND YOU FIND AN ENORMOUS, ANCIENT CITY THAT HAS EVIDENCE OF IPHONES."

—ASTRONOMER KEVIN HAINLINE, SPEAKING OF THE JADES-GS-Z14-0 GALAXY

A composite image of the Bullet cluster, the first observation humanity had of normal and dark matter together

CURRENT FACT COUNT 4185

GALAXIES: BY THE NUMBERS

515,000

miles an hour: the speed at which our solar system orbits the Milky Way's center

4,500,000,000

years: the amount of time until Andromeda and the Milky Way collide

100,000

years: the time it would take to travel at the speed of light from one end of the Milky Way to the other

230,000,000

years: the amount of time it takes for our solar system to orbit the Milky Way

103

the number of cosmic items listed in the original Messier catalog

3

the number of large galaxies in our Local Group: Milky Way, Andromeda, and Triangulum

80

the rough number of confirmed and suspected galaxies in the Local Group

10,000,000

the estimated number of stars in the globular cluster Omega Centauri

11

galaxy types in Hubble's tuning fork

40

Edwin Hubble's age when he proved the existence of galaxies beyond the Milky Way

6

The 17th century saw the telescope and microscope invented within mere decades of each other, throwing open new windows to the universe. How fitting that, 300 years later, we would develop the theories establishing an expanding universe and quantum physics within the same decade. Our macro- and micro-understanding of the cosmos expanded yet again.

An illustration depicting the spacetime continuum in grand strokes

SPACE

TIME

CURRENT FACT COUNT 4234

THE BEGINNING OF EVERYTHING

Which is harder to believe: that the universe came from nothing or that it always was and always will be? Astrophysicists grappled with this conundrum when confronted by the undeniable evidence of the Big Bang.

- **EINSTEIN INITIALLY BELIEVED THE UNIVERSE TO BE STATIC—ETERNAL AND UNCHANGING.**
- He believed the universe to be spatially finite but temporally infinite, meaning the universe had always existed and always would, a common assumption at the time.
- In 1931, Belgian astronomer and Catholic priest Georges Lemaître proposed the idea of an entire universe expanding from a "primeval atom" of extreme density from which all matter emerged.
- The term "Big Bang" was first uttered as a dismissive jab at Lemaître's theory by British astronomer Fred Hoyle during a radio interview. Hoyle's expression caught on, and he inadvertently coined the term for the theory he fiercely opposed.
- **HOYLE ADVOCATED HIS OWN RIVAL THEORY CALLED STEADY STATE.**
- The steady-state theory proposed that the universe expands at a constant rate, its density never fluctuating, and that new matter is continuously created along the way.
- Like the static universe, it presumed the universe had no beginning or end.
- In 1948, George Gamow and Ralph Alpher demonstrated how the Big Bang could have made all existing hydrogen, some helium, and a bit of lithium.
- Aside from accidentally coining the Big Bang theory, Hoyle was best known for pioneering the stellar nucleosynthesis theory (alongside several other scientists), which showed how stars forge elements heavier than helium.

✦ COSMIC AFTERGLOW

- The Big Bang left a remarkable afterglow that still radiates in the universe today, called the cosmic microwave background (CMB).
- When the CMB was first emitted, 380,000 years after the Big Bang, it was warm and bright and primarily fell into the visible and infrared portions of the electromagnetic spectrum.
- As space expanded, the CMB stretched into the microwave frequencies as they cooled—to just above absolute zero now.
- **THIS PRIMEVAL LIGHT SLOWLY LOSES ENERGY AS IT COOLS, MEANING ITS WAVELENGTHS ARE SLIPPING DOWN THE ELECTROMAGNETIC SPECTRUM.**
- Today, we detect the CMB as microwaves. In a distant future, many billions of years from now, it will register as radio waves.
- The discovery of cosmic background radiation in 1965 transformed the field of cosmology from one largely of pure theory to one rooted in observational evidence.

✦ EXPANSION AND ACCELERATION

- Long before he was honored with a space telescope, Edwin Hubble found evidence that confirmed the universe is expanding.
- Hubble showed that the most distant galaxies are receding from us more quickly than closer ones.
- **NOW CALLED THE HUBBLE CONSTANT, THE EXPANSION RATE OF THE UNIVERSE IS ONE OF THE KEYS TO UNDERSTANDING THE EVOLUTION OF THE COSMOS—AND, PERHAPS, ITS FATE.**
- The Hubble Space Telescope was named for Edwin Hubble because one of its primary missions was to determine the universe's rate of expansion.
- **THE TELESCOPE'S OBSERVATIONS SUCCEEDED IN THAT TASK, NARROWING THE AGE OF THE UNIVERSE TO 13.7 BILLION YEARS.**
- Prior to that, astrophysicists guessed that the age fell somewhere between 10 and 20 billion years.
- That figure was further revised to 13.799 billion years in 2015, thanks to data from the Planck space observatory.
- In 1998, astronomers discovered that the universe's expansion rate is actually accelerating.
- Scientists were surprised by that finding, because gravitational attraction suggests that expansion should slow down.

- The universe is engaged in a tug-of-war between gravity's attractive pull and a mysterious counterforce that appears to push things apart.

- **WE NOW CALL THIS COUNTERFORCE DARK ENERGY.**

- The ever expanding universe is now 1,100 times larger than it was at the time the cosmic microwave background was emitted.

- This perpetual expansion means that 94 percent of all galaxies we can see in the observable universe are already beyond our reach. We view them not as they are now, but as they were billions of years ago.

- If we sent radio signals to those distant galaxies at the speed of light right now, they could never receive the transmission—and we could never receive theirs.

- **EVERY SECOND, ABOUT 60,000 STARS DISAPPEAR ENTIRELY FROM VIEW.**

- The observable universe is about 93 billion light-years across.

- This means we can see about 46 billion light-years in any direction.

- What's beyond? Nobody knows. But likely more of the same.

✦ ANTIPARTICLES

- Albert Einstein's famous equation $E=mc^2$, which expresses the equivalence of mass and energy, was activated during the early universe.

- The early universe was so hot, photons had enough energy to spontaneously make particle-antiparticle pairs, which immediately "annihilated" back into photons on contact, turning into pure energy.

- Antiparticles are the antimatter versions of particles themselves.

- As the hot, expanding universe cooled, the energy of photons ultimately dropped low enough to no longer spontaneously make particles.

- For reasons that remain mysterious, during the hottest phase of the early universe, the symmetry of particle-antiparticle production broke, creating one excess particle in 10 billion annihilations.

- Because of this symmetry breaking, we live in a universe that contains some matter rather than all photons.

- One second after the Big Bang, the universe was 20 light-years across.

- After 380,000 years, the universe cooled enough for electrons, protons, and neutrons to assemble as atoms, clearing pathways for light to travel.

- In a universe that moves inexorably to disorder, pockets of order can arise from an infusion of energy.

- If it were a closed system, Earth would have long ago plunged into disorder. But Earth receives sunlight, and builds order at the expense of the Sun.

✦ DUAL THEORIES

- Quantum physics is the theory of the small. General relativity is the theory of the large.

- During the Big Bang, the large was small, forcing a marriage of the two theories in ways they wouldn't otherwise intersect.

- Merging quantum physics with general relativity tells us that our universe might be one of countless other universes created as part of a multiverse.

✦ ATOMS OF LIFE

- A common definition of life is something that has a metabolism and is capable of reproducing and evolving. By that definition of life, stars are alive.

- Atoms of life are traceable to the Big Bang and to the thermonuclear fusion in the cores of exploded stars. So, it's not just figuratively true, it's *literally* true that we are composed of stardust.

- We are not only alive in the universe, but the universe is alive within us. The human mind is a way for the universe to know itself, as Carl Sagan famously mused.

"ALL AROUND THE GALAXIES IS A COOL SEA OF MICROWAVES—THE ECHO OF THE BIG BANG SOME FIFTEEN BILLION YEARS AGO."

—JOHN D. BARROW, *THE ORIGIN OF THE UNIVERSE*

"GIVEN SHIPS OR SAILS ADAPTED TO THE BREEZES OF HEAVEN, THERE WILL BE THOSE WHO WILL NOT SHRINK FROM EVEN THAT VAST EXPANSE."

—JOHANNES KEPLER, *CONVERSATION WITH THE SIDEREAL MESSENGER*

A composite of 678 separate images of the Trifid and Lagoon Nebulae

CURRENT FACT COUNT 4255

A TIMELINE OF THE EVOLVING UNIVERSE

The further we turn our clocks back to the Big Bang, the more puzzling the physics becomes.

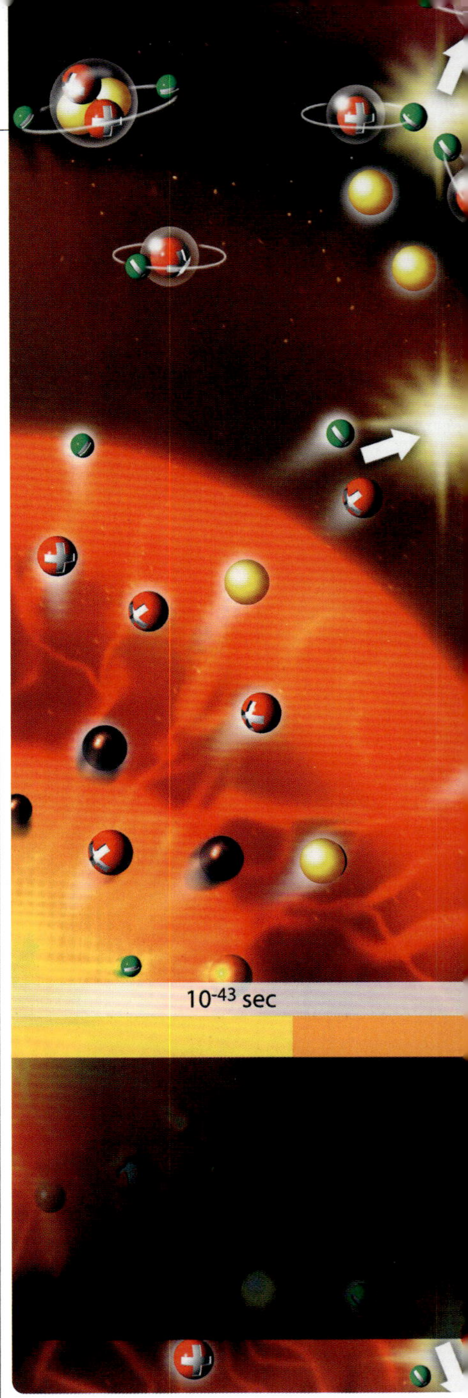

✦ THE PLANCK EPOCH

- In the beginning, everything was contained within an infinitely dense point, tinier than the period at the end of this sentence.
- From the beginning to 0.0000000000000000000000000000000000000000001 second after the Big Bang (collectively called the Planck epoch), the temperature of the universe was 100,000,000,000,000,000,000,000,000,000,000 degrees kelvin.
- These very first moments of birth rely on cosmologists' best formed conjectures, as they defy our current understanding of the laws of physics.
- During this epoch, all four fundamental forces (gravity, electromagnetism, and weak and strong nuclear forces) unified as one.
- **THAT FORCE IS OFTEN REFERRED TO AS QUANTUM GRAVITY.**
- A framework for quantum gravity is still under construction.

✦ AFTER 1 MICROSECOND ...

- As space expanded, it also cooled, enabling the first particles (neutrons, protons, and electrons) to form.
- This phase is called nucleosynthesis.

✦ AFTER 3 MINUTES ...

- The universe cooled enough for the first particles to combine.
- Ionized hydrogen, helium, and lithium began to form.
- **THE UNIVERSE EXISTED AS A HOT, DARK, AND MOSTLY FEATURELESS SOUP OF ELECTRICALLY CHARGED PARTICLES.**
- Light cannot permeate this soup, as any scattered photons are immediately captured and absorbed by roaming electrons.

✦ AFTER 380,000 YEARS ...

- The universe cooled just enough to form neutral atoms, freeing photons from their ionized shackles.
- The universe became transparent to light.
- This first dawn of light imbued the universe with what we now detect as the cosmic microwave background.
- For the next 150 million years or so, the universe entered the cosmic "dark ages"—the time before the first stars formed.

✦ AFTER ABOUT 150 MILLION YEARS ...

- Gas and dust condensed into the first stars, extinguishing the cosmic dark ages.

The evolution of the universe, from the Big Bang to present day

✦ AFTER 400 MILLION YEARS ...

- Influenced by the gravity of dark matter, small variations of density in this plasma began to clump together.
- These clumps seeded the galaxies that would emerge hundreds of millions of years later.

✦ AFTER 10 BILLION YEARS ...

- Dark energy accelerated the expansion of spacetime.

✦ AFTER 13.8 BILLION YEARS ...

- **YOU ARE HERE.**

> "IT IS ONLY IN THE MICROSCOPE THAT OUR LIFE LOOKS SO BIG."
>
> —ARTHUR SCHOPENHAUER, "THE VANITY OF EXISTENCE"

CURRENT FACT COUNT 4270

A COSMIC STORY

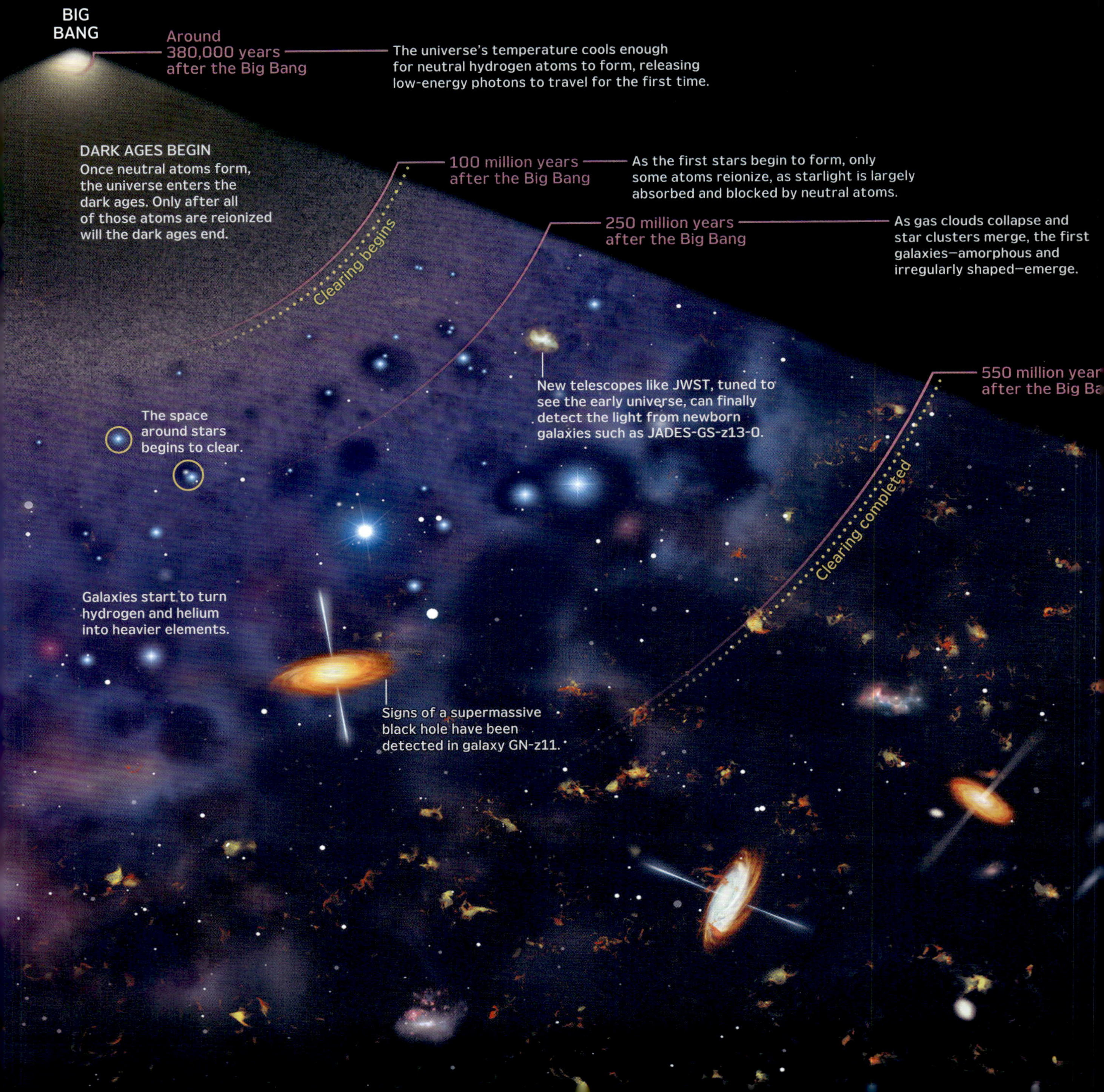

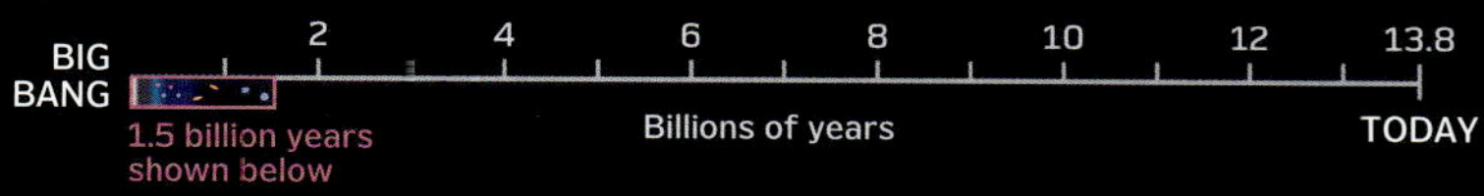

LOOKING TO LIGHT FOR CLUES

Scientists are using the James Webb Space Telescope (JWST) to explore questions about the earliest objects in the universe.

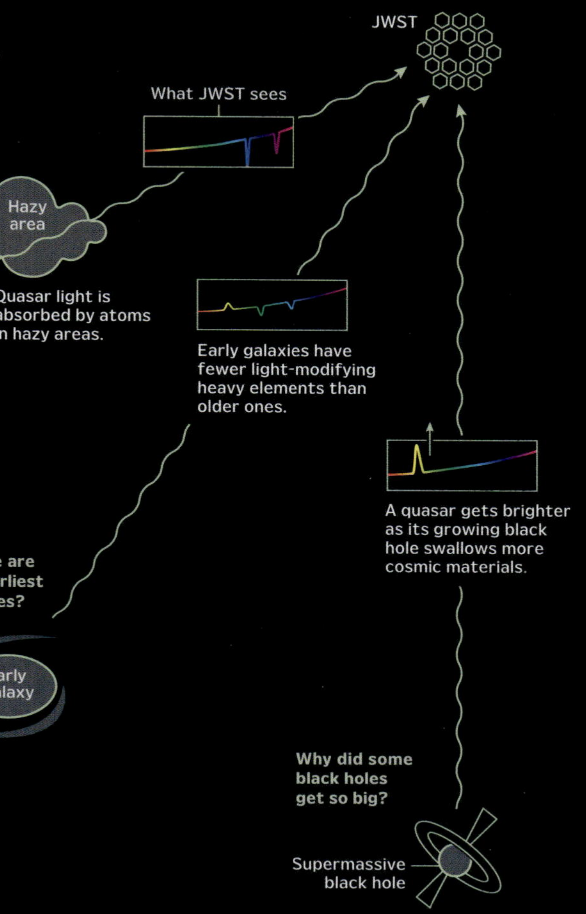

The dark ages end as neutral atoms become reionized, creating a universe transparent to visible starlight.

CURRENT FACT COUNT 4289

COSMIC CONCEPTS

Einstein's theories of special and general relativity provided the blueprints that would merge space and time into a single fabric.

- Our current model of the cosmos is quite young. Einstein revolutionized our perception of the universe by showing that time, space, gravity, and matter are inextricably intertwined, if not inseparable.

- **ACCORDING TO HIS GENERAL THEORY OF RELATIVITY, TIME AND SPACE ARE FUSED INTO A UNIFIED FABRIC: THE SPACETIME CONTINUUM.**

- He also posited that the gravity of massive objects can affect spacetime.

- We can think of gravity as curves in the fabric of spacetime, drawing objects toward each other.

- Any object with mass—you, me, the stars and planets—creates a dent or dimple, curving this fabric.

- The fabric may curve a little or it may curve a lot, depending on the mass.

- This concept is often visualized as a rubber sheet, with balls creating dimples, or gravity wells.

- The greater an object's mass, the deeper the dimple.

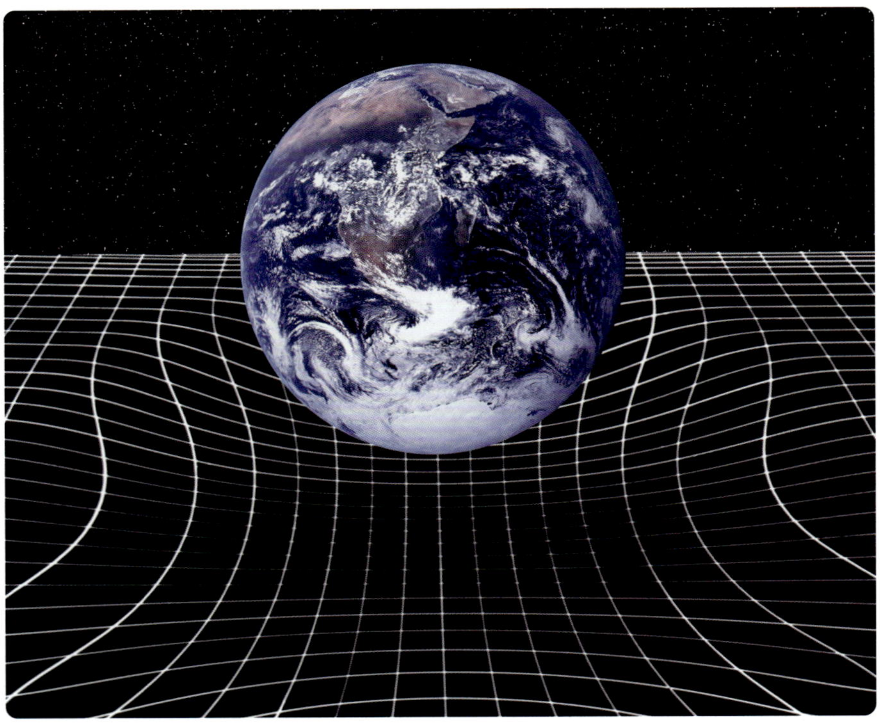

This bent grid visualizes the warping of spacetime due to gravity.

✦ GRAVITATIONAL LENSING

- Einstein proposed that the gravitational effect of extremely massive objects could bend light rays—an effect we now call gravitational lensing.

- Gravitational lensing can cause the illusion that a faraway star is in a different position than its actual location.

- This can happen when light from a distant object is influenced by the gravity of an object (or objects) closer to the observer—it's as if the observer sees the distant object through a distorted lens.

- Gravitational lensing also creates a magnifying effect, allowing us to see light from galaxies that are otherwise too distant.

- The effect can reveal objects that would otherwise be hidden behind something else in our line of sight.

- Sometimes, gravitational lensing creates a double image or even multiple images of a star or galaxy.

- The Hubble Space Telescope mastered images of gravitational lensing, like in the photo of Abell 370 at right.

- **SOME IMAGES FROM THE JAMES WEBB SPACE TELESCOPE (JWST) ARE POSSIBLE ONLY BECAUSE OF GRAVITATIONAL LENSING.**

- Thanks to gravitational lensing, JWST provided us an image of Earendel, among the most ancient and distant single stars ever viewed.

- Earendel is 28 billion light-years away.

- Its light peeked at us through a massive distortion in spacetime created by a galaxy cluster.

The arcs of blue light in this image of galaxy cluster Abell 370 are actually other galaxies hidden behind the cluster.

CURRENT FACT COUNT 4312

BREAKING STARLIGHT INTO SPECTRA

The invention of spectroscopy in the early 19th century heralded the branching of astronomy into a new field: astrophysics.

- The understanding of starlight's relationship with physical matter unlocked a new window to the cosmos.
- When a molecule of ordinary matter absorbs, emits, or reflects light, it imparts a unique signature onto that light.
- The analysis of these signatures is called spectroscopy.
- Spectroscopy can detect, among other useful measures, the temperature, rotation, density, and chemical composition of stars and even of the planets orbiting them.
- **THANKS TO SPECTROSCOPY, WE CAN STUDY NOT ONLY STARS, BUT ALSO PLANETS, BEYOND THE SOLAR SYSTEM.**
- Spectroscopy searches the atmosphere of planets for chemical clues suggesting the presence or possibility of life.
- Analyzing spectra allows scientists to peer into the distant past to the evolution of the universe itself.

✦ HOW IT WORKS

- Studies in the 1840s and '50s revealed the science behind spectroscopy: namely, that every atom or molecule has its own distinct spectrum.
- This can be visualized thanks to an instrument called the spectroscope and recorded with a spectrograph.
- A spectroscope works like a prism, splitting incoming light from across the electromagnetic spectrum into its component colors.
- Every chemical element leaves a characteristic pattern of spectral lines.
- Lines grouped together can form a spectral band.
- **SCIENTISTS STUDY DATA FROM BOTH EARTH-BASED AND SPACE-BASED SPECTROSCOPES.**

✦ REDSHIFT, BLUESHIFT

- Spectra also yield clues about an object's speed and direction.
- Just as a siren's sound changes as an ambulance approaches or recedes, light from distant stars also shifts as they move toward or away from Earth.
- **WE CALL THIS THE DOPPLER EFFECT, AND IT HAPPENS WHEN LIGHT WAVES EXPAND OR CONTRACT AS THEIR SOURCE MOVES RELATIVE TO THE RECEIVER.**
- If an object moves toward us, its light is blueshifted, moving toward the blue end of the color spectrum.
- If an object moves away from us, its light is redshifted, moving toward the red end of the color spectrum.
- **ASTRONOMERS USE REDSHIFT SURVEYS TO DETERMINE THE AGE AND DISTANCE OF STARS AND GALAXIES.**
- They also use redshift surveys to map the distribution of galaxies across the universe.
- As starlight traverses the expanding fabric of spacetime, the light is stretched along with it.
- The longer the light's journey, the more extreme its redshift.
- In general, the greater the redshift, the farther away an object is from us.

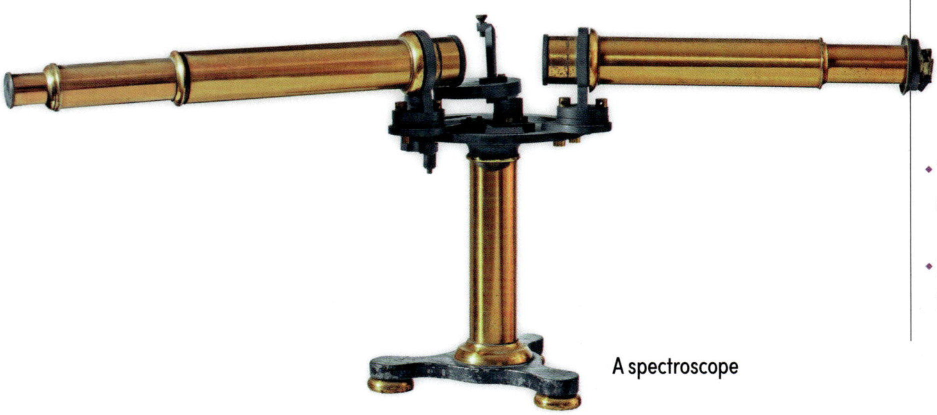

A spectroscope

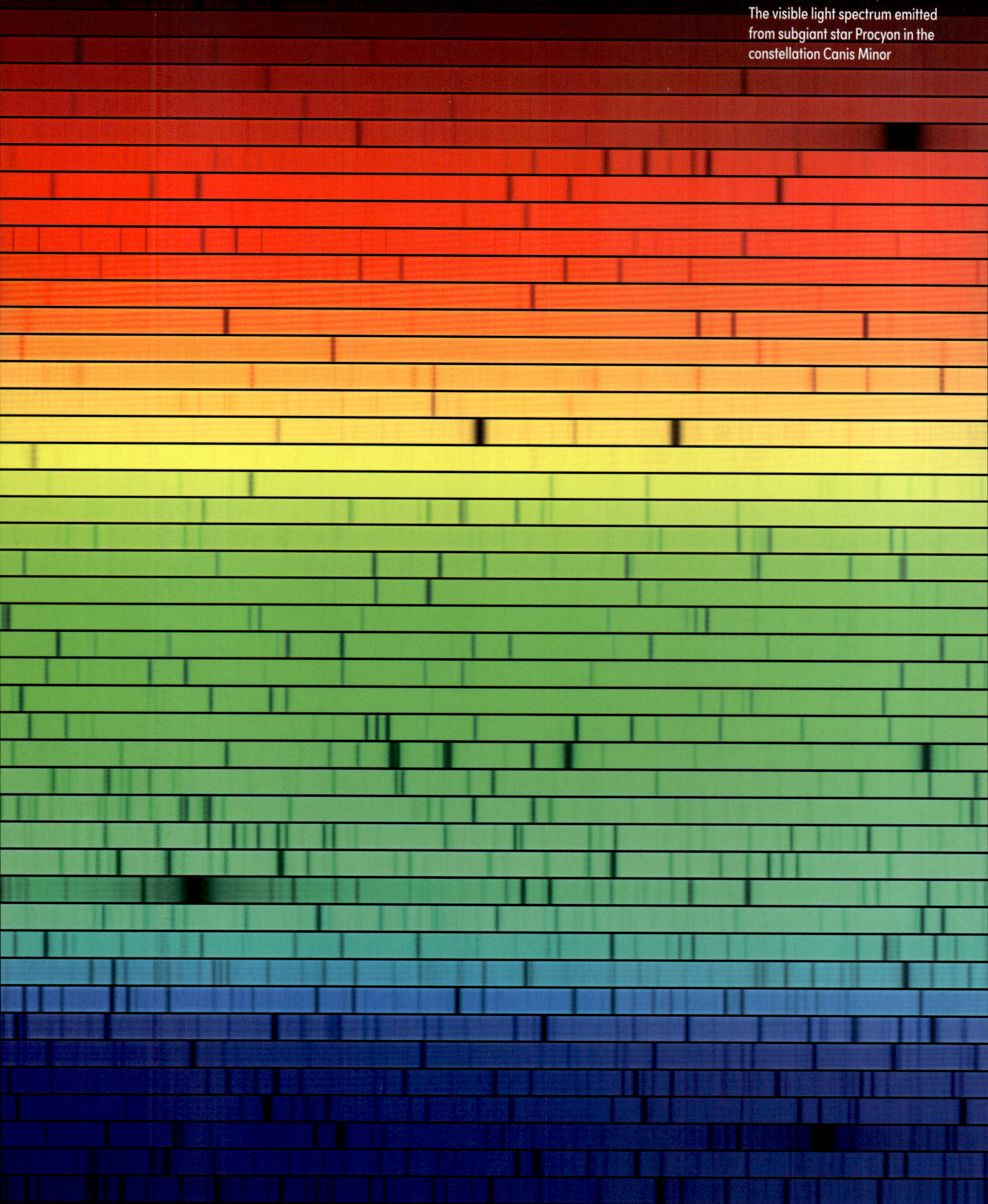

The visible light spectrum emitted from subgiant star Procyon in the constellation Canis Minor

CURRENT FACT COUNT 4378

INNER SPACE

The science of astrophysics connects outer space with inner space: the nature of the tiniest particles of matter.

✦ THE FUNDAMENTALS

- The structure of the universe and all the interactions within it are governed by four fundamental forces: gravity, electromagnetism, the strong nuclear force, and the weak nuclear force.
- These fundamental forces hold together the fundamental particles that make up our universe.
- An atom's nucleus sits at its core, where protons and neutrons pack together.
- The atom's nucleus makes up 99.95 percent of its total mass.
- Yet the nucleus makes up only 1/100,000 of the atom's total volume.
- **IF AN ATOM WERE THE SIZE OF A BASEBALL STADIUM, THE NUCLEUS WOULD BE THE SIZE OF A SINGLE CRACKER JACK.**
- Quarks are a fundamental particle and come in a variety of species, called flavors.
- There are six flavors of quarks: up, down, charmed, strange, top, and bottom.
- The quirky name "quark" comes from a line in James Joyce's novel *Finnegans Wake:* "Three quarks for Muster Mark!"

✦ GRAVITY

- Gravity attracts: planets to stars, galaxies to one another, material objects to Earth.
- Gravity is the weakest of all four forces, yet it works across an infinite distance.
- **IT IS THE PRIMARY FORCE SHAPING THE STRUCTURE OF OUR UNIVERSE.**
- At the subatomic level, however, the force of gravity is nearly negligible.

✦ ELECTROMAGNETISM

- Electromagnetism is the interrelated dynamic of electric and magnetic forces.
- This force is responsible for binding or repelling negatively charged electrons (another fundamental particle) to or from atomic nuclei.
- **ELECTROMAGNETISM, LIKE GRAVITY, ACTS ACROSS INFINITE DISTANCES.**

✦ STRONG NUCLEAR FORCE

- Physicists understood that an atom's nucleus contains positive protons and neutral neutrons.
- But they didn't understand why two or more protons don't repel each other and blow apart.
- A force even greater than the electromagnetic urge to repel must be acting on the structure.
- It was aptly named the strong nuclear force.
- The strong force holds together quarks, the fundamental particles that make up protons and neutrons.
- **THE STRONG NUCLEAR FORCE IS, THEREFORE, THE MIGHTIEST OF THE FOUR FORCES.**
- The strong nuclear force operates at a subatomic level, binding quarks into the clusters that form protons and neutrons and holding atomic nuclei together.
- It is immensely stronger than gravity—on the order of trillions of trillions times.

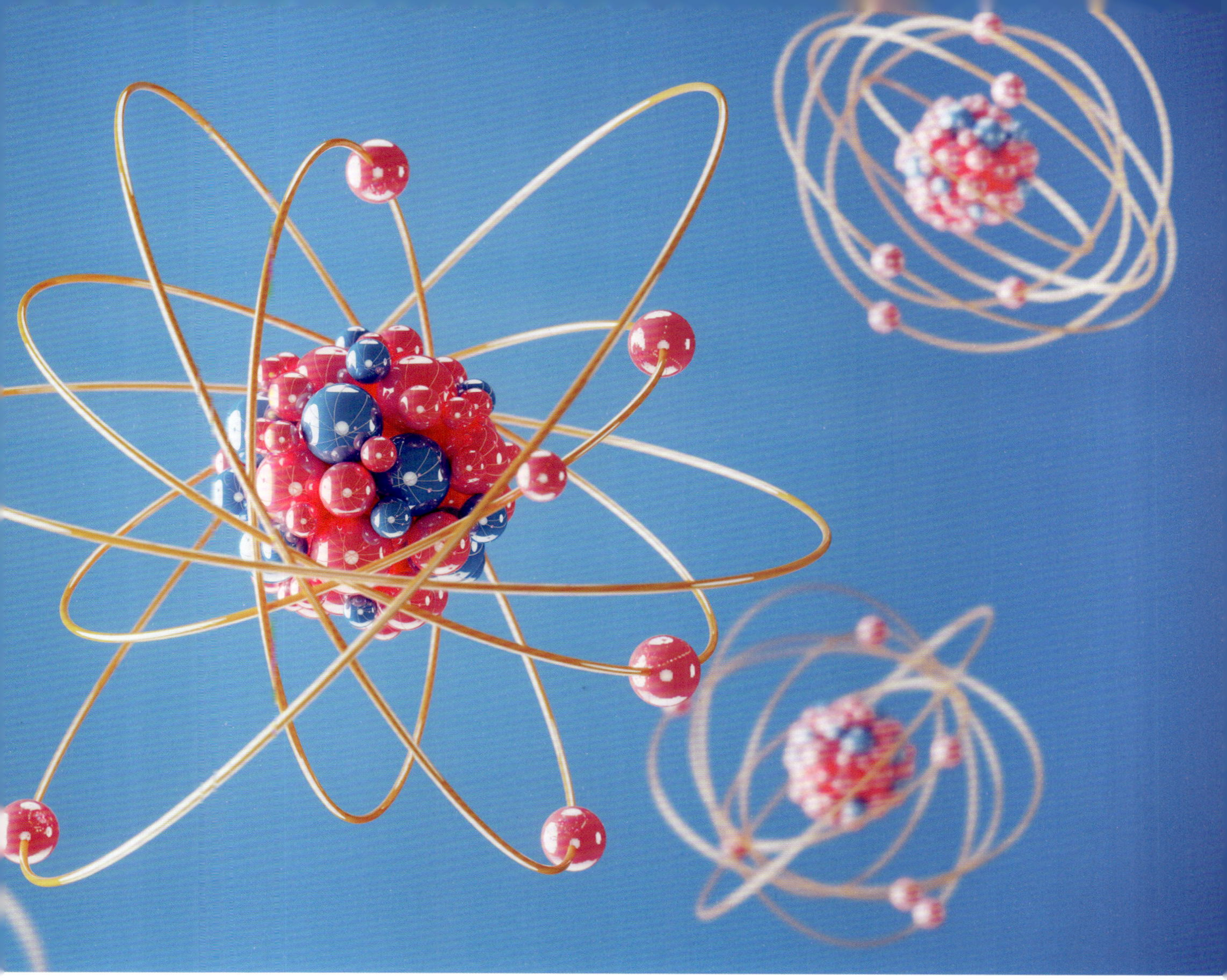

Electrons orbit a colorful nucleus in this 3D illustrated atom model.

✦ WEAK NUCLEAR FORCE

- The weak force is the only one of the four fundamental forces that can alter the flavors of quarks—which can alter an atom's atomic number and, thus, its elemental identity.

- Over time, the weak nuclear force turns carbon-14 into nitrogen-14 at a steady and predictable rate.

- **THE OLDER AN ORGANIC OBJECT IS, THE LESS CARBON-14 IT CONTAINS—A SIMPLE YET PROFOUND CALCULATION THAT ALLOWS CARBON DATING OF ANCIENT MATERIAL.**

- This weak nuclear force enables nuclear fusion in stars.

- It is about a million times weaker than the strong nuclear force, yet still much stronger than gravity.

✦ PHOTONS, GLUONS, AND MORE

- The photon, a fundamental particle, carries the electromagnetic force from particle to particle.

- Similarly, gluons propagate the strong nuclear force from particle to particle.

- These particles that carry fundamental forces are called bosons.

- Photons and gluons are both massless, so they both move at the speed of light.

- The graviton is a hypothetical particle that theorists have suggested could propagate the force of gravity.

- W and Z bosons are propagators of the weak force. They carry mass, and therefore do not move at light speed, but they are responsible for altering a proton's charge by altering the quarks that compose it.

- Because gravity appears to affect matter at the speed of light, then gravitons should be massless and able to move at light speed, like photons or gluons. But none have been observed to prove this hypothesis.

✦ THE SPACE INSIDE

- You may have heard that an atom is 99.9999 percent empty space, but it's more accurate to say an atom is 99.9999 percent *invisible* space.

- **TRUE EMPTINESS DOES NOT EXIST ANYWHERE IN OUR UNIVERSE, NOT EVEN WITHIN ATOMS. INVISIBLE SPACE IS SUFFUSE WITH ELECTRON CLOUDS AND FORCE FIELDS.**

- If the space within atoms were truly empty, then our hands could not clap, but would instead permeate and pass through one another like a ghost through a wall.

- The electromagnetic force repels atoms from one other, creating the illusion of a solid barrier. You have never touched another atom because of this repulsion.

DC Universe's The Flash defies quantum physics.

- So, when you clap your hands, you are colliding the force fields of your left hand's atoms with your right hand's atoms.

- When you sit in a chair, you are really hovering about four-billionths of an inch over it, separated by the electromagnetic force of your chair's atoms against your butt's atoms.

- DC Comics' The Flash can traverse any barrier at will.

- This ability is explained with loose interpretations of quantum jargon that make for great stories, but there's no real way a human could bypass solid objects.

✦ QUANTUM PARTICLES

- Quantum particles actually do possess this superpower in the real world through a phenomenon called quantum tunneling.

- Almost like magic, an electron can appear to move from one side of a barrier to the other.

- Unlike everyday objects we're familiar with, quantum particles do not exist at one point in time, but rather as a probability of existing at one point and at another point, and at another, spread across a field of possibilities.

- **ELECTRONS, THEREFORE, DO NOT EXIST IN ONE SINGULAR POINT OF AN ATOM'S SHELL, BUT ACROSS A WAVE OF POSSIBLE POINTS.**

- Since there's a small chance that part of an electron's wave exists on the other side of a barrier, that electron could appear on that other side.

- Heisenberg's uncertainty principle states that you cannot know both a quantum particle's speed and its position at the same time.

- If you know its speed, then its position is a wave of possibilities.

- And if you know its position, then you can't be certain of its speed.

✦ THE SCIENCE OF ANT-MAN

- Marvel Studios' Ant-Man can shrink to the size of a subatomic particle to enter the Quantum Realm.

- But a fundamental law of physics governing the universe—the conservation of mass—states that mass cannot be created or destroyed.

- To avoid violating this law, Ant-Man utilizes a fictional "Pym Particle" to reduce the space between atoms, allowing him to shrink at will while maintaining his mass.

- If he retains mass, however, his density must fluctuate, making Ant-Man in his diminutive form the densest object on Earth.

- **SHRINK ANT-MAN DOWN TO THE SIZE OF A FUNDAMENTAL PARTICLE, AND HE BECOMES A MICRO-BLACK HOLE, INSTANTANEOUSLY EVAPORATING FROM EXISTENCE.**

- The square-cube law poses another danger to Ant-Man.

- The law states that an object's volume grows faster than its surface area.

- For example, as mammals, we lose heat through our skin. Small mammals have a far larger surface area to volume ratio. Thus, small mammals lose heat more rapidly than large mammals do and must eat multiple times their weight every day.

- This suggests that Ant-Man's suit must be the most insulating material in the world, or he'd need to incessantly munch on snacks.

Ant-Man hitches a ride while in miniature in Marvel Studios' 2015 film *Ant-Man*.

✦ THE HUMAN MIND

- Each human brain has 100 times more synaptic connections than there are stars in our galaxy, but we don't yet know whether it's possible for the human mind to figure out how itself works, or if it's possible for the universe to create something more complex than itself.

- The size of the solar system is to the human brain as the size of the human brain is to the size of an atomic nucleus.

- **SCIENCE DID NOT ACHIEVE MATURITY UNTIL IT CREATED METHODS, TOOLS, AND MACHINES TO REPLACE OUR SENSES WITH ACCURATE AND PRECISE MEASUREMENTS OF THE WORLD AROUND US.**

- Understanding the universe may require a level of intelligence beyond our reach, much as the smartest chimpanzee is likely incapable of doing long division.

- Chimpanzees, our closest genetic relatives, cannot meaningfully communicate with us, yet we remain convinced that we will be able to communicate with intelligent aliens.

"OF COURSE, MANKIND HAS MADE GIANT STEPS FORWARD. HOWEVER, WHAT WE KNOW IS REALLY VERY, VERY LITTLE COMPARED TO WHAT WE STILL HAVE TO KNOW."

—PARTICLE PHYSICIST FABIOLA GIANOTTI

CURRENT FACT COUNT 4409

THE LARGE HADRON COLLIDER

Sometimes, the best way to understand something is to smash it apart.

- Studying the insides of atoms reveals principles of the universe that help us understand outer space.

- A primary research facility dedicated to this quest is the Large Hadron Collider (LHC), the world's most powerful particle accelerator.

- Built about 300 feet underground near Geneva, Switzerland, the LHC started running in 2008.

- **IN ESSENCE, THE LHC IS DESIGNED TO UNDERSTAND THE BASICS OF MATTER BY REPLICATING THE EARLIEST CONDITIONS OF THE UNIVERSE.**

A visual of the four detectors around CERN's Large Hadron Collider: ALICE, ATLAS, CMS, and LHCb

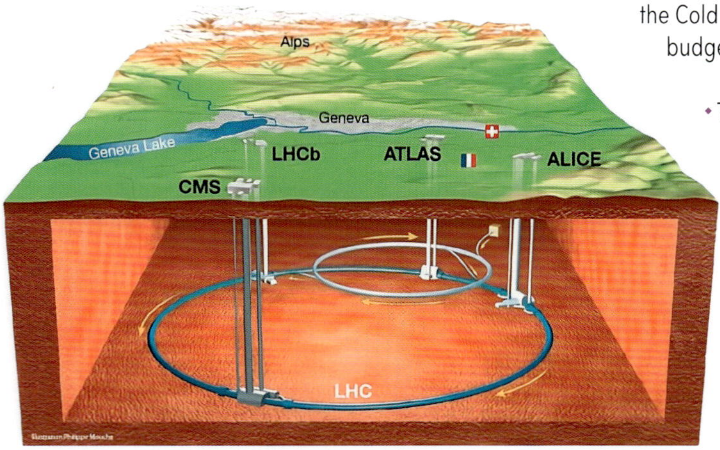

✦ HOW THE LARGE HADRON COLLIDER CAME TO BE

- In the 1980s, the U.S. allocated funds to build the world's largest collider.

- That collider would benefit from new magnet technologies that would greatly enhance the machine's energy.

- It was (unimaginatively) called the Superconducting Super Collider (SSC).

- The SSC's funds were eliminated in the early 1990s due to management problems and cost overruns, with projected expenses rising from $5.3 billion to more than $11 billion.

- Simultaneously, the end of the Cold War shifted U.S. priorities away from physics-leading endeavors to less lofty goals. No other collider built during the Cold War had its budget cut.

- The center of research in particle physics swung to Europe and the European Organization for Nuclear Research (CERN) in Switzerland.

- There, they built the Large Hadron Collider, usurping America's lead in the study of particle physics.

- The unbuilt American Superconducting Super Collider would have been at least three times more powerful than the LHC.

✦ LHC BASICS

- Hadrons are subatomic particles made of quarks, including mesons (built with a quark and an antiquark) and baryons (made up of three quarks).

- The more energy a collider can reach, the closer it gets to simulating conditions during the Big Bang itself.

- In the LHC, a pair of high-energy particle beams are sent in opposite directions through a ring of superconducting electromagnets nearly 17 miles in circumference.

- Inside the ring is an ultrahigh-vacuum tube kept at minus 456°F—colder than outer space.

- **THE LHC CAN PRODUCE ONE BILLION COLLISIONS PER SECOND.**

- These particle collisions continue to reveal new information about the nature of atoms and forces during the birth of the universe.

✦ FEARS ALLAYED

- The construction of the Large Hadron Collider was halted due to public outcry over fears it could create black holes through particle collisions.

- **IF THE LHC COULD CAUSE BLACK HOLES TO FORM, THEY WOULD BE MICRO-SIZE—AND THEY WOULD EVAPORATE ALMOST INSTANTANEOUSLY.**

CERN's ATLAS detector, which measures the particles and processes that emerge during proton collision within the Large Hadron Collider

- The LHC uses supercooled semiconductor magnets to accelerate particles to near light speed.

- Cosmic rays—charged particles streaming from the Sun—bombard Earth constantly. It's the original, albeit less controlled, version of a particle accelerator.

- Those bombardments carry far greater energy than within the LHC—enough to create micro-black holes upon collision with Earth's atmosphere.

- So, if cosmic ray collisions have yet to end civilization, you can rest assured the LHC won't either.

✦ THE MYTHICAL HIGGS BOSON

- European scientists used the LHC to search for the Higgs boson, a particle whose field grants mass to other particles when it interacts with them.

- **IN 2012, THE LHC INDEED DISCOVERED THE HIGGS BOSON IN A FIELD OF SMASHED PARTICLES CREATED BY THE COLLIDER ITSELF.**

- To ensure a detection wouldn't be spurious or a fluke, two separate groups researched two beams, created independent of each other.

- The discovery was announced only after both teams confirmed the same results.

- The existence of the Higgs boson was predicted by the "standard model of particle physics," which establishes the existence of, and interrelationships among, fundamental particles and forces of nature.

- A year after the Higgs boson discovery, Peter Higgs, who proposed the idea back in 1964, shared the Nobel Prize in Physics with François Englert, who independently proposed the same idea.

- The LHC continues to be upgraded, in search of even more elusive particles. Perhaps it may one day discover a weakly interacting particle responsible for dark matter in the universe.

CURRENT FACT COUNT 4441

NEUTRINOS

The discovery of elusive neutrinos—the ghosts of the cosmos—unlocked a new vista on the universe.

- There exists one singular particle capable of escaping the Sun's thermonuclear core and traveling through its surface into space, past Earth's atmosphere, and through Earth's surface: the neutrino.
- Before we detected neutrinos, we could not prove for certain that the Sun's thermonuclear furnace was powered by the fusion of hydrogen with helium atoms.
- To account for conservation of energy, the reaction should emit a neutral particle.
- **PHYSICISTS CALLED THE HYPOTHETICAL PARTICLE A NEUTRINO.**
- These once hypothetical particles became of utmost interest to astronomers seeking to understand the nature of our Sun and other stars.
- In the blockbuster film *2012*, apocalypse ensues after myriad natural disasters wreak havoc on Earth, caused by a blitz of neutrinos heating Earth's core. But neutrinos are fundamental particles that barely interact with matter at all.
- **100 TRILLION NEUTRINOS PASS HARMLESSLY THROUGH YOUR BODY EVERY SINGLE SECOND.**
- Throughout an entire average human lifetime, there is a 25 percent chance of one single neutrino interacting with one atomic particle in your body.
- As their name might imply, neutrinos carry no charge and very little mass.
- They are nearly 500,000 times lighter than electrons, the next heaviest particle.

✦ THE SUPERNOVA NEUTRINO BACKGROUND

- The universe is awash in neutrinos emitted by supernovae. Scientists have dubbed this the cosmic neutrino background.
- Like the cosmic microwave background, these elusive particles may open an entire new window into the universe and its formation.
- One supernova can produce an estimated 10 octodecillion neutrinos—10^{58}. That's a one followed by 58 zeros: 10,000,000,000,000,000,000,000,000,000,000,000,000,000,000,000,000,000,000,000.

High-energy particles and radiation bombard Earth's atmosphere in this artist's rendering.

"THERE WAS A LOT OF DISBELIEF THAT THIS COULD BE REAL. IN A DECADE OF OBSERVATIONS, WE HAVEN'T SEEN ANYTHING QUITE LIKE THIS."

—PHYSICIST ERIK BLAUFUSS ON THE 2023 NEUTRINO DISCOVERY

A bubble chamber photograph reveals the presence of a neutron in its V-shaped tracks.

✦ THE SOLAR NEUTRINO MYSTERY

- Neutrinos are the most abundant particles in the universe (not including dark matter and dark energy), likely outnumbering even photons.
- Yet they are impossible to detect directly.
- **AS WITH DARK MATTER AND DARK ENERGY, WE CAN ONLY STUDY THE INTERACTIONS OF NEUTRINOS WITH SURROUNDING MATTER RATHER THAN THE PARTICLES THEMSELVES.**
- Because neutrinos interact so rarely and weakly with other particles, the only way to detect them is by creating a behemoth vat of potential particles for them to interact with and then observe that vat carefully.
- In the late 1960s, chemist Raymond Davis, Jr., did just that. He built a 100,000-gallon drum deep underground in a South Dakota gold mine and filled it with chlorine.
- If neutrinos from the Sun penetrated the tank, they would change chlorine atoms into argon atoms.
- Davis detected about 4 to 11 collisions per day, proving that the Sun emitted neutrinos.
- But only a third of the neutrinos that were predicted to have been emitted actually showed up in Davis's experiments. Either the standard model of physics was wrong, or Davis's experiment was flawed.
- **FOR 30 YEARS, THE MYSTERY WENT UNSOLVED, AND DAVIS HELD FIRM THAT HIS EXPERIMENT WAS SOUND.**
- By the turn of the 21st century, Davis's experiment methods were exonerated when scientists discovered that neutrinos oscillate—they change species when their quarks change flavors—on their journey from the Sun.
- Davis's experiment was equipped to measure only one of the three neutrino species.
- His experiment proved the Sun's mechanism of nuclear fusion.
- Davis shared the Nobel Prize in 2002 for proving the existence of cosmic neutrinos.

✦ JUST SHY OF THE SPEED OF LIGHT

- Neutrinos are the fastest traveling particle after photons.
- They can travel more than 1,000 times faster than particles observed in Earth-bound particle accelerators, earning them the title "fastest moving particles with mass."
- **ONLY LIGHT ITSELF OUTPACES NEUTRINOS.**
- They travel so fast that they're nearly impossible to clock at slower-than-light speed.
- But we know if a neutrino can become another species of neutrino, they must carry an internal clock.
- If they experience time, then they cannot travel at the speed of light.

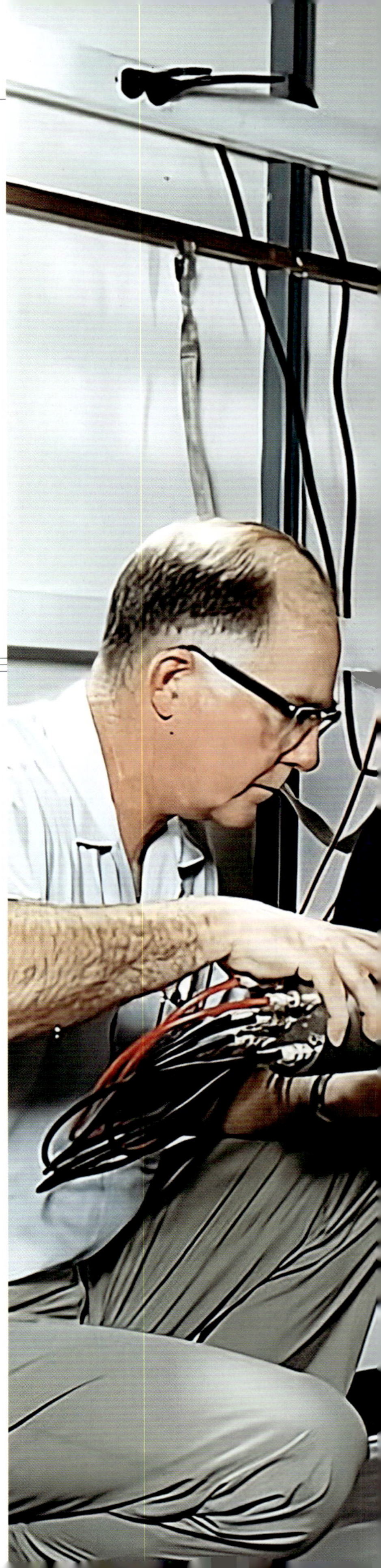

Shields at Raymond Davis's Brookhaven National Laboratory were built from battleship gun barrels.

CURRENT FACT COUNT 4460

NEUTRINO DETECTORS: THE REAL GHOST HUNTERS

Neutrinos may hold clues to our unanswered questions about the very origin of the universe and the nature of dark matter.

✦ SUPER-KAMIOKANDE

- The hit sci-fi TV series *3 Body Problem*, based on the novel by Cixin Liu, opens by featuring the ominous and spectacular interior of the Super-Kamiokande neutrino observatory.
- Japan's Super-Kamiokande neutrino observatory utilizes a type of detection method called a Cherenkov tank.
- The tank sits more than 3,000 feet belowground and contains 50 tons of ultrapure water and 13,000 optical sensors specially designed to detect the aftermath of a rare neutrino collision.
- When a neutrino collides with an electron, proton, or neutron, it produces another high-energy particle that zips through the water at speeds faster than light can travel within water.
- **THAT NEW PARTICLE EMITS A TINY SHOCK WAVE THAT CREATES A BLUE GLOW CALLED CHERENKOV RADIATION—EVIDENCE OF A NEUTRINO COLLISION.**
- The enormous observation vat detects about 30 of these rare, teensy disturbances per day.

Inside the tank of the Super-Kamiokande neutrino observatory in 2018

✦ ICECUBE NEUTRINO OBSERVATORY

- Like Super-Kamiokande, the IceCube Neutrino Observatory in Antarctica searches for Cherenkov radiation.
- **RATHER THAN A MASSIVE VAT OF WATER, HOWEVER, THE OBSERVATORY USES A SOLID BLOCK OF CRYSTAL CLEAR, ANCIENT ANTARCTIC ICE AS THE DETECTION MEDIUM.**
- Eighty-six vertical tethers containing more than 5,000 optical sensors sink a mile and a half beneath the ice, down to the bedrock.
- Project engineers had to drill through solid ice using high-pressure hot-water hoses.
- The technique melted through the Antarctic ice at about seven feet per minute.
- We may one day use similar thermal drilling technology to dig through the thick icy surfaces of the Europa and Enceladus moons.

The IceCube Neutrino Observatory in 2017

✦ DEEP-SEA NEUTRINO DETECTORS

- Until 2022, the first deep-sea neutrino detector, ANTARES, operated 1.5 miles beneath the surface on the ocean floor, 25 miles off the coast of France.

- The deep, dark Mediterranean waters offer the quiet stillness required to observe the ephemeral blips of Cherenkov radiation.

- At these depths, sensors can detect neutrino bombardments coming both from above and below—neutrinos actually passing through Earth from the opposite side.

- China has plans to implement the world's largest neutrino detector, called TRIDENT, at the bottom of the tropical South China Sea, two miles below the surface.

- **AT A NEAR-EQUATORIAL LATITUDE, TRIDENT WILL DETECT NEUTRINOS FROM ALL ANGLES OF THE COSMOS AS EARTH ROTATES.**

- In February 2025, scientists announced the discovery in 2023 of an ultrahigh-energy neutrino with 220 million billion electron volts of energy—tens of thousands of times more energy than the Large Hadron Collider can produce.

- It was observed using the Cubic Kilometre Neutrino Telescope (KM3NeT), a neutrino telescope located in the Mediterranean Sea.

CURRENT FACT COUNT 4482

THE FABRIC OF SPACETIME

Neutrinos may hold clues to our unanswered questions about the very origin of the universe and the nature of dark matter.

✦ THEORIES OF RELATIVITY

- Special relativity in a nutshell: The faster you move relative to something else, the slower your clock ticks.
- General relativity in a nutshell: The stronger the force of gravity acts on you relative to something else, the slower your clock ticks.
- **ALBERT EINSTEIN ALSO SHOWED THAT OBJECTS MOVE IN STRAIGHT LINES OVER CURVED SPACETIME, AND THE CURVATURE ITSELF CREATES THE "FORCE" OF GRAVITY THAT'S FELT.**
- If an object is massive enough, the space around it bends so radically, it can drag light and other objects toward it.

Albert Einstein, photographed while giving a lecture in 1921

- These phenomena are called gravitational waves, and they point to ripples in spacetime tracing back to the Big Bang. They could allow us to detect the "dark ages" of our universe, when light had not yet escaped.

✦ EINSTEIN'S "BLUNDER"

- Einstein's general theory of relativity predicted the existence of something working in opposition to gravity.
- Why? Because gravity pulls things together, so some opposing force was needed to explain why gravity wasn't pulling the universe into a giant clump.
- He called this the "cosmological constant" and inserted the Greek letter *lambda* into his equations of general relativity.
- He originally plugged it in because scientists at the time assumed the universe was static.
- In 1929, when Edwin Hubble proved that the universe was not static, Einstein removed lambda from his equations, calling it the "greatest blunder" of his life.
- Lambda came and went over the decades, as astronomers played with various models of the universe.
- In 1998, lambda made a comeback when two separate teams of astrophysicists reported that more than 50 distant supernovae were noticeably dim, as if they were 15 percent farther away than ordinary cosmological models predicted.
- To resolve this finding, the scientists put lambda back into Einstein's equations for general relativity—and all was well.
- The three scientists (physicist Saul Perlmutter and astronomers Brian Schmidt and Adam Riess) won the Nobel Prize in Physics for this discovery in 2011.
- We now call this counterforce dark energy.

✦ THREADS OF SPACETIME

- Quantum physics predicts that virtual particle pairs pop in and out of existence in the vacuum of space.
- While they exist, the particle pairs are quantum entangled with each other.
- One speculation is that the entangled particle pairs are connected via wormholes through the fabric of space and time.
- **THESE WORMHOLES MAY BE THE LITERAL THREADS THAT WEAVE TOGETHER THE FABRIC OF SPACETIME.**
- "Matter tells space-time how to curve; space-time tells matter how to move," Charles Misner, Kip Thorne, and John A. Wheeler once explained.
- The shape of spacetime in the vicinity of black holes is maximally distorted, harming life and limb.
- Light from objects far behind a massive object will be "lensed," or distorted, as the light bends around the object en route to the observer. Though curved spacetime will bend the path of light, as far as the light knows, it's traveling in a straight line.

The merging of two black holes could look like the simulated dance of light shown here.

CURRENT FACT COUNT 4534

BLACK HOLES

What happens in a black hole, stays in a black hole.

- The idea of black holes first surfaced in the 18th century with England's Rev. John Michell and French astronomer Pierre-Simon Laplace.

- They independently suggested that there might be objects in the universe with gravity strong enough to trap light.

- Michell called them "dark stars," and Laplace referred to them as "invisible bodies." In time, the idea evolved and gained scientific support. Today, we call them black holes.

- **AT THEIR MOST BASIC, BLACK HOLES ARE MONSTROUS GRAVITY SINKS FROM WHICH NOTHING, NOT EVEN LIGHT, CAN ESCAPE.**

One of NASA's "Galaxy of Horrors" posters

- Black holes don't suck. They're not giant vacuum cleaners. Instead, think of them like playground slides molded by the force of gravity—slides so curved and steep that objects fall toward and into them.

- A black hole holds a phenomenal amount of matter in an infinitesimal amount of space.

✦ ESCAPE VELOCITY

- Escape velocity is the minimum velocity (or speed) that any moving object must obtain to break free of the force of gravity and enter into space.

- The higher the density of a celestial object, the higher the velocity required to escape its gravity.

- An object's density is its mass divided by its volume.

- So, a planet could be much larger than Earth by total volume, but unless it's much denser, the escape velocity won't necessarily vary.

- VY Canis Majoris, one of the largest stars known to humanity, is 1,500 times larger than our Sun. But its escape velocity is less.

- **THE ESCAPE VELOCITY OF ...**
  - Earth = eight miles a second
  - Jupiter = 37 miles a second
  - The Sun = 400 miles a second
  - Neutron stars (the second densest objects in the universe after black holes) = around 100,000 miles a second
  - Escape velocity of a black hole = There is no escape.

- Once your escape velocity equals the speed of light (186,282 miles a second), you are, by definition, a black hole.

✦ ANATOMY OF A BLACK HOLE

- At a black hole's center is what's called the singularity: an infinitely dense point.

- **AROUND THE SINGULARITY IS THE EVENT HORIZON: THE BOUNDARY PAST WHICH NOTHING ESCAPES THE BLACK HOLE'S GRAVITY, NOT EVEN LIGHT.**

- The event horizon represents a point of no return.

A star leaves a glowing stream of light in its wake as it's sucked into a black hole in this illustration.

- More than a physical barrier, it is a literal causal separation: Any event beyond the event horizon will never be known to anyone on the other side.
- A photon cresting the exact precipice of the event horizon would appear to hover motionlessly, despite moving at the speed of light.
- The accretion disk is a whirling disk of gas, dust, and anything else caught in the black hole's grip, including stars and planets.
- **A BLACK HOLE GROWS BY DEVOURING THE MATERIAL THAT APPROACHES IT.**
- The energetic activity in the accretion disk enables us to see the otherwise invisible black hole.
- The friction and heat from all that material spinning around emits radiation across many wavelengths.
- Sometimes, large black holes emit enormous spurts of plasma, called relativistic jets, reminiscent of the super-blaster of the planet-destroying Death Star in *Star Wars.*
- **THESE JETS CAN SPAN THOUSANDS TO MILLIONS OF LIGHT-YEARS ACROSS.**
- The jets form when a black hole gobbles down too much matter too quickly, causing a shock wave that throws some of the ionized particles and energy back out into space.
- One black hole spewed jets that spanned 23 million light-years—about 140 times longer than the Milky Way galaxy.
- This black hole was named Porphyrion, after one of the giants of ancient Greek mythology who waged war against the gods.
- Porphyrion's jets are estimated to carry the energy of trillions of stars and heat surrounding particles to millions of degrees.
- If Porphyrion's jets were the size of Earth, then Porphyrion itself would be about the width of a human hair (about 0.2 millimeter).
- Astrophysicists hypothesize that these ancient, violent outgushings may have helped shape the very structure of our universe.

Cygnus X-1, a black hole about
15 times the mass of our Sun,
photographed in 2011

✦ NOTHING WAGERED, NOTHING GAINED

- The first black hole that was discovered was Cygnus X-1, formed from the collapse of a massive star.
- It was discovered as a bright x-ray source in 1964.
- In 1975, physicists Stephen Hawking and Kip Thorne made a bet: Hawking wagered that Cygnus X-1 was not a black hole (despite being pretty certain that it was).
- In Hawking's landmark book, *A Brief History of Time*, he explained, "This was a form of insurance policy for me. I have done a lot of work on black holes, and it would all be wasted if it turned out that black holes do not exist. But in that case, I would have the consolation of winning my bet."
- **BY 1990, HAWKING CONCEDED THAT IT MUST INDEED BE A BLACK HOLE.**
- Cygnus X-1 turns out to be a black hole orbiting with a companion star.
- Cygnus X-1 lies about 7,200 light-years from Earth.
- The approximate mass of the Cygnus X-1 black hole is 20 times that of the Sun.

✦ TYPES OF BLACK HOLES

- Primordial, or mini, black holes may have existed shortly after the universe formed.
- Most have dissipated since then.
- Stellar black holes form during the death and collapse of very massive stars—20 times the Sun's mass or more.
- **SUPERMASSIVE BLACK HOLES ARE MILLIONS OR BILLIONS OF TIMES MORE MASSIVE THAN THE SUN.**
- As far as we know, all large galaxies have supermassive black holes in their centers.
- Supermassive black holes can contain hundreds of thousands to billions of times worth of solar masses inside themselves.
- **THE ORIGINS OF SUPERMASSIVE BLACK HOLES ARE STILL A MYSTERY.**

✦ THE NATURE OF BLACK HOLES

- German physicist Karl Schwarzschild took Einstein's general theory of relativity a step further by showing precisely how spacetime curved in the area around a given mass.
- Scientists now calculate the size of a black hole for a specified mass using the Schwarzschild solution.
- A black hole with Earth's mass, for instance, would have a radius of just nine millimeters (a third of an inch).

A Brief History of Time **by Stephen W. Hawking**

CURRENT FACT COUNT 4537

SILHOUETTES OF A BLACK HOLE

Machine learning helped create a sharper image (taken originally with the Event Horizon Telescope using more than 30,000 simulations, at right) of a black hole at the center of the Messier 87 (M87) galaxy. The method, dubbed PRIMO, shows a thinner ring around the black hole: matter pulled inward by the sheer force of gravity.

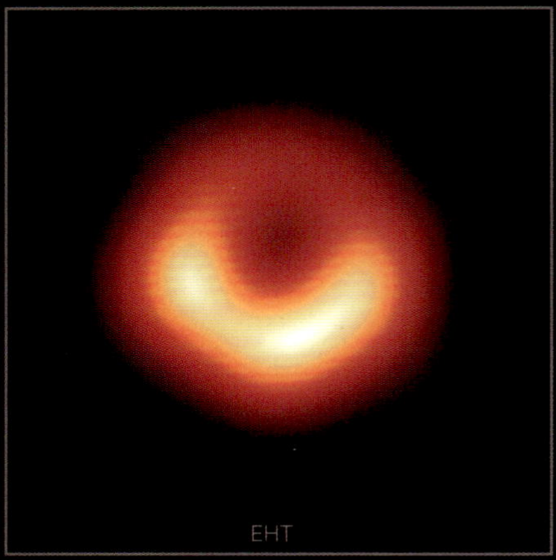

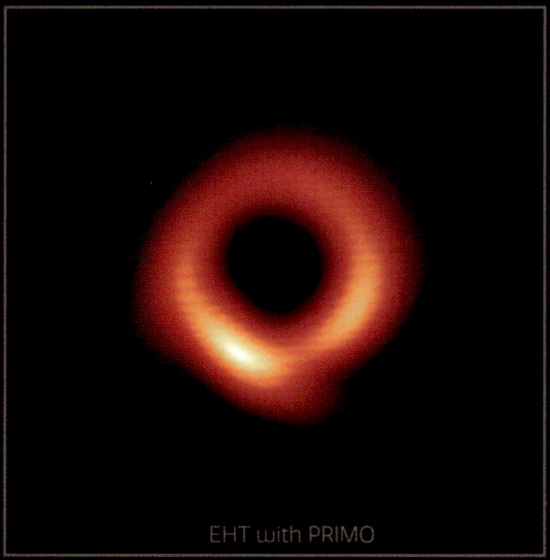

The top photograph shows the first image of M87's black hole, originally published in 2019. Below is the PRIMO image, released in 2023.

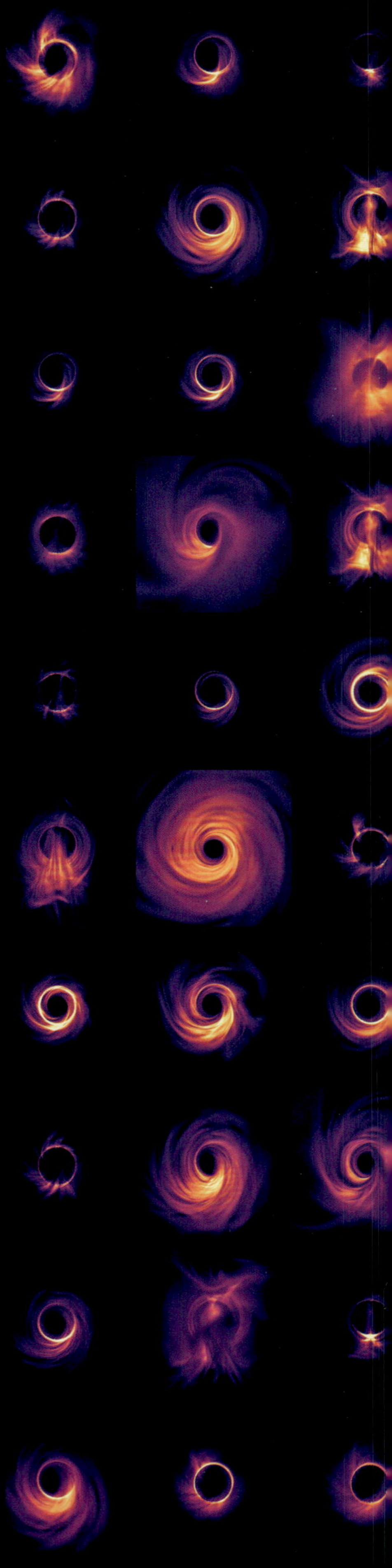

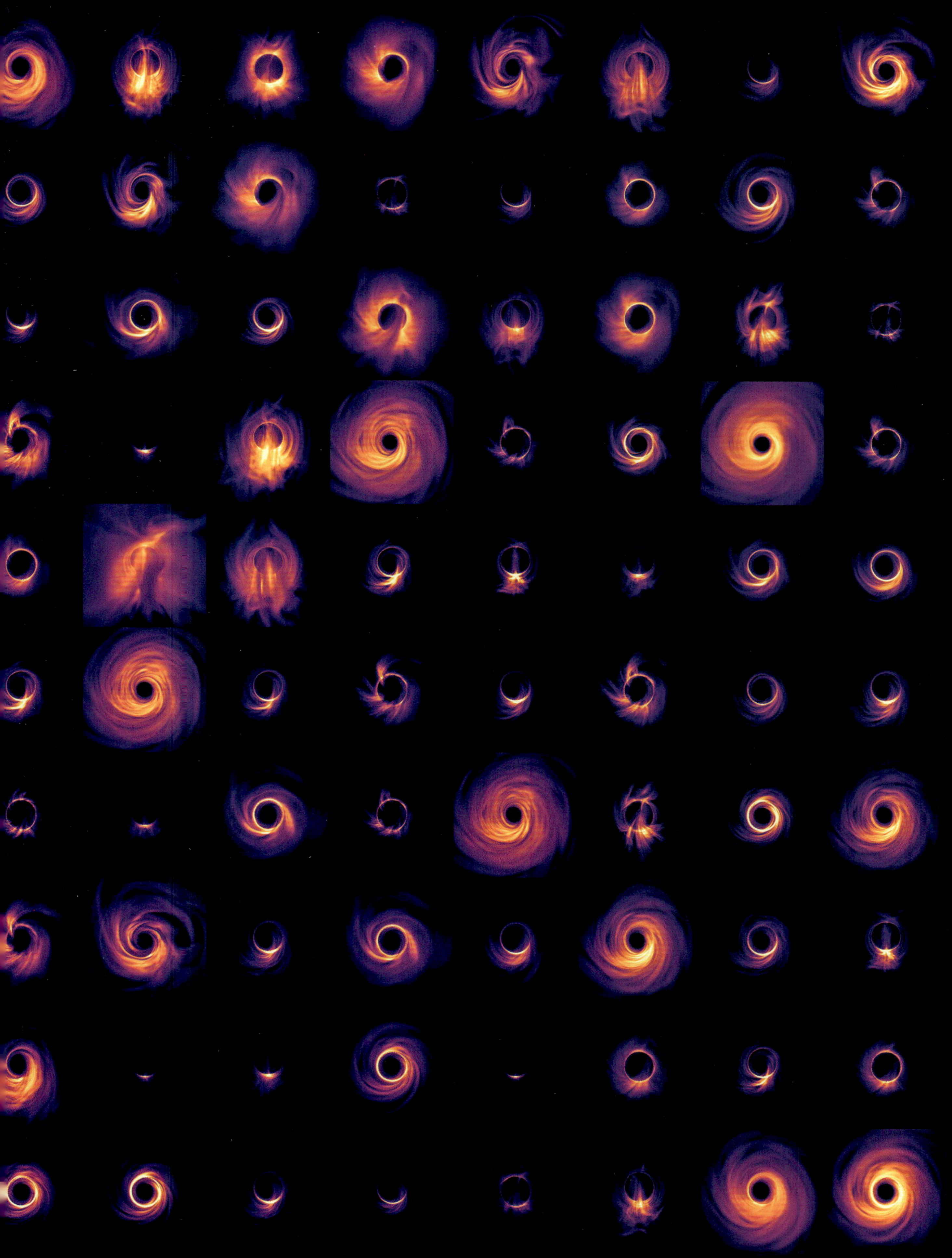

CURRENT FACT COUNT 4565

THEORETICAL NO LONGER

Once merely conjecture, black holes are now discovered with increasing frequency, thanks to groundbreaking science and powerful telescopes.

- Einstein's general theory of relativity predicted the existence of black holes—but because they are invisible, they were nearly impossible to observe and prove objectively. Until now.
- **IN 2019, HUMANITY CAPTURED THE FIRST DIRECT VISUAL EVIDENCE OF A BLACK HOLE.**
- The feat required an array of Earth-based radio telescopes across continents and synchronizing data through atomic clocks.
- Together, the telescope system is called the Event Horizon Telescope (EHT), an international collaboration designed to detect and observe black holes in the universe.
- The telescopes are located in Hawaii and Arizona in the U.S. and Antarctica, Mexico, Spain, Chile, France, and Greenland.
- EHT captured the image of a supermassive black hole in the Messier 87 galaxy, 54 million light-years away in the constellation Virgo.
- **IT'S COMPARABLE TO PHOTOGRAPHING A LEMON ON THE SURFACE OF THE MOON ALL THE WAY FROM EARTH.**
- Or reading the date on a quarter on a Los Angeles sidewalk from New York City.
- The image shows a dark region (the signature shadow of a black hole) surrounded by a bright ring of light bent by intense gravity.
- The mass of M87's black hole is around 6.5 billion times that of our Sun.
- Black holes also give off strong radio and x-ray signals—energy from the heat of all that material whirling around.
- The amount of x-ray radiation emitted is a good measure of a black hole's mass.
- Roughly one out of every 1,000 stars is massive enough to become a black hole.

✦ JWST'S DISCOVERIES

- In 2023, the James Webb Space Telescope (JWST) found a supermassive black hole in the young universe, this time hosted by a galaxy that formed 570 million years after the Big Bang.
- Then in March 2024, JWST detected a distant, extremely luminous supermassive black hole located in the protogalaxy GN-z11—a whopping 13.4 billion light-years from Earth.
- GN-z11 existed just 430 million years after the Big Bang, and its black hole has a mass of two million Suns.
- In 2025, scientists utilized JWST to confirm another black hole, this time in the galaxy CAPERS-LRD-z9. Our view of the black hole is as it was 13.3 billion years ago, or 500 million years after the Big Bang.

✦ BLACK HOLE WHAT-IFS

- If you crammed all Earth's mass—including everyone and everything on its surface, beneath its oceans, and within its atmosphere—into the volume of a plum, Earth would become a black hole.
- If that happened, our Moon and every other planet in the solar system would continue to orbit as usual, unencumbered.
- What would happen if you fell into a black hole? Your body would spaghettify—it would stretch apart, molecule by molecule, atom by atom, until you were nothing but a long stream of fundamental particles.
- One thing's for sure: If you survived the descent, you could never report back.

✦ GALAXIES AND THEIR BLACK HOLES

- Sagittarius A* (pronounced "A-star") is the supermassive black hole at the center of the Milky Way galaxy.
- It is four million times more massive than the Sun.
- As with other black holes, a whirling disk of hot gas and dust surrounds Sagittarius A*, generating radio waves and x-rays.
- **OUR FIRST PICTURE OF SAGITTARIUS A* WAS TAKEN IN MAY 2022 BY THE EVENT HORIZON TELESCOPE.**
- To get to this black hole, travel about 26,000 light-years from Earth toward the constellation Sagittarius.
- The Andromeda galaxy hosts a black hole at least 100 million times heavier than the Sun.
- An ultraheavy black hole lies in the distant galaxy TON 618—a whopping 66 billion solar masses.

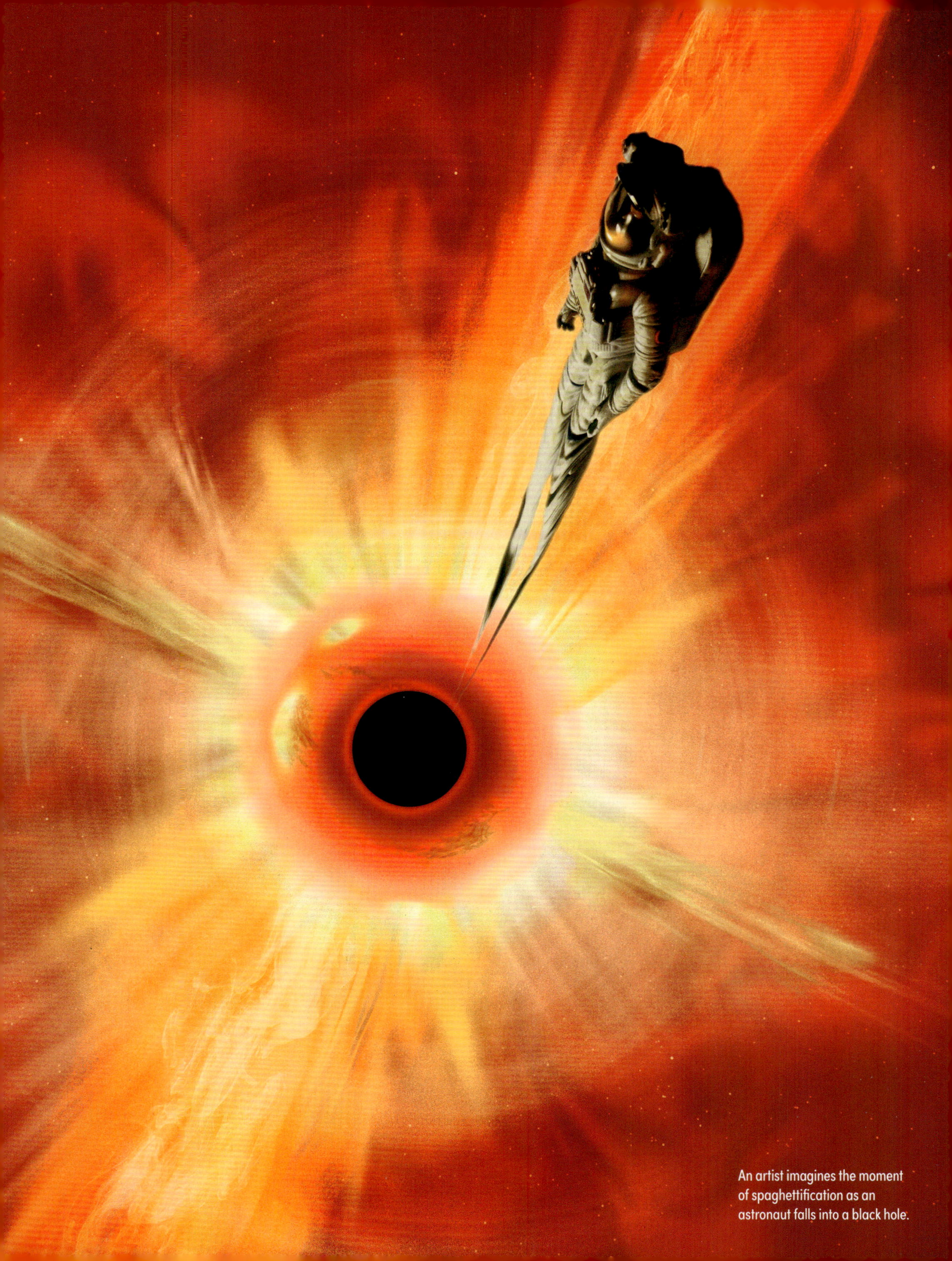

An artist imagines the moment of spaghettification as an astronaut falls into a black hole.

"BLACK HOLES AREN'T AS BLACK AS THEY ARE PAINTED. THEY ARE NOT THE ETERNAL PRISONS ONCE ENVISAGED . . . SO, IF YOU FEEL YOU ARE IN A BLACK HOLE, DON'T GIVE UP: THERE IS A WAY OUT!"

—STEPHEN HAWKING, *BLACK HOLES: THE BBC REITH LECTURES*

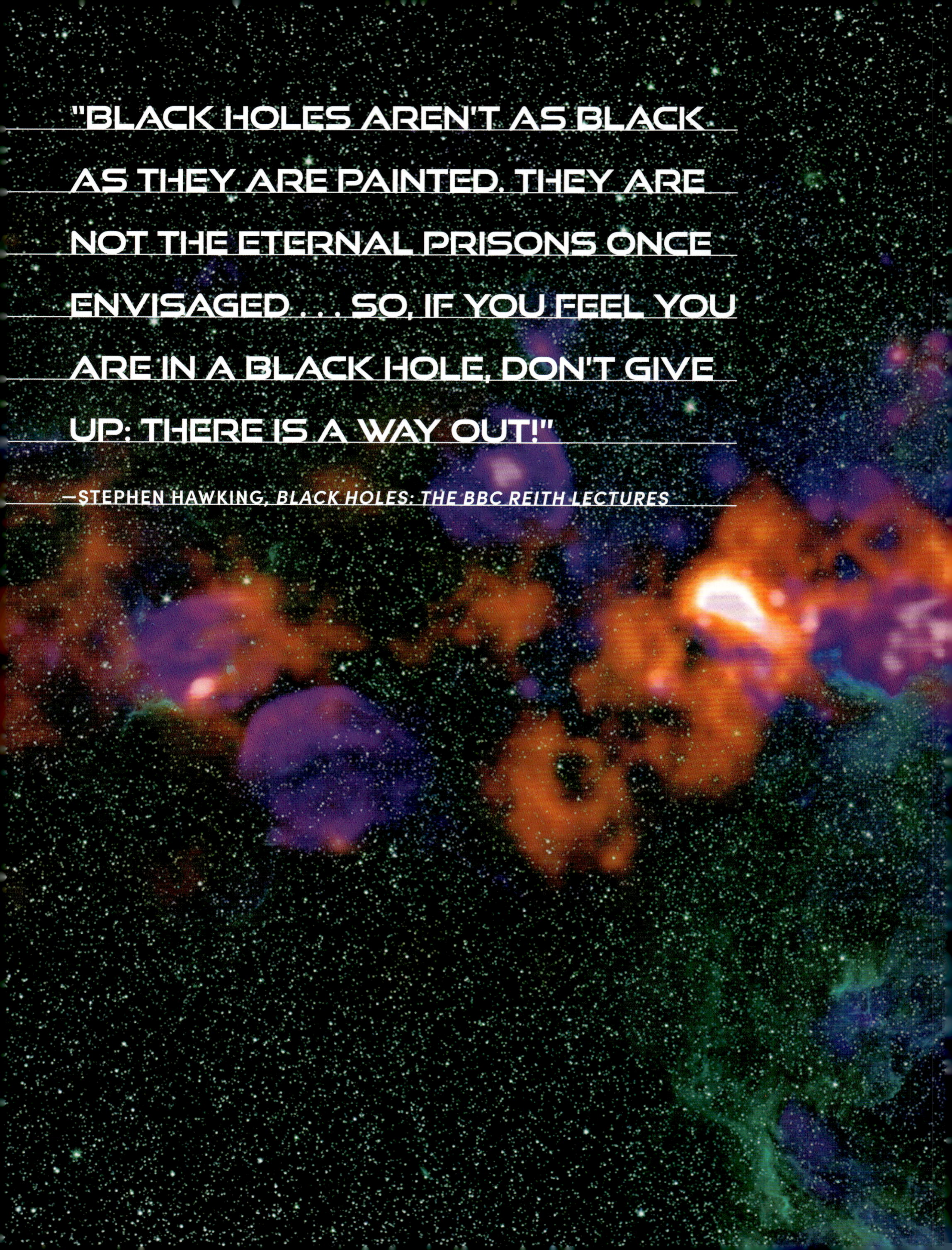

A close-up view of Sagittarius A* and the center of the Milky Way galaxy

CURRENT FACT COUNT 4578

BIG BLACK HOLES NEED BIG NUMBERS

The human brain did not evolve to intuit large numbers or deep time. But we must tap the very limits of our mathematical vocabulary to describe the most massive and ancient objects in our universe.

- The supermassive black holes at the centers of galaxies will require more than a googol years to evaporate completely.

Mathematician Edward Kasner

- One googol is represented as 10 raised to the 100th power: 10^{100}.
- It's written as a one followed by 100 zeros: 10,000,000,000,000,000,000,000,000,000,000,000,000,000,000,000,000,000,000,000,000,000,000,000,000,000,000,000,000,000,000,000,000,000,000,000,000,000,000,000.
- One googol is one quintillion times larger than the number of atoms in the universe.
- **THE NAME WAS INVENTED BY NINE-YEAR-OLD MILTON SIROTTA, THE NEPHEW OF MATHEMATICIAN EDWARD KASNER.**
- In the late 1990s, a search engine website borrowed a misspelled version of the name, propelling it into the zeitgeist: Google.
- Today, "google" has become synonymous with looking something up on the Internet.
- What does 10 raised to the googol power look like? 10^{googol}.
- That's 10 raised to the 100 billion billion billion billion billion billion billion billion billion billion billion power.
- Behold the googolplex—a number with more zeros than particles in the observable universe.
- **THE GOOGOLPLEX MUST BE NAMED: IT CANNOT BE WRITTEN ANY OTHER WAY.**
- There are literally not enough atoms in the universe to form the ink to write out the required number of zeros.
- Even if we could magically borrow enough atoms to write out the number for googolplex, and assuming we wrote one numeral per second, then the time required to do so would far exceed the age of the universe.

"I WAS WALKING IN THE WOODS WITH MY NEPHEW ONE DAY, AND I ASKED THE BOY TO THINK UP ANY NAME FOR THE NUMBER, ANY AMUSING NAME THAT ENTERED HIS HEAD."

—EDWARD KASNER

NASA's WISE captured galaxies clumped together in the Fornax cluster, 60 million light-years from Earth.

CURRENT FACT COUNT 4618

EVEN MORE BIG NUMBERS

• **Million**

1,000,000 (10^6)

• **Billion**

1,000,000,000 (10^9)

• **Trillion**

1,000,000,000,000 (10^{12})

• **Quadrillion**

1,000,000,000,000,000 (10^{15})

- An estimated 20 quadrillion ants inhabit Earth.
- That's 2.5 million ants for every individual human.

• **Quintillion**

1,000,000,000,000,000,000 (10^{18})

- There are an estimated 7.5 quintillion grains of sand on all of Earth's beaches and deserts.

• **Sextillion**

1,000,000,000,000,000,000,000 (10^{21})

- All the oceans on Earth combined hold about six sextillion cups of water.

• **Septillion**

1,000,000,000,000,000,000,000,000 (10^{24})

- This is the estimated upper limit of the number of stars in the universe.

• **Octillion**

1,000,000,000,000,000,000,000,000,000 (10^{27})

- There are about seven octillion atoms in the average human body.

- **Nonillion**

1,000,000,000,000,000,000,000,000,000,000 (10^{30})

  - An estimated one nonillion microbes inhabit Earth.
  - That's about 50 trillion microbes for every one ant.
  - Numbers larger than this are rarely discussed, even among microbiologists and astrophysicists.

- **Decillion**

1,000,000,000,000,000,000,000,000,000,000,000 (10^{33})

- **Undecillion**

1,000,000,000,000,000,000,000,000,000,000,000,000 (10^{36})

- **Duodecillion**

1,000,000,000,000,000,000,000,000,000,000,000,000,000 (10^{39})

- **Tredecillion**

1,000,000,000,000,000,000,000,000,000,000,000,000,000,000 (10^{42})

- **Quattuordecillion**

1,000,000,000,000,000,000,000,000,000,000,000,000,000,000,000 (10^{45})

- **Quindecillion**

1,000,000,000,000,000,000,000,000,000,000,000,000,000,000,000,000 (10^{48})

- **Sexdecillion**

1,000,000,000,000,000,000,000,000,000,000,000,000,000,000,000,000,
000 (10^{51})

- **Septendecillion**

1,000,000,000,000,000,000,000,000,000,000,000,000,000,000,000,000,
000,000 (10^{54})

- **Octodecillion**

1,000,000,000,000,000,000,000,000,000,000,000,000,000,000,000,000,000,000,
000 (10^{57})

- **Novemdecillion**

1,000,000,000,000,000,000,000,000,000,000,000,000,000,000,000,000,000,000,
000,000 (10^{60})

- **Vigintillion**

1,000,000,000,000,000,000,000,000,000,000,000,000,000,000,000,000,000,000,
000,000,000 (10^{63})

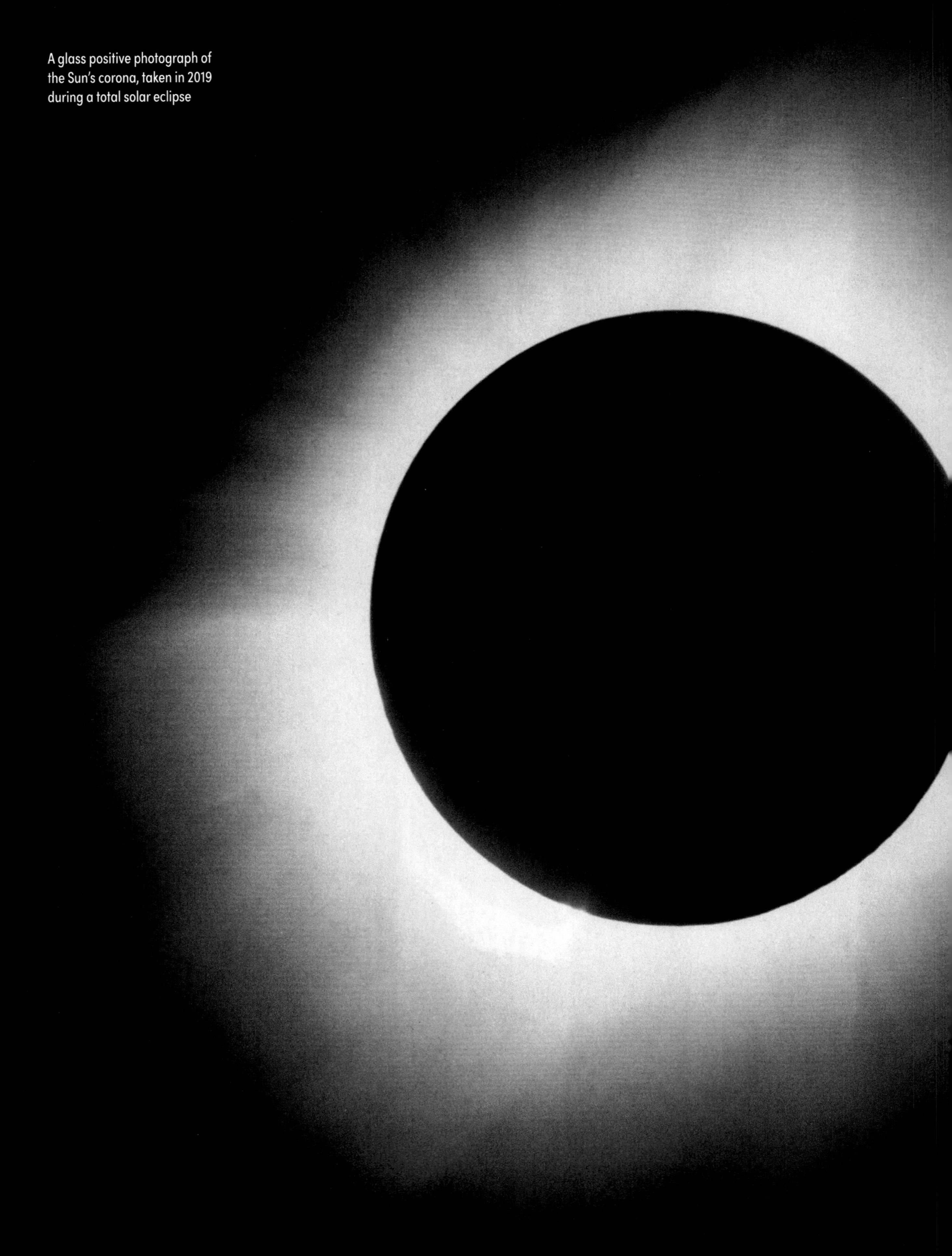

A glass positive photograph of the Sun's corona, taken in 2019 during a total solar eclipse

✦ THE FINAL "ILLION"

- The estimated number of atoms in the observable universe is 10^{80}.
- 10^{80}: 100,000,000,000,000,000,000,000,000,000,000,000,000,000,000,000,000,000,000,000,000,000,000,000,000,000,000,000,000,000.
- **THIS NUMBER IS SO HUGE THAT IT HAS NO NAME.**
- It is sometimes referred to as the Eddington number, named for physicist Sir Arthur Eddington, who was the first to speculate on how to precisely estimate the number of atoms in existence.
- This number does not include dark matter or dark energy, which makes up about 95 percent of our universe.
- Physicists arrived at this ballpark estimate by combining a few known facts with a few estimated guesses.
- We know that hydrogen and helium atoms make up 98 percent of ordinary visible matter, so we use their known masses as a base.
- Next, we can calculate the number of hydrogen and helium atoms in an average star.

Sir Arthur Eddington, circa 1930

- Since the Sun comprises 99.8 percent of the mass of the solar system, we can disregard planets, moons, and asteroids as insignificant contributors to the total number of atoms.
- Now, we multiply that number by the average number of stars in an average-size galaxy.
- **FINALLY, WE MULTIPLY THAT NUMBER BY THE PREDICTED NUMBER OF GALAXIES IN THE UNIVERSE, AND VOILÀ: EDDINGTON'S NUMBER.**

"WE USED TO THINK THAT IF WE KNEW ONE, WE KNEW TWO, BECAUSE ONE AND ONE ARE TWO. WE ARE FINDING THAT WE MUST LEARN A GREAT DEAL MORE ABOUT 'AND'."

—SIR ARTHUR EDDINGTON

CURRENT FACT COUNT 4649

QUASARS

Abject mysteries for decades, quasars have been an object of astrophysical fascination for more than a century. Their secret energy source? Black holes.

- During the early universe, supermassive black holes in the centers of galaxies voraciously devoured stars and gas clouds that wandered too close. We call these galaxies quasars.
- **WHEN A BLACK HOLE DINES UPON STARS AND GAS CLOUDS, THEY TYPICALLY SPIRAL INWARD.**
- Friction within this accretion disk renders the surrounding area of the black hole aglow, radiating stupendous quantities of electromagnetic energy that spirals into the abyss.
- The glow emanates from a tiny region in an otherwise large galaxy.
- Billions of years later, once all the nearby stars and gas have been fully consumed by the central black hole, the quasar mechanism shuts down.
- When this happens, all that's left is an ordinary galaxy with a silent supermassive black hole in its center.
- Quasar formation peaked around 10 to 11 billion years ago.

✦ DISCOVERY

- The first quasar was discovered in 1963, when astronomer Maarten Schmidt realized a bright radio source thought to be a star was

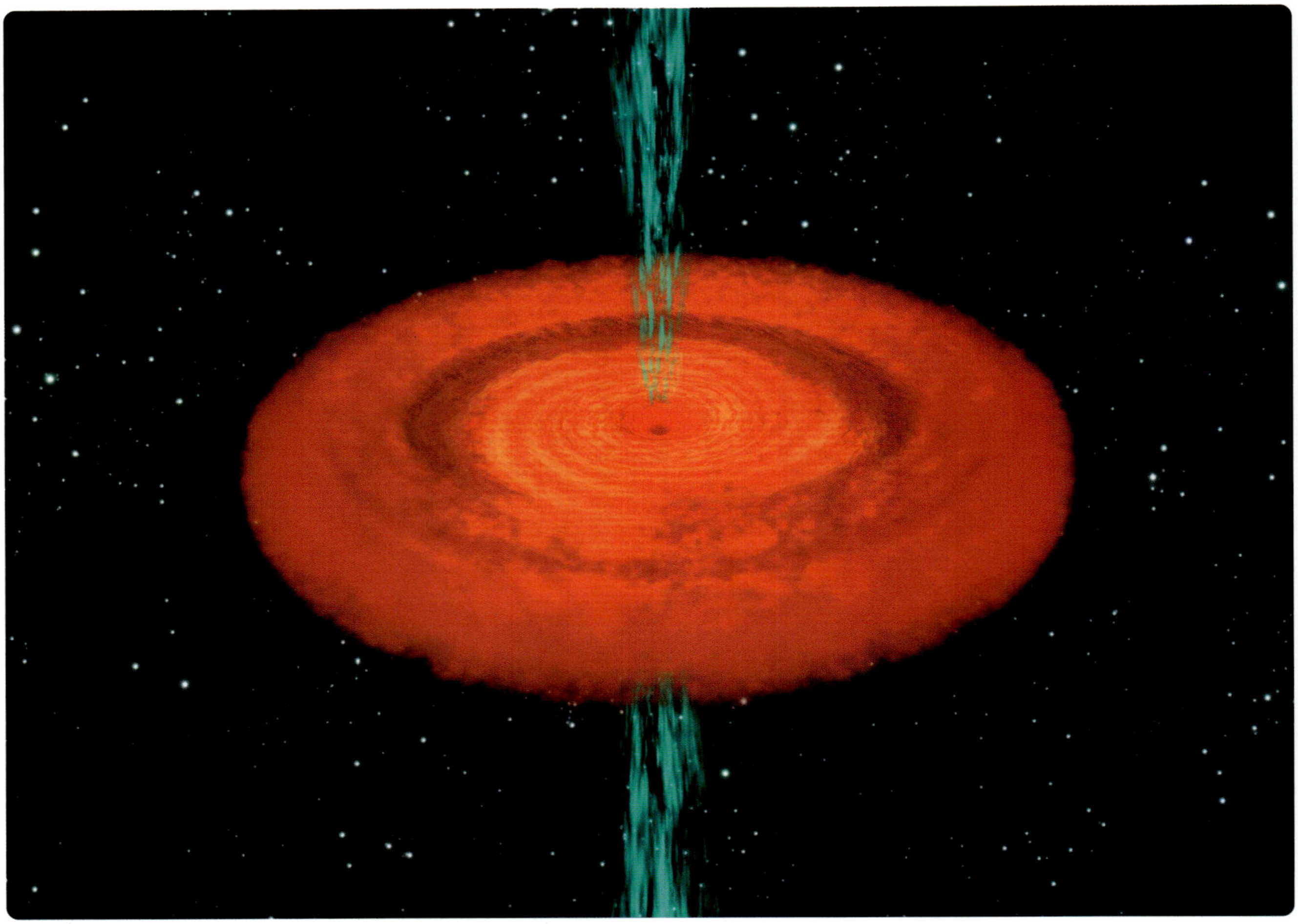

An artist's impression of a rare broad absorption line (BAL) quasar, where gas flows away from the accretion disk in fact an extremely distant and luminous object that lay beyond the Milky Way galaxy. People were surprised something so far away could appear so bright.

- This first quasar is called 3C 273.
- It is so named because it was the 273rd object in the Third Cambridge Catalogue of Radio Sources.
- Quasars give off strong radio emissions, which is why they were first detected in 1950s radio surveys of the skies.
- The name "quasar" is actually a loose acronym for quasi-stellar radio source.
- In 2009, astronomers reported the discovery of Clowes-Campusano, a large quasar group (LQG) containing a cluster of 34 quasars, the largest structure known in the universe at the time.
- In 2013, astronomers found an even bigger LQG, appropriately named Huge-LQG, which bumped Clowes-Campusano from top place as largest in the universe.
- **HUGE-LQG CONTAINS 73 QUASARS.**
- Its size is estimated at four billion light-years across.
- Might something even bigger lurk beyond? Many believe the Hercules-Corona Borealis Great Wall, a gamma-ray burst cluster, might top the list at 10 billion light-years across.

✦ LUMINOUS BEINGS

- Quasars are among the most distant known objects in the universe.
- When observing galaxies far away with a telescope, the central emitting region is so small, it's almost starlike.

Captain Marvel speeds into space.

- The brightest quasars can radiate as much light as 500 trillion Suns.
- The light must travel for billions of years before reaching Earth.
- **QUASARS CAN EMIT THOUSANDS OF TIMES THE LUMINOSITY OF THE ENTIRE MILKY WAY COMBINED.**
- The high luminosity of black holes renders quasars visible across the universe, at a time when quasars ruled spacetime.
- If there were any quasars in our galaxy, the high-energy radiation would make life on Earth impossible.

✦ WHITE HOLES

- If we could hit rewind on a black hole, it would appear to us instead as a white hole.
- A white hole is the hypothetical mathematical opposite of a black hole.
- If real, it would be a region where matter and light may only exit, never enter.
- The concept of a white hole is a paradox, since any matter drawn to its immense gravity would cause the white hole to collapse on itself and transform into a black hole.
- While there is no observational evidence for the existence of white holes, and most physicists doubt their existence, they do make for fun science fiction plot devices.
- **IN THE MARVEL COMICS, WHEN CAPTAIN MARVEL ACCIDENTALLY SIPHONED THE ENERGY OF A WHITE HOLE, SHE BECAME ONE OF THE MOST POWERFUL SUPER HEROES OF ALL TIME.**
- If white holes did exist, they would be the most energetic phenomena in the universe.

CURRENT FACT COUNT 4683

DARK THINGS

We call them "dark matter" and "dark energy," but we might as well just call them "Fred" for how much we understand their composition.

- The universe is constructed of matter and energy, most of it invisible to us.
- Normal matter—the stuff of Earth, planets, and stars—makes up less than 5 percent of the universe.
- **BUT THERE ISN'T ENOUGH ORDINARY MATTER IN THE VISIBLE UNIVERSE TO EXPLAIN THE GRAVITATIONAL EFFECTS WE OBSERVE.**
- That's how astrophysicists know that dark energy and dark matter dominate the universe.

Vera Rubin, circa 1985

- Like black holes, dark matter and dark energy are invisible except for their very noticeable gravitational effects.

✦ DARK ENERGY

- Two-thirds of the universe consists of dark energy.
- **DARK ENERGY IS A KIND OF ANTI-GRAVITY THAT'S PUSHING GALAXIES FARTHER APART, ACCELERATING THE EXPANSION RATE OF THE UNIVERSE.**
- If dark energy continues unchecked, then in 200 billion years, the expanding universe will stretch beyond its own capacity to hold itself together.
- That hypothetical ultimate moment in the universe's evolution is called the Big Rip.

✦ DESI

- A quest to create the largest 3D map of the universe occurred with the Dark Energy Spectroscopic Instrument (DESI).
- It provided the most precise measurements to date of our cosmic expansion history.
- These measurements offered further insight into the nature of dark energy and its role in the evolving structure of the universe.
- To map the universe, DESI gathered light from millions of galaxies encompassing more than a third of the entire sky.
- It observed more than 100,000 galaxies every single night.

✦ DARK MATTER

- Discovered in 1933 by Swiss American astrophysicist Fritz Zwicky, dark matter remains the longest-unsolved problem in astrophysics.
- The first hint of dark matter surfaced in the early 1930s, when Dutch astronomer Jan Oort (for whom the Oort cloud is named) found that stars in the Milky Way travel faster than expected, given their visible mass.
- He deduced that the gravity of other, unseen matter must keep them from flying off into space.
- Not long after, in 1933, Fritz Zwicky concluded the same about matter in galaxy clusters.
- In 1976, the work of Vera Rubin confirmed that galaxies must contain about 10 times as much dark matter as all their visible stars and nebulae.
- **WE NOW KNOW THAT 85 PERCENT OF THE GRAVITY OF THE UNIVERSE COMES FROM DARK MATTER.**
- Dark matter exerts about six times the total gravity of all visible matter.
- It doesn't emit, reflect, or absorb light.
- It also doesn't interact with ordinary matter at all, other than through the force of gravity. So like black holes, we can only *observe* the powerful influence of its gravitational effect. Truth in advertising: Dark matter should instead be called dark gravity.

The Mayall Telescope, where DESI operates, at Kitt Peak National Observatory in Arizona

- Dark matter doesn't appear to interact through any of the other three fundamental forces of the universe: the strong, weak, or electromagnetic forces.

- **THE PHENOMENON OF GRAVITATIONAL LENSING, BY WHICH LIGHT CURVES AS IT PASSES BY A MASSIVE BODY, HELPS TO IDENTIFY AND MAP LOCATIONS OF DARK MATTER IN THE UNIVERSE.**

- Ordinary matter attaches to itself to make molecules using electromagnetic forces within the atom. Dark matter has no capacity to attach to itself.

- Because dark matter does not interact with itself, it does not coalesce into solid physical objects.

- Thus, there are no dark matter planets.

- Astrophysicists have suggested that dark matter makes stars on the outer edges of spiral galaxies, including the Milky Way, move much faster than expected.

- The rotation speeds within a galaxy appear more or less the same, regardless of the distance from the center—a fact that suggests the presence of dark matter.

- Other indications of dark matter include the distribution of galaxies in space.

- The gravitational influence of dark matter may be what distorts the light of a galaxy cluster like MACS J1206, making the galaxies appear to have a warped shape.

- **WERE IT NOT FOR DARK MATTER, THE ORBITAL SPEEDS OF ALL THE STARS IN THE MILKY WAY WOULD EXCEED THE GRAVITY OF THE MILKY WAY ITSELF AND FLY OFF INTO INTERGALACTIC SPACE.**

- In the DC Universe, there's a villain called Dark Matter whose civilian name is Fred Fleming.

CURRENT FACT COUNT 4708

OBSERVING DARK MATTER

The best observatories attempting to study dark matter are the same ones built to study neutrinos. They require the most cosmically isolated conditions possible on Earth.

✦ WIMPs

- One candidate for the composition of dark matter are weakly interacting massive particles, affectionately, if irreverently, nicknamed WIMPs.

- Just being weak doesn't make them wimpy. Here, "weak" refers to their interactions with matter via the weak nuclear force—the same force responsible for radioactive decay.

- The weak force can alter a neutron's quark, transforming it into a proton and emitting tiny amounts of energy. This phenomenon changes the atom into an entirely different element.

- If hypothetical WIMPs exist, then they might occasionally interact with atoms on Earth and kick out tiny, yet detectable, bursts of energy.

- **TO WITNESS THIS HYPOTHETICAL EVENT REQUIRES EXORBITANT FUNDING, ADVANCED TECHNOLOGIES, AND FEATS OF ENGINEERING.**

- Dark matter detectors are generally built underground, where they encounter the least interference from standard charged particles, like muons, which can penetrate nearly a mile through Earth's crust before being absorbed.

- The world's largest and deepest lab can be found a mile and a half below China's Jinping Mountains.

- Encased in layers of the mountain's dense rock, the China Jinping Underground Laboratory (CJPL) is the most shielded laboratory in the world.

- Its protection results in near-negligible bombardment by the constant streams of cosmic rays from space.

- CJPL and other deep-underground research laboratories hunt tirelessly for one elusive moment: the interaction of a dark matter particle with any singular atom. The hunt continues ...

✦ AXIONS

- Axions are another hypothetical particle that may account for dark matter.

- Unlike WIMPs, axions interact using the strong nuclear force—the same force that binds protons and neutrons into a nucleus.

- **IN THEORY, WHEN AXIONS ENCOUNTER STRONG MAGNETIC FIELDS, SOME OF THEM WOULD CONVERT INTO PHOTONS, AND PHOTONS INTO AXIONS.**

- If axions exist, they could be detected by introducing light to specially magnetized chambers and watching for anomalies that only axions could account for.

- Those anomalies are so teensy, they amount to only one-octillionth of the power of an average light bulb. (An octillion is a one followed by 27 zeros.)

- A ground-based telescope, such as CERN's Axion Solar Telescope (CAST), can find axions by aiming its magnetized detector directly at the Sun.

- **AN AXION-PHOTON INTERACTION FROM COSMIC RAYS SHOULD RESULT IN EASILY DETECTABLE X-RAYS.**

- The Sanford Underground Research Facility (SURF), nearly a mile below the hills of South Dakota, was converted from an old gold mine into a dark matter observatory.

- The same properties that make an observatory excellent for dark matter detection also make it excellent for neutrino detection.

- In fact, SURF occupies the same gold mine as the one Raymond Davis, Jr., used to detect neutrinos emitted by the Sun in 1968.

- Scientists at SURF search for dark matter particle collisions within a massive vat containing 10 tons of liquid xenon.

- **THE COST OF THAT MUCH XENON ALONE COULD AMOUNT TO $30,000,000.**

- But even with the most advanced, sensitive, and isolated observatories we have today, we've yet to understand the nature of dark matter—or if it's even composed of particles at all.

✦ DARK STARS

- Dark stars are hypothetical, named for their power source: dark matter.

- Despite the name, they would shine billions of times brighter than the Sun, with millions of times more mass.

Liquid nitrogen clouds an experiment at the China Jinping Underground Laboratory in 2023.

CURRENT FACT COUNT 4754

TRAVELING THROUGH SPACE AND TIME

Both scientists and sci-fi writers conjure new methods of transportation that could carry us far beyond the capabilities of modern technologies.

✦ FASTER THAN LIGHT

- According to *Star Trek,* the warp drive was invented on April 5, 2063.
- Warp drives enabled the USS *Enterprise* to explore the universe at speeds faster than light (FTL), contradicting Einstein's theory.
- ***STAR TREK* VESSELS USE MATTER-ANTIMATTER ENGINES TO POWER THEIR WARP DRIVES.**
- Antimatter was predicted in 1928 by physicist Paul Dirac and discovered in 1932 by physicist Carl Anderson.
- Matter-antimatter annihilation is the most efficient conversion of matter to energy known.
- Inspired by *Star Trek,* physicist Miguel Alcubierre proposed a hypothetical warp drive model that complied with Einstein's general theory of relativity.
- The Alcubierre drive takes advantage of a loophole in the equations: An object cannot traverse spacetime at FTL speeds, but spacetime itself need not adhere to any speed limit.
- Einstein proved that nothing can travel faster than light—except for spacetime itself. We know that space itself is expanding faster than the speed of light.
- An Alcubierre drive would form a bubble of spacetime around it to travel in, and thereby break no physical laws.
- By shrinking spacetime in front of the ship and stretching it behind the ship, an Alcubierre drive travels at FTL through its own spacetime bubble.
- If it were nonfiction and not space fantasy, the *Millennium Falcon,* piloted by Han Solo in the *Star Wars* galaxy, technically would enter a separate dimension called "hyperspace" to achieve "hyperspeed" and travel at the speed of light to distant planets. But as far as we can tell, the same laws of physics preside over every dimension.
- The Guild Heighliner spacecraft in the world of *Dune* can fold up space to jump over vast distances with a tiny hop—much like the concept of wormholes.
- Imagine an ant on one end of a sheet of paper. The fastest way for that ant to traverse to the other side (short of true teleportation) is to fold the entire paper in half and join the two sides together. The ant simply steps onto its destination. Wormholes (and Guild Heighliners) scrunch up spacetime similarly.
- Sci-fi books, TV shows, and films like to depict spacecraft abruptly emerging from hyperspace, warp speed, subspace, or whatever word they use to describe FTL. One problem: The law of conservation of momentum dictates that everyone within that spacecraft just careened against the front of the ship or against their seat belts at that FTL speed.
- Also, an abrupt jump to FTL speed from a complete standstill imparts an opposite and equally deadly force upon anyone aboard. To remedy this, *Star Trek*'s starships enable fantastical "inertial dampeners" to absorb the shock from sudden acceleration or deceleration.
- **IF FTL TRAVEL COULD EXIST, IT WOULD UNDO CAUSALITY AND LEAD TO TIME PARADOXES.**

✦ THE TACHYON

- In the late 1950s, physicist Arnold Sommerfeld proposed a concept for a FTL particle that could obey the laws of physics by exploiting a mathematical loophole in Einstein's theories.
- In 1967, physicist Gerald Feinberg refined the concept and proposed the hypothetical subatomic particle called the tachyon.
- The word "tachyon" comes from Greek *tachus,* meaning "swift."
- Einstein's theories mandate that nothing can *accelerate* to faster-than-light speed from slower-than-light speed. This implies that crossing that barrier is what prevents FTL.
- The hypothetical tachyon outmaneuvers the law because it is *birthed* at FTL speed. It never crosses that illegal barrier between slower than and faster than the speed of light.

✦ SPECIAL RELATIVITY

- For a photon traveling at light speed, time ceases to elapse entirely.
- In *Star Wars:* Episode IV *A New Hope,* Han Solo boasts that the *Millennium Falcon* can make the Kessel Run in less than 12 parsecs. But the parsec is a unit of distance, not time.
- The word "parsec" is a contraction of "parallax-second" and, based on the trigonometry of parallax astronomy, is used to measure the distance to and between stars.
- **A PARSEC EQUALS 3.26 LIGHT-YEARS.**
- The 2018 movie *Solo: A Star Wars Story* justified Han's earlier use of "parsec" by showing how only powerful ships can navigate close to gravity wells without spaghettification.
- Most pilots take an 18-parsec-long route to pass around the black hole cluster of the Kessel Run, but by skimming dangerously close, Solo cut the run to only 12 parsecs.

✦ WORMHOLES

- Wormholes aren't just fantastic plot devices, they're theoretically possible—although they have yet to be observed.
- Wormholes are shortcuts forged through the very fabric of spacetime, connecting two distant, and otherwise untraversable, points in space.
- **A WORMHOLE REQUIRES ENORMOUS GRAVITATIONAL PULL TO "FOLD" SPACE BETWEEN TWO POINTS.**
- The concept arose in the 1930s when Albert Einstein and Nathan Rosen proposed a theory to bridge two points of spacetime.
- The scientific name for a wormhole is the Einstein-Rosen bridge.
- Renowned physicist John Wheeler coined the term "wormhole" and popularized the term "black hole."
- In the film *Contact* (1997), Jodie Foster's Ellie Arroway is whisked to the distant star Vega, some 26 light-years from Earth, via a wormhole.
- According to Hollywood, passing through a wormhole would look like gliding through a glittering, variegated, tubular slide. But physics tells us that stepping through a wormhole would be a visually unspectacular experience. One would simply step through the portal and instantaneously emerge on the other side.
- In Norse mythology, the glimmering rainbow Bifröst bridge connects the Earthly realm with Asgard, home of the gods.
- **IN THE MARVEL CINEMATIC UNIVERSE, THE BIFROST BRIDGE THOR AND THE ASGARDIANS TRAVERSE IS DESCRIBED AS AN EINSTEIN-ROSEN BRIDGE.**

✦ TIME TRAVEL

- Time travel to the future is not just a sci-fi phenomenon. It is a real, measurable occurrence predicted by Einstein's theories of general and special relativity.
- International Space Station astronauts do it all the time: They age more slowly in space.
- **THE CONCEPT OF TIME TRAVEL TO THE PAST, THOUGH, VIOLATES A FUNDAMENTAL PRINCIPLE GOVERNING THE UNIVERSE: CAUSE AND EFFECT.**
- In 1992, Stephen Hawking wrote a seminal paper arguing that time travel to the past would violate causality and was, therefore, impossible.
- Hawking offered as circumstantial evidence "the fact that we have not been invaded by hordes of tourists from the future."
- He wryly suggested that there seems to be a chronology protection agency in place that makes the universe safe for historians.
- The Time Variance Authority in the Marvel Cinematic Universe monitors and eliminates timelines that threaten the larger Multiverse.
- The Time Lords of *Doctor Who* are advanced extraterrestrials from the planet Gallifrey who can monitor—and interfere—with timelines to maintain order.
- The Temps Commission in Netflix's *The Umbrella Academy* are authorized to assassinate anyone who might cause a temporal paradox.

"FOR THOSE OF US WHO BELIEVE IN PHYSICS, THE DISTINCTION BETWEEN PAST, PRESENT, AND FUTURE IS ONLY A STUBBORNLY PERSISTENT ILLUSION."

—ALBERT EINSTEIN

CURRENT FACT COUNT 4786

E.T., WHERE'S HOME?

The search for life in the universe—a bustling, serious, scientific pursuit—often leaves us with more questions than answers. Examining our origins is a good place to begin.

✦ SEEKING THE GOLDILOCKS ZONE

- Astrobiology explores the origin, evolution, and possibilities of life across the universe.
- This field of study blends chemistry, physics, astronomy, planetary science, molecular biology, and geology.
- **ONE STRATEGIC APPROACH IN ASTROBIOLOGY IS TO FIND PLANETS IN THE GOLDILOCKS ZONE OF A STAR.**
- Within a Goldilocks Zone, a planet is not too close, not too far, but at just the right distance from its host star to support liquid water on the surface—the best (and only) conditions known to be capable of supporting life.

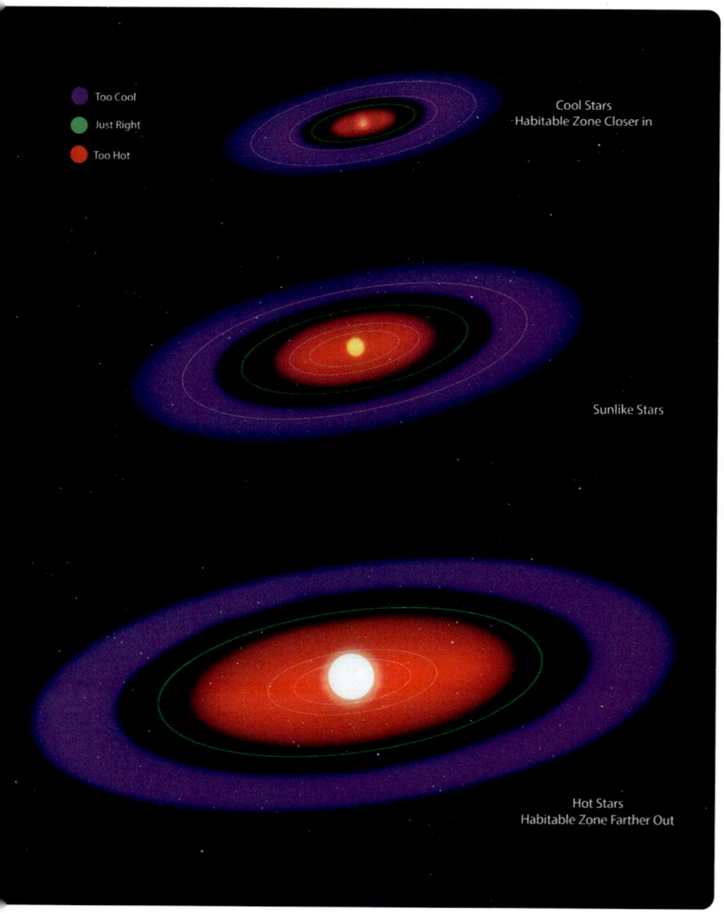

- The Goldilocks Zone can become larger, depending on a planet's ability to retain heat, such as by way of a greenhouse effect in its atmosphere.
- For cooler stars, the Goldilocks Zone is narrower than it is for hotter stars. But cool stars outnumber hot stars by a factor of a hundred, boosting the chances of finding a planet orbiting there.
- The host star's temperament is another critical factor. How frequent and how violent are its weather tantrums?

Categories of the Goldilocks Zone

- Because a red giant is much cooler than sunlike stars, for example, its Goldilocks Zone hugs the star more closely.
- A planet residing within that red giant's habitable zone is thus much more exposed to the star's solar weather. If the star spews violent or frequent flares and coronal mass ejections, conditions may prove too hostile for life as we know it to persist.
- The dynamics of planetary gravity limit the total number of planets that can sustain a stable orbit within a star's habitable zone. That number is about six.
- The pudgy, lovable, microscopic, mildly creepy tardigrade, or water bear, is one of the few Earth animals capable of living outside of a temperate Goldilocks Zone.
- NASA's Kepler telescope was specifically designed to discover Earthlike planets orbiting sunlike stars.
- With two releases of data, K1 and K2, the Kepler telescope showed that it has discovered more than 2,700 such planets.
- The Transiting Exoplanet Survey Satellite, also a telescope tuned to find planets, has 400 times the search area that Kepler had, generating nearly 8,000 exoplanet candidates.
- **THE LATEST ESTIMATES PLACE THE NUMBER OF PLANETS ORBITING WITHIN THEIR HOST STAR'S HABITABLE ZONE ACROSS THE MILKY WAY GALAXY AT HALF A BILLION.**
- Searching only for life as we know it may bias and blind us from finding life as we *don't* know it on a planet we don't think is habitable.
- Thus, the search for intelligent life in the universe makes the big assumption that *we* are intelligent, so we look for life-forms just like us.

E.T. and Elliott cycle to freedom in *E.T. the Extra-Terrestrial* (1982).

✦ LIFE AS WE (DON'T) KNOW IT

- In the 1958 film *The Blob*, the alien was ... a blob. No body parts. No skeleton, exo- or otherwise. No face. No mouth. An excellent display of creativity when representing an alien from another planet.

- In the 1982 film *E.T. the Extra Terrestrial*, the director Steven Spielberg originally conceived of the main character, E.T., as a sentient plant.

- Showing hardly any creativity at all, in the 1987 film *Predator*, the alien has a head, two eyes, a nose, mouth, neck, torso, two arms, fingers, two legs, and feet. But an ugly face.

✦ EARTHLIKE, YES OR NO?

- The Earth Similarity Index (ESI) scale measures how Earthlike an exoplanet is on a scale of zero to one, with Earth ranked as one.

- Teegarden's Star b, for example, ranks around 0.9 on the ESI scale.

- Discovered in 2019, Star b is among the most Earthlike planets yet discovered.

- In the search for habitable exoplanets, astronomers study a planet's atmosphere for elements that might support life.

- The recipe for life as we know it requires the right set of chemicals, a liquid water medium where said chemicals can react, and an energy source to fuel those reactions.

- **WATER, HEAT, LIGHT, AND OXYGEN ARE SOME OF THE KEY LIFE-SUSTAINING COMPONENTS THAT ANOTHER PLANET WOULD NEED TO SUPPORT LIFE.**

- Since carbon is the foundation for life on Earth, astrobiologists focus their search on carbon-based life-forms elsewhere in the universe.

- NASA's modern mantra is to "follow the water" in the search for alien life in the universe and especially in our cosmic backyard, such as on Mars, Jupiter's moon Europa, and Saturn's moon Enceladus.

- Everywhere on Earth where there's liquid water, there's life—even in the Dead Sea, evidence that whoever named it did not have access to a microscope.

- Carbon is abundant in the universe and is highly chemically active, making it the ideal element upon which to base the diversity of life on any planet.

- The four most abundant chemically active ingredients in the universe match one for one, in sequence, the abundance of chemical elements in life on Earth: hydrogen, oxygen, carbon, nitrogen.

- The animal with the closest DNA to humans—sharing about 98 to 99 percent—is the chimpanzee. Imagine a life-form one percent beyond us in intelligence, just as humans are one percent beyond chimps. What would our greatest intellectual achievements look like to them?

CURRENT FACT COUNT 4813

LIFE FINDS A WAY

Our searches for extraterrestrials are impeded by the fact that we know of only one form of life: life as we know it. What might life look like as we *don't* know it?

- We know that life on Earth flourishes in even the most hostile of environments, even in the deepest reaches of the darkest ocean, where sunlight cannot penetrate.
- In those realms, seawater exerts thousands of pounds of pressure per square inch—conditions that would crush any terrestrial animal.
- **THESE DEEP-SEA REGIONS HOLD CLUES TO HOW LIFE ORIGINATED ON EARTH, AND HOW WE MIGHT FIND IT BEYOND.**

✦ LIFE AT THE VENTS

- Hydrothermal vents were discovered in 1977 by marine geologist Robert Ballard.
- Most edges of Earth's seven major tectonic plates lie within an ocean.
- Where two plates separate on land, such as in Africa's Great Rift Valley, you find yawning chasms, deep lakes, and volcanic activity of every kind.
- Where two plates separate along the ocean floor, near-freezing ocean brine seeps down into deep fissures and meets Earth's upwelling molten magma.
- One tectonic boundary is the Mid-Atlantic Ridge running between the Americas to the west and Europe and Africa to the east.
- **AT 10,000 MILES LONG, THIS SUBSURFACE RIDGE FORMS THE LONGEST MOUNTAIN RANGE ON EARTH.**
- A similar boundary, the Pacific-Antarctic Ridge, runs from New Zealand to South America.
- Another mid-ocean ridge, the East Pacific Rise, runs from South America to the Gulf of California.
- These boundaries expose Earth's mantle, the thick layer of solid rock otherwise hidden beneath the 10 or so miles of crust.
- The ridges are dynamic. The East Pacific Rise, for example, is spreading about six inches a year.
- Mid-ocean ridges are crowned with chimneys containing hydrothermal vents: fissures in the ocean floor that eject plumes of jet-black, superheated seawater laden with minerals.
- **THE SEAWATER AT THESE VENTS CAN REACH A SCALDING 750°F, HOT ENOUGH TO MELT LEAD OR FORGE STEEL.**
- Minerals separate from the water and form monoliths up to 200 feet tall—taller than the Leaning Tower of Pisa.
- One 200-foot-tall vent, named Poseidon, stands in an Atlantic vent field called the Lost City.
- These bizarre monstrosities of the deep look like they more likely belong in the fantastical realm of *The Lord of the Rings'* Mordor than within Earth's oceans.
- Flourishing all around and atop these dark chimneys are entire ecosystems of flora and fauna that do not require sunlight and can survive in hellish heat.
- **THIS FREAKISH DISCOVERY UPSET A FUNDAMENTAL UNDERSTANDING OF LIFE.**
- Before we discovered hydrothermal vent ecosystems, we believed all organisms required sunlight to live—either by consuming it through photosynthesis or by eating organisms that already consumed it.
- The creatures living around hydrothermal vents are chemosynthetic: To make food, they utilize the energy released by chemical reactions and not energy from the Sun.
- Since 1977, we have uncovered hundreds of species unique to hydrothermal vents, including crabs, shrimp, and even one octopus called the vulcanoctopus.
- A vulcanoctopus can live long and prosper near the hostile hydrothermal vents.
- **NOW SOME SCIENTISTS BELIEVE LIFE MAY HAVE ORIGINATED AT HYDROTHERMAL VENTS BILLIONS OF YEARS AGO.**
- Some believe that early in our planet's evolution, hot thermal springs propelled chemical reactions that led to more complex biomolecules.
- Simulation models of Earth's early geology indicate that the chemical interactions between seawater and the molten, mineral-rich crust around the vents could have formed amino acids: the building blocks of DNA.

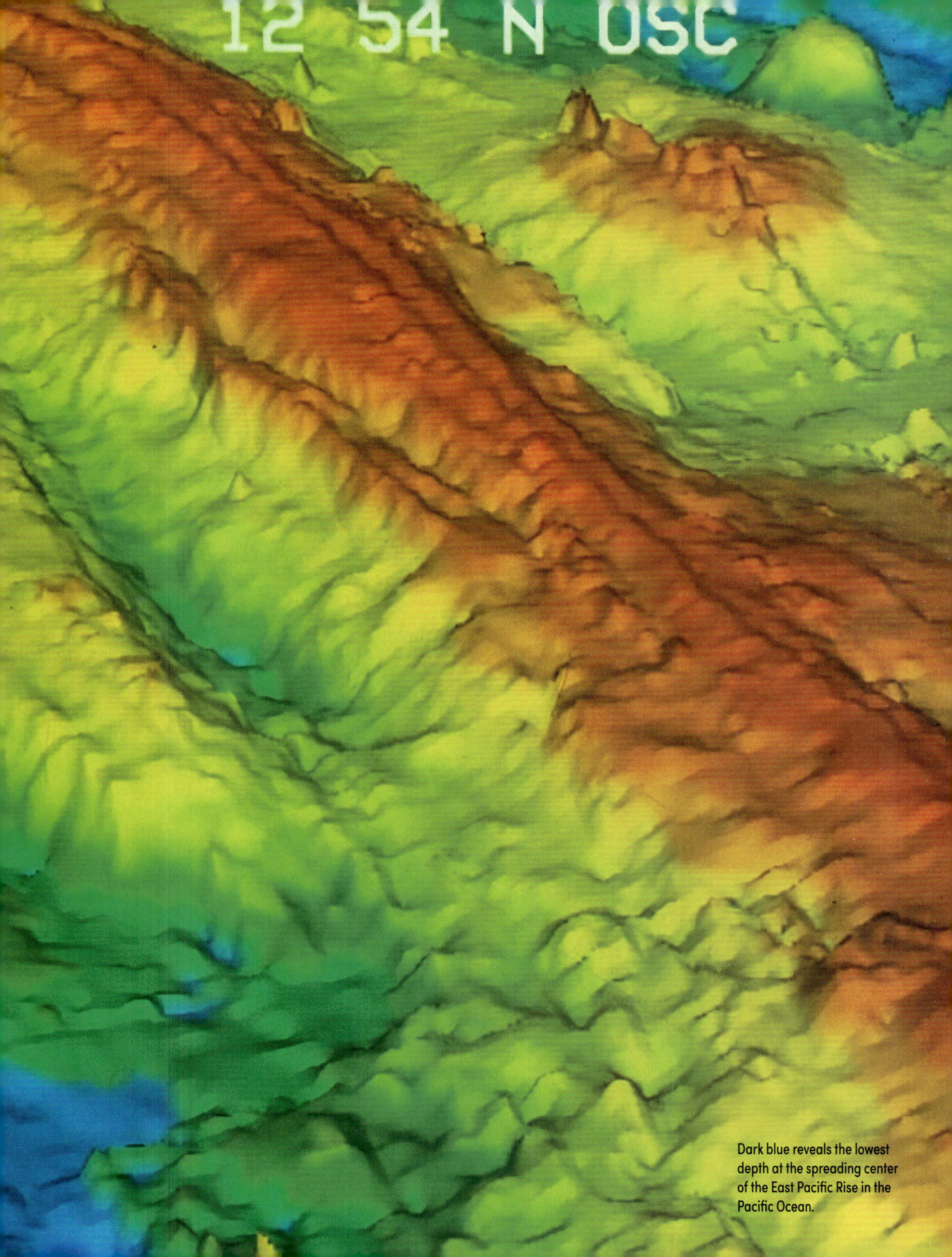

Dark blue reveals the lowest depth at the spreading center of the East Pacific Rise in the Pacific Ocean.

CURRENT FACT COUNT 4835

SEEKING HABITABLE PLANETS

Innovating telescopes and spacecraft continuously probe and uncover new potential habitable worlds.

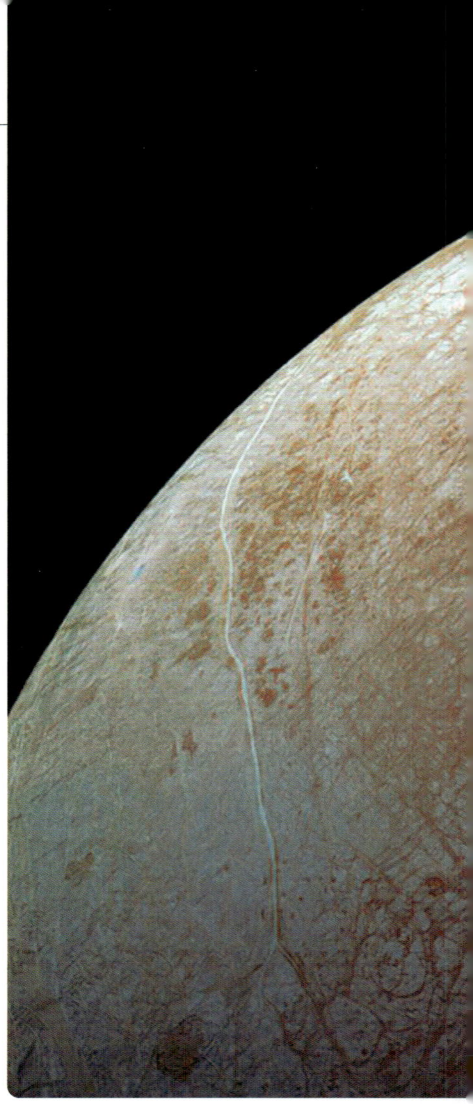

✦ EXTRATERRESTRIAL VENTS

- Ocean worlds like Jupiter's moon Europa or Saturn's moon Enceladus may harbor deep-sea hydrothermal vents of their own.
- **IN FACT, RESEARCH SUGGESTS THAT HYDROTHERMAL VENTS MAY BE EVEN MORE ABUNDANT ON OCEAN WORLDS THAN ON EARTH.**
- Scientists believe hydrothermal vents could have endured on Enceladus or Europa for millions or billions of years.
- This would allow plenty of time for at least one form of life as we know it to have taken hold.
- Massive plumes spewed by Enceladus were sampled by the Cassini spacecraft and shown to contain chemical markers that strongly suggest hydrothermal activity.
- Europa may be spewing similar geysers.
- In the early 2030s, two probes are scheduled to arrive at Jupiter's icy ocean moons to investigate habitability and signs of life.

A depiction of what Europa's internal structure could look like (not to scale)

- NASA's Europa Clipper should arrive in 2030, carrying sophisticated instruments designed to investigate the ocean moon.
- Clipper carries a small triangular metal plate etched with the waveforms for the word "water" in 103 languages.
- In 2031, the JUICE mission (JUpiter ICy moons Explorer) will inspect three of the four Galilean moons of Jupiter: Callisto, Ganymede, and Europa.
- Unlike the other three Galilean moons, the volcanic world of Io does not harbor any water or water ice.
- JUICE will be the first probe to orbit the moon of another planet beyond Earth.

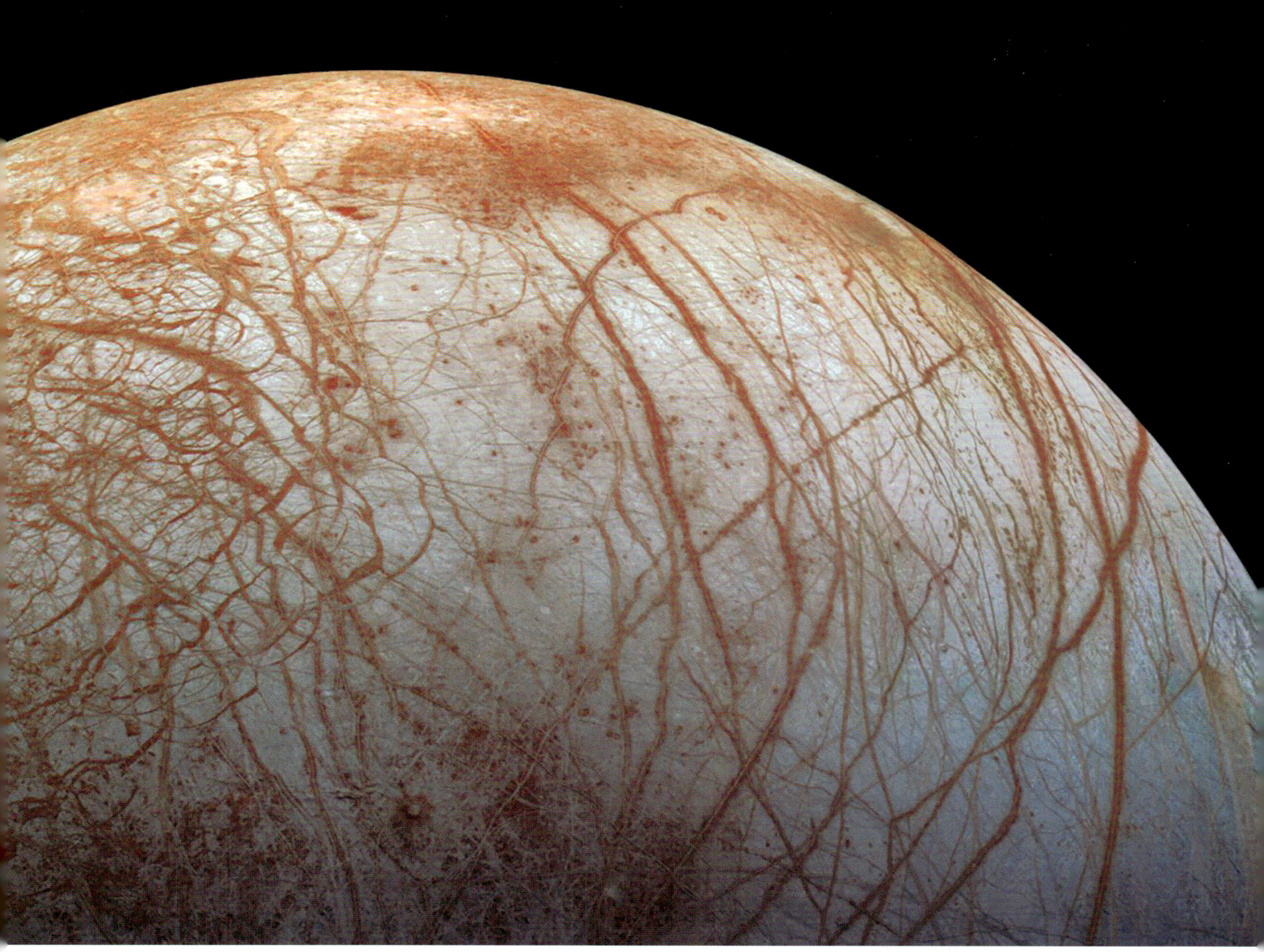

NASA's Galileo spacecraft took this photo of Jupiter's Europa moon in the late 1990s; reprocessing it now reveals even more detail.

✦ CANDIDATE FOR LIFE: WASP-39 B

- In 2022, scientists using the James Webb Space Telescope (JWST) discovered the first evidence of photochemistry—chemical reactions necessary for life on Earth—in the atmosphere of an exoplanet.
- WASP-39 b was first discovered in 2011.
- It is located about 700 light-years away from Earth, so conditions we observe today are actually 700 years old.
- It is a gas giant orbiting a G-type star (like our Sun).
- WASP-39 b's mass is similar to Saturn's and just over a quarter of Jupiter's.
- It orbits eight times closer to its star than Mercury does to the Sun.
- WASP-39 b is fast: It circles its star in just 4.1 Earth days.
- Observations by both the Hubble and Spitzer telescopes have found water vapor, sodium, and potassium in the exoplanet's atmosphere.
- **JWST IDENTIFIED CARBON DIOXIDE IN WASP-39 B'S ATMOSPHERE—THE FIRST DISCOVERY OF CARBON ON AN EXOPLANET.**
- Sulfur dioxide was also detected, likely produced by photochemical reactions involving high-energy light from the exoplanet's star.

"IT'S LIKE GETTING TO OPEN 15 DOORS ALL AT ONCE. THE HARD PART IS KNOWING WHICH ONE TO GO THROUGH FIRST."

—SCIENTIST DAVID SING ON THE POWER OF STUDYING EXOPLANETS WITH THE JAMES WEBB SPACE TELESCOPE

CURRENT FACT COUNT 4854

REACHING OUT

Humans have sent multiple messages into space, via spacecraft and radio signals, in hopes of broadcasting our existence to potential alien species.

- We first attached a golden plaque etched with symbols to both Pioneer 10 and 11, launched in 1972 and 1973.
- Then we included a golden record with sounds, images, and symbols from Earth on both Voyager spacecraft, launched in 1977.

Enrico Fermi (1946)

- When it comes to sending messages to potential alien species, however, radio transmissions are orders of magnitude faster and cheaper than attaching plaques and golden records to long-distance spacecraft.
- In 1974, we beamed a three-minute radio signal in binary code toward globular cluster Messier 13. It included information about human DNA and our solar system.
- Our messages, whether visual or coded, presume science and mathematics to be a common, decipherable language for any technically advanced civilization.

✦ THE QUESTION OF THE FERMI PARADOX

- Astronomer Enrico Fermi was known as the architect of the nuclear age and the inventor of the first nuclear reactor.
- In 1950, Fermi posed a conundrum to his colleagues during their lunch break at the Los Alamos National Laboratory in New Mexico.
- **HE ASKED A SIMPLE QUESTION, NOW CALLED THE FERMI PARADOX: "WHERE IS EVERYBODY?"**
- In other words, "Where are all the aliens, and why have we found zero evidence of their existence?"
- **THE FACT THAT WE HAVE YET TO DISCOVER EXTRATERRESTRIAL LIFE IS A STAGGERING STATEMENT WHEN YOU COLLECTIVELY CONSIDER CERTAIN FACTS, NAMELY:**
    - The observable universe holds an estimated two trillion galaxies.
    - Each galaxy contains an estimated 100 billion stars on average.
    - Each star might host one planet, on average.
    - Chemistry and physics govern the entire universe with the same laws that breathed life into human civilization.
    - The universe is about 14 billion years old—ample time for incalculable planets to have formed and for civilizations to have developed.
- Humanity evolved from multicellular microorganisms into a tool-wielding, war-waging, cave-painting species in two billion years.
- We progressed from an agricultural revolution to an industrial revolution in 12,000 years.
- We then progressed from steam power to nuclear power in 150 years.
- **UNDERLYING THE FERMI PARADOX IS THIS QUESTION: WHAT MIGHT AN INTELLIGENT CIVILIZATION BE CAPABLE OF IF GIVEN THOUSANDS, MILLIONS, OR EVEN BILLIONS MORE YEARS THAN HUMANITY TO DEVELOP?**
- Surely, given such timescales and opportunities, we would at least be aware of some evidence of extraterrestrial intelligence, if not bombarded by their presence.

The Milky Way brightens the sky above Lake Tahoe, Nevada.

CURRENT FACT COUNT 4915

IMAGINING THE POSSIBILITIES

Modern technologies would appear as magic to Earthlings a mere few hundred years ago. What about on timescales of the universe—what would technology look like in hundreds of thousands of years? Millions of years? What about a society that has existed for billions of years?

✦ THE DRAKE EQUATION

- One theoretical estimate of the number of advanced societies in the Milky Way galaxy was proposed by astronomer Frank Drake in 1961. The Drake equation is: $N = R_* \times f_p \times n_e \times f_l \times f_i \times f_c \times L$.

- **THE DRAKE EQUATION'S VARIABLES ARE AS FOLLOWS:**
  - N: the number of civilizations in the Milky Way galaxy with technology to send detectable signals
  - R_*: the rate of formation of stars suitable for developing intelligent life
  - F_p: the fraction of those stars with planetary systems
  - N_e: the number of planets, per star system, with an environment suitable for life
  - F_l: the fraction of suitable planets on which life actually appears
  - F_i: the fraction of life-bearing planets on which intelligent life emerges
  - F_c: the fraction of civilizations that develop a technology that produces detectable signs of their existence
  - L: the average length of time civilizations produce such signs

- If we solve for the Drake equation with our current best guesses about the formation and prevalence of worlds in the universe, then there should exist plenty of alien civilizations with technologies so sufficiently advanced that they are indistinguishable from magic to us.

American astronomer Frank Drake demonstrates his Drake equation.

✦ VON NEUMANN THEORY

- Von Neumann devices, or self-replicating robots, would be the most efficient way to explore and colonize a galaxy.

- They are named for physicist John von Neumann who, among many other feats, analyzed mathematics models of self-replication in nature.

- The concept is that a space probe sent into space with the capability of building a clone of itself using in situ materials (materials found on a visited celestial body) would populate that celestial body at an exponential rate with new clones that would then depart on their own journey to discover more celestial bodies and repeat the same process.

- At the exponential rate of growth and discovery that von Neumann probes are hypothetically capable of, any probes sent by other civilizations should have already been discovered by humanity.

- Some astronomers have taken this lack of discovery as evidence for the lack of extraterrestrial life.

A first edition of *Ozma of Oz* (1907)

The 85-foot Tatel radio telescope at the National Radio Astronomy Observatory in West Virginia

✦ THE WIZARD OF OZ

- Project Ozma—the first official search for extraterrestrial intelligence using radio wave signals—heralded a new field of astronomy.

- It was led by Frank Drake in 1960.

- **DRAKE NAMED PROJECT OZMA AFTER THE FICTIONAL WORLD OF L. FRANK BAUM'S *WIZARD OF OZ* BOOKS.**

- Baum kept the fantastical whimsy of his fictional world alive through the playful claim that he was not an author of novels, but rather the first Royal Historian of Oz, tasked with documenting the accounts of the realm.

- As Royal Historian, so Baum claimed, he communicated with Oz via radio waves.

- Drake described a possible real Oz as "a land far away, difficult to reach, and populated by strange and exotic beings," much like a hypothetical alien planet.

- Princess Ozma, the rightful ruler of the land of Oz in the series, was a source of inspiration for Queen Padmé Amidala of the *Star Wars* prequel trilogy.

"THERE COULD WELL BE MESSAGES FROM THE STARS FLYING RIGHT THROUGH THIS ROOM. THROUGH YOU AND ME."

—ASTRONOMER FRANK DRAKE

Atomic weapons and other destructive forces may prevent aliens from revealing themselves to us.

✦ SOLVING THE FERMI PARADOX

- The first searches for extraterrestrial intelligence coincided with the height of the Cold War, an era defined by fear of nuclear weapons and the apocalyptic consequences of nuclear war.

- **EARTHLINGS HAVE NOW AMASSED ENOUGH NUCLEAR WEAPONS TO ERADICATE OURSELVES MANY TIMES OVER.**

- With these self-annihilating contemplations coloring the cultural landscape, astronomers wondered if intelligent civilizations eradicate themselves soon after inventing radio communication.

✦ THE KARDASHEV SCALE

- In 1964, Soviet astrophysicist Nikolai Kardashev proposed a scale to measure a civilization's level of technological advancement based on how efficiently they can harness energy.

- Level 1: Planetary civilizations can harvest the entirety of the energy provided by their home planet.

- Level 2: Stellar civilizations can harvest all the energy provided by their star.

- Level 3: Galactic civilizations can harvest the energy of an entire galaxy.

- **ON THE KARDASHEV SCALE, EARTHLINGS ARE AT LEVEL 0.**

- The United Federation of Planets in the *Star Trek* franchise has achieved level 2.

- A level 2 civilization might utilize a megastructure called a Dyson sphere, named for the physicist Freeman Dyson, to fully exploit their star.

- This hypothetical structure is made of a network of orbiting crafts engulfing the star. The crafts work together to absorb and reemit nearly all the star's energy output.

✦ THE COPERNICAN PRINCIPLE

- In 1993, J. Richard Gott III, a Princeton astrophysicist, approached questions of extraterrestrial life by applying what he dubbed the Copernican principle: If we are not special in the universe, then we are representative of other civilizations, and since we can't (yet) travel to the stars, then neither can aliens.

✦ THE GREAT FILTER HYPOTHESIS

- In 1998, Robin Hanson, an economist at George Mason University, proposed the Great Filter hypothesis: If advanced civilizations exist within our own galaxy, then surely we would witness evidence of their technological artifacts, be they robots or spaceships—yet we've seen no such thing.

- There must then be some great filter beyond which few, if any civilizations, can pass.

- It might arise early in the evolution of life, preventing life from forming at all or from becoming complex. Or it might arise much later, preventing an already intelligent civilization from colonizing space. Might such a great filter ultimately limit humanity?

✦ THE RARE EARTH HYPOTHESIS

- Perhaps complex, intelligent life is exceedingly rare in the universe. The conditions contributing to life's proliferation on our planet are so profoundly unique that similar circumstances rarely, if ever, occur anywhere else.

✦ THE BERSERKER OR DEADLY PROBES HYPOTHESIS

- We have detected no alien civilizations simply because none remain. In their efforts to colonize the galaxy, von Neumann–type devices destroyed every other civilization.

- Civilizations may inevitably destroy themselves too early to achieve galactic dominance.
- Or they may destroy one another too soon to achieve galactic dominance.
- Because we have not yet encountered a von Neumann device, either the Berserker hypothesis is wrong or Earth is next.

✦ THE DARK FOREST HYPOTHESIS

- Perhaps civilizations hide from detection from fear of discovery by other advanced—and hostile—civilizations.
- The name for this theory comes from the novel *The Dark Forest* by Cixin Liu.
- To counter this hypothesis, consider that civilizations don't venture far from their home worlds. Based on what we know of humanity's innate urge to explore, and considering the four probes we have already dispatched into interstellar space, this solution seems improbable.

✦ THE ZOO HYPOTHESIS

- Intelligent aliens exist and are aware of Earthlings, but we have not detected them because they are silently observing us.
- **WE ARE LIKE ANIMALS IN A ZOO, LIVING UNDER THEIR WATCH WITHOUT OUR KNOWLEDGE.**
- These advanced civilizations operate under a prime directive to avoid interference with our development, much like an ecologist studying the behavior of animals without interfering.

✦ THE PANSPERMIA HYPOTHESIS

- The fifth century B.C. Greek philosopher Anaxagoras wondered whether the first seeds of life fell from the heavens to Earth.
- The name "panspermia" is derived from the ancient Greek word *pan*, meaning "all," and *sperma*, meaning "seed."
- The idea was modernized in 1903 by Svante Arrhenius, who proposed that microscopic spores travel through space thanks to the Sun's radiation pressure.
- The hypothesis suggests that life was delivered to Earth billions of years ago as microbial organisms aboard one or more comets or meteorites, like cosmic hitchhikers.
- According to the hypothesis, life existed before Earth began and proliferated throughout the galaxy through the same means.
- **IF THE PANSPERMIA HYPOTHESIS IS CORRECT, THEN WE SHOULD FIND TRACES OF MICROBIAL LIFE ON OTHER PLANETS OR MOONS IN OUR SOLAR SYSTEM.**
- Experiments on the International Space Station and other orbiting craft have confirmed that microbes can survive in space.
- This new iteration of the panspermia hypothesis guides some studies of Mars and the moons of Jupiter and Saturn.

✦ THE WOW! SIGNAL

- The most compelling indication that aliens have attempted communication with Earth is known as the Wow! signal.
- In 1977, astrophysicist Jerry Ehman, a volunteer at the Big Ear Radio Observatory in Ohio, was analyzing transmission data when he noticed something extraordinary.
- The observatory picked up a 72-second-long signal that aligned with all our expectations of how extraterrestrials might successfully communicate with Earth.
- Ehman circled the data in red pen and wrote "Wow!"
- Despite many theories to explain its origin, none has been widely accepted.

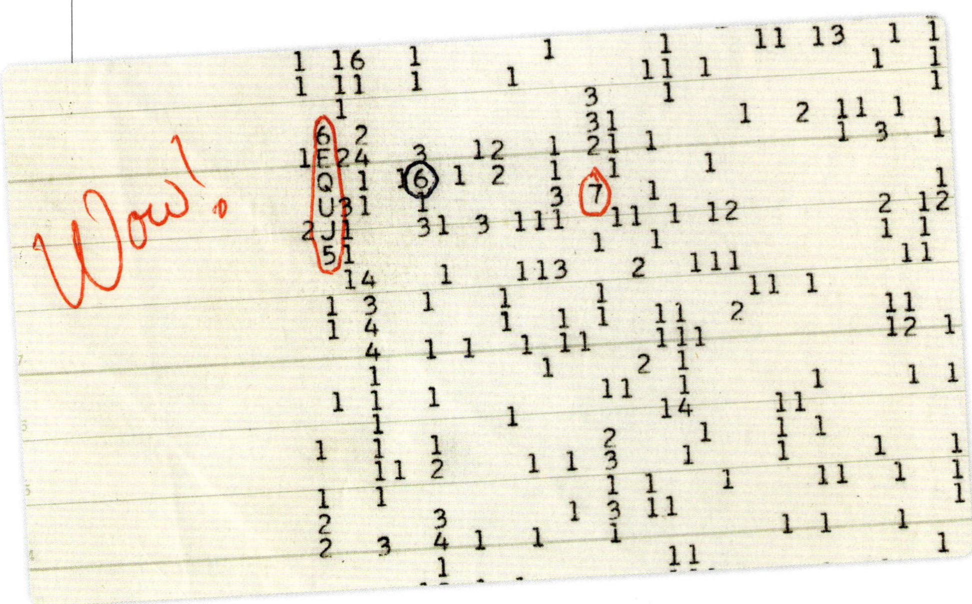

Ehman's note, which inspired the name for the Wow! signal theory

"THE COPERNICAN REVOLUTION TAUGHT US THAT IT WAS A MISTAKE TO ASSUME, WITHOUT SUFFICIENT REASON, THAT WE OCCUPY A PRIVILEGED POSITION IN THE UNIVERSE."

—J. RICHARD GOTT III, "IMPLICATIONS OF THE COPERNICAN PRINCIPLE FOR OUR FUTURE PROSPECTS"

The young star cluster NGC 602
in the Small Magellanic Cloud

CURRENT FACT COUNT 4950

HOLLYWOOD HYPOTHESES

Humans have yet to bridge communication with our closest evolutionary ancestors, the great apes, with whom we share about 98 percent of our DNA. Can we realistically hope to communicate with an intergalactic alien? What's more, can we realistically assume they would care what we have to say?

- An unsolved mystery of Hollywood sci-fi: How is it that intelligent aliens can traverse thousands or millions of light-years yet always manage to crash-land on the third rock from the Sun?

- Sci-fi writers like to think of silicon-based life as an alternative to carbon. But that's not a random choice. Silicon sits just below carbon on the periodic table, indicating that silicon forms the same families of molecules that carbon does.

Linguists and scientists work to decode the nonlinear language in *Arrival*.

- Following the same pattern, tin lies directly below silicon. Maybe the Tin Man isn't the most far-fetched creature in the land of Oz.

- **CARBON CAN COMBINE WITH OTHER ATOMS TO MAKE MORE KINDS OF MOLECULES THAN ALL OTHER KINDS OF MOLECULES COMBINED.**

- The sticky nature of carbon makes it ideal as the foundation of chemical experiments that would lead to self-replicating life.

- There's 10 times as much carbon as silicon in the universe, so there's no need to imagine extraterrestrial life based on silicon. There's plenty of carbon for this purpose.

✦ MYSTERIOUS *ARRIVAL*

- If aliens appeared tomorrow, as they do in the sci-fi film *Arrival*, based on the novella "Story of Your Life" by Ted Chiang, how might we communicate? In the story, a linguist and physicist are recruited to crack the aliens' language. But if aliens do arrive, let's consult the foremost astrobiologists *first*.

- *Arrival* explored how we might translate the language of a species we know nothing about. The peaceful aliens in *Arrival* communicate with circular symbols representing not words or letters but entire ideas.

- Unlike humans, the aliens, called heptapods, experience time without beginning or end, as a circle rather than as a sequence of events. To write in the heptapod language, one must remember the future and foresee the past.

✦ MAKING *CONTACT*

- The 1997 film *Contact*, based on Carl Sagan's novel, explores the Fermi paradox.

- In *Contact*, an advanced alien civilization has attempted to communicate with Earth via signals transmitted on an electromagnetic frequency we could not detect.

- In our current Search for ExtraTerrestrial Intelligence (SETI) efforts, we transmit signals in radio waves, as longer frequencies can penetrate our atmosphere to travel vaster distances without losing much energy or information along the way.

✦ E.T., WHAT ARE YOU?

- In Steven Spielberg's 1982 sci-fi film *E.T. the Extra-terrestrial*, the charmingly creepy alien protagonist is classified, according to Spielberg, as a vegetable, not an animal.

- E.T.'s existence as an animate vegetable accounts in part for its ability to rejuvenate wilting plants with its finger.

- A new species of sponge discovered in the Pacific, *Advhena magnifica*, was given the common name "E.T. sponge" because it looks so much like the film's protagonist.

✦ *STAR TREK*'S ALIENS

- The *Star Trek* universe spans myriad galaxies and explores alien civilizations of all kinds.
- Most *Star Trek* aliens, including the Vulcans, Klingons, and Cardassians, appear as bipedal humanoids with faces like ours.
- But *Star Trek* offers a creative explanation for their otherwise unimaginative aliens: their own take on the panspermia hypothesis.
- A federation archaeologist discovers all humanoid aliens share a common ancestor: the highly advanced, intelligent Progenitors, who seeded their DNA across many worlds.

Leonard Nimoy (left) as Mr. Spock and John Colicos (right) as Kor in a 1967 episode of *Star Trek.*

✦ MAYBE IT'S THE MATRIX

- It would take tens of thousands of years to reach the nearest exoplanet with current technology.
- Assuming aliens face the same biological constraints humans do (eating, sleeping, finite life spans), they will face the same challenges of long-term spaceflight.
- Galaxy-traversing aliens either know how to make wormholes, travel near light speed, and keep their flesh on ice and consciousness intact for indeterminate lengths of time ... or they're not made of flesh at all.
- If intelligent aliens wished to conquer Earth, there'd be little we could do to stop them. After all, an ant can do little to prevent your foot from smooshing it if you want to.
- If a conquering extraterrestrial civilization did consist of artificially intelligent creatures, who's to say we're not living in a *Matrix*-like simulation even as you read this?
- Our grasp of artificial intelligence (AI) remains nearly as rudimentary as our understanding of consciousness. Manipulating a human mind might be as easy for an advanced civilization as it is for gamers to manipulate their favorite virtual characters.
- **THE SIMULATION ARGUMENT BY SWEDISH PHILOSOPHER NICK BOSTROM POSITS THAT AT LEAST ONE OF THREE REALITIES IS TRUE.**
  - Reality #1: Nearly all civilizations go extinct before becoming technologically advanced enough to create true simulated realities.
  - Reality #2: If and when civilizations reach advanced technological maturity, they lose interest in creating simulated realities, or lose interest in creating simulations of their own histories.
  - Reality #3: We live in a simulation.
- If the first two are false, Bostrom claims, the third must be true.
- If some nontrivial number of civilizations reach technological advancement, and some nontrivial proportion of those civilizations run simulations of their own histories, then billions of simulations should exist for every one original biological reality.
- If so, a rational assumption would be that we are a simulation—just one among trillions—rather than that we beat the odds of billions (or more) to one.
- The stark lack of probability and statistics courses offered in schools, coupled with the human brain's innately inadequate grasp of these concepts, leaves us susceptible to irrational thinking, a phenomenon that casinos prey upon. Your odds of guessing the number a roulette wheel lands on are one in 38 (just 2.63 percent) every time.
- AI is progressing at an exponential rate. Some worry we'll soon be unable to distinguish AI from human thought or likeness. Should we consider, or assume, that an advanced alien civilization might also be indistinguishable from AI?
- Futurists have proposed the possibility of a post-biological society that has evolved beyond biological constraints. Post-biologics achieve immortality through AI. This might look like human consciousness uploaded into cyborgs—or the annihilation of human consciousness entirely in favor of a new AI species.

CURRENT FACT COUNT 5000

ENDURING MYSTERIES

As the area of our knowledge grows, so too does the perimeter of our ignorance.

- Robert Boyle, the 17th-century scientist foundational to modern chemistry and the scientific method, once wrote one of the most ambitious to-do lists in history.

- The nameless, undated list consisted of 24 unsolved problems or yet-to-be-invented technologies that he believed science would one day accomplish.

- It begins with the simple and enduring problem of prolonged life and ends with a perfume activated by rubbing. The full list is as follows:

- ***1. "The Prolongation of Life":*** In mid-17th-century Europe, when and where this list was written, the average life expectancy was about 35 years. Today, an average European's life expectancy is about 75 years.

- ***2. "The Recovery of Youth, or at least some of the Marks of it, as new Teeth, new Hair colour'd as in youth":*** Veneers, hair implants, and face lifts? Check.

- ***3. "The Art of Flying":*** Airplanes, paragliders, and rockets? Check.

- ***4. "The Art of Continuing long under water, and exercising functions freely there":*** SCUBA diving? Check. Deep-sea submersibles? Check.

- ***5. "The Cure of Wounds at a Distance":*** This desire was rooted in the alchemical and metaphysical beliefs of Boyle's time.

- ***6. "The Cure of Diseases at a distance or at least by Transplantation":*** Organ transplants? Check.

- ***7. "The Attaining Gigantick Dimensions":*** Boyle may have been referencing architecture, although it's unclear. If so, we can check off towering skyscrapers.

- ***8. "The Emulating of Fish without Engines by Custome and Education only":*** Free diving? We checked that off thousands of years before Boyle's time.

- ***9. "The Acceleration of the Production of things out of Seed":*** Genetic crop modifications to accelerate growth? Check.

- ***10. "The Transmutation of Metalls":*** Boyle was referencing the alchemical belief in changing one element to another. We may not be able to turn lead into gold, but radioactive decay does fundamentally alter an element. Check.

- ***11. "The makeing of Glass Malleable":*** This is a reference to a legend of "unbreakable glass." We're still working on this.

- ***12. "The Transmutation of Species in Mineralls, Animals, and Vegetables":*** Boyle may have been hoping for a magical, alchemical answer, but Darwin provided the closest solution: evolution by natural selection.

- ***13. "The Liquid Alkaest and Other dissolving Menstruums":*** Another alchemical reference, this time to a universal solvent that could dissolve anything. Sorry, Boyle. No check. And how could one store it, anyway?

- ***14. "The making of Parabolicall and Hyperbolicall Glasses":*** Curved lenses, such as those used in Galileo's telescopes, were grainy and flawed. We have long since perfected them. Check.

- ***15. "The making Armor light and extremely hard":*** It's no Elvish mithril of Middle-earth, but we do have bulletproof vests and glass. Check.

- ***16. "The practicable and certain way of finding Longitudes":*** The challenge of calculating accurate longitude was finally solved in 1773. Today, we simply rely on GPS satellites. Double check.

- ***17. "The use of Pendulums at Sea and in Journeys, and the Application of it to watches":*** Physicist Christiaan Huygens invented the first pendulum clock around the time this list was written. Although it was the best way to tell time until the invention of quartz clocks in 1927, pendulum clocks were and are useless on a rocking ship. Partial check.

- ***18. "Potent Druggs to alter or Exalt Imagination, Waking, Memory, and other functions, and appease pain, procure innocent sleep, harmless dreams, etc.":*** Prescription and over-the-counter painkillers and sleep aids are widely available. Check.

- ***19. "A Ship to saile with All Winds, and A Ship not to be Sunk":*** A motorized boat requires zero wind. But all boats are vulnerable to sinking. Partial check.

- ***20. "Freedom from Necessity of much Sleeping exemplify'd by the Operations of Tea and what happens in Mad-Men":*** Humans still need around eight to 10 hours of sleep per night, but we can temporarily stave off sleep with the help of potent stimulants. Check.

- ***21. "Pleasing Dreams and physicall Exercises exemplify'd by the Egyptian Electuary and by the Fungus mentioned by the French Author":*** This is likely a reference to hallucinogenic plants, but the exact meaning is unclear. This merits a partial check.

- ***22. "Great Strength and Agility of Body exemplify'd by that of Frantick Epileptick and Hystericall persons":*** While performance-enhancing drugs can aid in building muscle, they come at a high risk and possible long-term detriment. A partial check.

- ***23. "A perpetuall Light":*** The electric light bulb fulfills this one. Boyle would have delighted in the garlands of LED bulbs festooning many houses and trees throughout the month of December. Check.

- ***24. "Varnishes perfumable by Rubbing":*** Scratch and sniff? Check.

- Many of Boyle's wishful tech has emerged, while we're still in pursuit of others. What questions might appear on our list today, questions Boyle and his contemporaries might never have thought to ask?

✦ STUFF WE DON'T KNOW ... YET

- No one has ever measured the size of electrons. As far as we know, they could be infinitely small.

- We have no idea how large our universe extends beyond our cosmic horizon. It could be infinite.

- We measure the existence of what we call dark matter, which makes up 85 percent of the gravity in the universe. But we have no idea what's causing it.

- **WE DON'T KNOW FOR SURE WHAT'S ON THE OTHER SIDE OF A BLACK HOLE'S EVENT HORIZON.**

- We have no idea how Earth went from organic molecules to self-replicating life.

- Quarks always come in pairs. We've never seen an isolated quark. When we try to pull them apart, the energy required spontaneously re-pairs them.

- We have no idea why the fundamental particles of the universe have the masses they do.

- The expansion of the universe is accelerating. We can measure this "dark energy" but have no idea what's causing it.

- We still don't know what broke the symmetry between matter and antimatter in the early universe, creating a cosmos of matter and photons.

- The temperature of the universe, before stars and galaxies formed, varied by no more than one part in 100,000. We are still testing the theory that predicts why this is so.

- Most tested predictions of quantum physics offer no intuitive understanding for why particles behave the way that they do.

- There remains no agreement on how the planet Uranus got its rotation axis tilted 98 degrees from the vertical—far more than the tilt of any other planet.

- **IF THE EFFECTS OF DARK ENERGY CONTINUE UNABATED, IN 10^{22} YEARS, THE UNIVERSE WILL EXPAND SO FAST THE FABRIC OF SPACETIME WILL RIP. WE DO NOT KNOW IF THAT FATE AWAITS FUTURE GENERATIONS OF LIFE IN THE UNIVERSE.**

✦ THE ULTIMATE QUESTION

- How did life begin?

- Are we alone?

- What preceded the Big Bang?

- **IS THE ANSWER TO THE ULTIMATE QUESTION OF LIFE REALLY 42? ACCORDING TO THE FICTIONAL SUPERCOMPUTER IN THE NOVEL *THE HITCHHIKER'S GUIDE TO THE GALAXY* BY DOUGLAS ADAMS, IT IS.**
  - 42: the number of times you could fold a piece of paper before it reached the Moon
  - 42: the degrees of the angle that light reflects off a rainbow toward a viewer's eyeball
  - 42: the atomic number of the element molybdenum, a common additive to make alloy metals stronger and more heat resistant, favored by the aerospace industry to make rocket nozzles and spaceship nose cones
  - 42: the number of minutes you'd spend falling through the center of Earth to the opposite end (negating air resistance and your inevitable vaporization within the core)
  - 42nd: the street in New York City that's home to the main branch of the New York Public Library. Perhaps its cumulative contents actually hold the answer to the ultimate question of life.

"ONE CANNOT HELP BUT BE IN AWE WHEN HE CONTEMPLATES THE MYSTERIES OF ETERNITY, OF LIFE, OF THE MARVELOUS STRUCTURE OF REALITY."

—ALBERT EINSTEIN, *LIFE* MAGAZINE

ACKNOWLEDGMENTS

Every page of this book reflects a collaborative effort of creativity and careful craftsmanship.

We are grateful, once again, to the team at National Geographic Books, who work tirelessly on our shared and ongoing mission to bring the universe down to Earth. We thank executive editor Hilary Black, our flight director keeping us on course and ever ensuring that failure is not an option, and senior editor Susan Tyler Hitchcock, for her invaluable creative and scientific instincts and command of the written word. We also thank editorial project manager Ashley Leath, whose meticulous eye ensured no detail went overlooked.

A hard-earned thank-you to the publishing team at National Geographic Books: Nicole Miller Roberts, Sanaa Akkach, TJ Tucker, Meredith Wilcox, Jerome Cookson, Michael O'Connor, and Heather McElwain. And thank you to Yehrin Tong for the fabulous cover illustration.

Thanks to George Chudley and Lane Unsworth for their painstaking fact-checking (no simple feat for a book of 5,000 facts).

And finally, we thank the lifelong learners—the readers, viewers, and listeners of *StarTalk*—whose ignited curiosity kindles the flame of our enduring mission.

ABOUT THE AUTHORS

Neil deGrasse Tyson is an astrophysicist and Frederick P. Rose Director of the Hayden Planetarium at New York's American Museum of Natural History. He is the author of more than a dozen books, many of them international bestsellers, and numerous articles, both scholarly and for the general public. He is the host of *StarTalk,* a podcast, and of two seasons of *Cosmos,* televised by Fox and National Geographic. He has received 27 honorary doctorates as well as NASA's Distinguished Public Service Medal. He and his wife live in New York City.

Lindsey Nyx Walker is the senior producer and head writer at *StarTalk,* a popular science podcast and a four-time Emmy-nominated television show on National Geographic. She collaborated on both *Cosmic Queries* and the *New York Times* bestseller *To Infinity and Beyond,* the second and third books, respectively, in the publishing collaboration between *StarTalk* and National Geographic Books. Walker lives with her partner in Manhattan. Visit her on TikTok @spacetok.

CONTRIBUTORS

Named for the first Apollo moon landing, **Luna Shyr** has been a staff editor and writer at *National Geographic,* the *Wall Street Journal,* and Associated Press. With curiosity and the joy of exploration as her fuel, she covers a multitude of topics, from space travel, meteor craters, whale skeletons, and Antarctic vegetation to cathedral architecture, British gardening, and artistic collaboration in the performing arts. She studied art history and premedical science at Princeton University before earning a master's degree in broadcast journalism at New York University.

James Trefil is the Clarence J. Robinson Professor of Physics at George Mason University. He is the author of more than 60 books about science for general readers, including *National Geographic Space Atlas: Mapping the Universe and Beyond,* and co-author of *Cosmic Queries: StarTalk's Guide to Who We Are, How We Got Here, and Where We're Going* and *The Next Big Thing: Innovations for a Better, Smarter, Stronger Tomorrow.* He and his wife live in Fairfax, Virginia.

Maya Wei-Haas is a freelance science writer with a particular affection for rocks and reactions. She's published in a variety of outlets, including the *New York Times, National Geographic,* and Smithsonianmag.com. Maya holds a bachelor's degree in geology from Smith College and a Ph.D. in earth science from the Ohio State University. She made the jump to journalism as a fellow for the AAAS Mass Media program. She is the author of a children's book, *What a Rock Can Reveal,* and was a contributing writer to *National Geographic Stargazer's Atlas.*

MAPS AND ILLUSTRATIONS CREDITS

Maps

80–84 The Northern Sky/The Southern Sky: Sources: NASA/JPL; International Astronomical Union (IAU); Space Telescope Science Institute; ESA/GAIA/DPAC; Tycho Catalog Skymap, NASA/Goddard Space Flight Center Scientific Visualization Studio. **132–33 Cosmic Journeys:** Images: NASA Goddard Space Flight Center; NASA/JPL-Caltech; Johns Hopkins University Applied Physics Laboratory; Carnegie Institution of Washington; ESA; JAXA. **158–59 You Are Here:** Art: Antoine Collignon. Sources: Guillermo Gonzalez, Ball State University; Michael Gowanlock, Northern Arizona University; Icarus; *Astrobiology;* NASA/JPL; *International Journal of Astrobiology.* **196–97 Map of the Moon:** Global mosaic: Lunar Reconnaissance Orbiter, NASA, Arizona State University. Lunar landings source: NASA Goddard Space Flight Center. **218–19 Europa and Its Ocean:** Art: Dana Berry. Sources: Steven D. Vance, NASA/JPL; NASA. **226–27 Enceladus and the E Ring:** Art: Dana Berry. Sources: Steven D. Vance, NASA/JPL; NASA. **228–29 Titan and the Octocopter:** Art: Dana Berry. Sources: Steven D. Vance, NASA/JPL; Steve Gribben, Johns Hopkins University Applied Physics Laboratory; NASA. **230–31 Oceans Beyond Earth:** Art: Dana Berry. Sources: Steven D. Vance, NASA/JPL; NASA. **302–303 Expanding Galaxies:** Art: Ken Edward. **320–21 A Cosmic Story:** Art: Mark Garlick. Sources: Enrico Garaldi, Max Planck Institute for Astrophysics; Kevin Hainline, University of Arizona; Erica Nelson, University of Colorado Boulder.

Illustrations

2–3, NASA, ESA, CSA, STScI; 4, ESO/Y. Beletsky (LCO; 6, NASA/Bill Ingalls; 8–9, Image: NASA, ESA, CSA, STScI, Yu Cheng (NAOJ); Image Processing: Joseph DePasquale (STScI); 10–1, Rogelio Bernal (deepskycolors); 12, NASA Earth Observatory image by Joshua Stevens, using data from NASA/METI/AIST/Japan Space Systems, and the U.S./Japan ASTER Science Team; 13, Chris Butle/Science Source; 14, mantisdesign/Shutterstock; 15, Science & Society Picture Library/Getty Images; 16, Image taken by Luke Thompson from Chisholm Lab and Nikki Watson from Whitehead, MIT, 2007; 17 (UP), Hans Wismeijer/iStock/Getty Images; 17 (LO), Herb Ball/NBCU Photo Bank/NBCUniversal via Getty Images; 20–1, NASA; 22, Maxar via Getty Images; 23, Alan Dyer/Stocktrek Images/Getty Images; 24–5, Caryn Hill Photography/Science Source; 28, zabanski/Shutterstock; 29, NASA via AP; 30, Visual China Group via Getty Images; 32–3, Jill Schneider/National Geographic Image Collection; 36, Phil Watson Photography/Shutterstock; 37 (UP), Carlos Mamani/AFP via Getty Images; 37 (LO), AP Photo/Thanassis Stavrakis; 38–9, Rogelio Bernal (deepskycolors); 40, Galilei, Galileo (1564–1642)/Italian/Biblioteca Nazionale Centrale, Florence, Italy/© Iberfoto/Bridgeman Images; 41, Donald Smith/Alamy Stock Photo; 42, Biblioteca Medicea Laurenziana, Florence/© Raffaello Bencini/Bridgeman Images; 43, SheraleeS/iStock/Getty Images; 44–5, Stephen Chung/Alamy Live News; 46,

Milamai/Moment/Getty Images; 47, illpaxphotomatic/Shutterstock; 48, W. & F. Langenheim/Metropolitan Museum of Art/Gilman Collection, Gift of The Howard Gilman Foundation, 2005; 49, Brian Cassella/Chicago Tribune/Tribune News Service via Getty Images; 50–1, Colleen Pinski; 52, NASA/JSC/Stanford University/Science Source; 53, NASA via Alamy Stock Photo; 54–5, Detlev van Ravenswaay/Science Source; 58–9, Framalicious/Shutterstock; 60–1, Robert E. Peary Collection/National Geographic Image Collection; 62, Science Museum, London, UK/SSPL/UIG/Bridgeman Images; 63, Yuri Beletsky; 64–5, Wally Pacholka/Astropics; 67, Science Source; 68–9, Wally Pacholka/Astropics; 70, piemags/Alamy Stock Photo; 71, Gogh, Vincent van (1853–90)/Dutch/Museum of Modern Art (MoMA), New York, USA/Bridgeman Images; 72, Jim Patterson/TandemStock; 75, Alan Dyer/VW Pics//Science Source; 78, John R. Foster/Science Source; 79, *Emu Sky*, New South Wales, 2008/Bates, Badger (William Brian) (b.1947)/Art Gallery of South Australia, Adelaide, Acquisition through Tarnanthi: Festival of Contemporary Aboriginal & Torres Strait Islander Art supported by BHP 2019/Photo © Art Gallery of South Australia/Bridgeman Images; 84, NASA; 86, Jean-Leon Huens/National Geographic Image Collection; 87, World History Archive/Alamy Stock Photo; 88, Akademie/Alamy Stock Photo; 89 (UP), Ron Miller/Stocktrek Images/Science Source; 89 (LO), CBS Photo Archive via Getty Images; 90, Bettymaya Foott; 91, IMAGO/Ou Dongqu/Alamy Stock Photo; 92, prosot-photography/iStock/Getty Images; 93, peace portal photo/Alamy Stock Photo; 96–7, NASA, ESA, CSA, and STScI; 98, NASA; 99, NASA/MSFC/David Higginbotham; 102–3, Babak Tafreshi/National Geographic Image Collection; 104, NASA; 105, NASA, ESA, STScI; 106–7, NASA, ESA, and S. Beckwith (STScI) and the HUDF Team; 108, Venus Angel/Shutterstock; 109, Craig P. Burrows; 110 (UP), Willhelm Röntgen/CBW/Alamy Stock Photo; 110 (LO), Henri Manuel, 1908/IanDagnall Computing/Alamy Stock Photo; 111, Chandra X-ray Observatory; 113, Lucasfilm/Disney; 114, Sovfoto/Universal Images Group via Getty Images; 115, near infinity studios/Alamy Stock Photo; 118, NASA; 119 (UP), Pictorial Press/Alamy Stock Photo; 119 (LO), BerS-089/iStock/Getty Images; 120–1, NASA; 122, NASA/Rick Guidice; 123, NASA/ABMA/JPL-Caltech; 124–5, japatino/Moment/Getty Images; 126, NASA/John W. Young; 127, Mark Thiessen, NGP; 128, Gilmanshin/Shutterstock; 129, NASA/Josh Valcarcel; 130 (LE), NASA/JPL-Caltech; 130 (RT), NASA/digitaleye/J Marshall – Tribaleye Images/Alamy Stock Photo; 131, NASA/JPL; 134, United States Patent and Trademark Office; 134–5, NASA/Marshall Space Flight Center; 136, Alan Dyer/Stocktrek Images/Alamy Stock Photo; 137, Photon Illustration/Stocktrek Images/Getty Images; 138, NASA/GSFC/Arizona State University; 139, NASA/Joel Kowsky; 140, NASA; 141, Anadolu via Getty Images; 142, NASA; 143, Paolo Nespoli and Roland Miller, courtesy of NASA and ASI; 144–5, NASA; 148 (UP), Smithsonian National Air and Space Museum, photograph by Mark Avino; 148 (LO), Stringer/AFP via Getty Images; 149, NASA; 150, Sovfoto/Universal Images Group via Getty Images; 151, Marka/Universal Images Group via Getty Images; 152, Nicolle R. Fuller/National Science Foundation via Wikimedia Commons; 153, Eye of Science/Science Source; 156–7, buradaki/Shutterstock; 160–1, Ron Miller/Stocktrek Images/Alamy Stock Photo; 163, JAXA/NASA; 164, James P. Blair/National Geographic Image Collection; 165, Museum of Fine Arts, Boston, Massachusetts, USA, Gift of Jeannette and Jonathan Rosen in honor of John J. Herrmann, Jr., Curator of Classical Art, 1976–2004/Photograph © 2025

Museum of Fine Arts, Boston. All rights reserved./Bridgeman Images; 170, Giambologna (Giovanni da Bologna), (1529–1608)/Flemish/Fitzwilliam Museum, University of Cambridge, UK/© Fitzwilliam Museum/Bridgeman Images; 171, NASA/Johns Hopkins University Applied Physics Laboratory/Carnegie Institution of Washington; 172–3, Alan Dyer/VW Pics/Science Source; 175, Dennis Mammana; 176, NASA/JPL; 177, NASA/JPL/USGS; 178–9, NASA/Goddard Space Flight Center/SDO; 181, Wally Pacholka/Astropics; 183, Tamara Merino/National Geographic Image Collection; 184, Patrick Kelley/National Geographic Image Collection; 185, Kateryna Kon/Science Source; 187, Barrett & MacKay/All Canada Photos/Alamy Stock Photo; 188–9, ESA/Eumetsat; 190, PRISMA ARCHIVO/Alamy Stock Photo; 191, The Planetary Society; 192–3, Stefan Seip; 198 (UP), Alexandria, Greek-Roman Museum/G. Dagli Orti/© NPL – DeA Picture Library/Bridgeman Images; 198 (LO), SilverScreen/Alamy Stock Photo; 199, NASA, ESA, and The Hubble Heritage Team (STScI/AURA); 200, Janny2/Shutterstock; 200–1, NASA/JPL-Caltech; 203, Gulliver's Travels by Jonathan Swift. Illustrated by Harry G Theaker. Published Ward, Lock, London (1928)/Lebrecht Authors/Bridgeman Images; 204, Alex McDougall-Page, Astroll; 206–7, Richard Becker/Alamy Stock Photo; 210, NASA/JPL; 210–1, Jacopin/BSIP/Alamy Stock Photo; 212–3, NASA, ESA, A. Simon (Goddard Space Flight Center), and M.H. Wong (University of California, Berkeley) and the OPAL team; 214, House of Dionysus, Kato Paphos, Cyprus/Sonia Halliday Photographs/Bridgeman Images; 215, NASA/JPL/DLR; 221, NASA/JPL/Space Science Institute; 222, NASA/JPL-Caltech/Space Science Institute; 232 (UP), Harry Taylor © Dorling Kindersley/Alamy Stock Photo; 232 (LO), Private Collection/Bridgeman Images; 233, NASA/Science Source; 234–5, NASA, ESA, CSA, STScI, J. DePasquale (STScI); 238, Image: NASA, ESA, CSA, STScI; Image Processing: Joseph DePasquale (STScI), Naomi Rowe-Gurney (NASA-GSFC); 239 and 240–1, NASA; 241, Fahad Sulehria/Stocktrek Images/Science Source; 244–5, Ahmad Sarem/iStock/Getty Images; 246–7, NASA/JPL-Caltech/UCLA/MPS/DLR/IDA; 248 (UP), Everett/Shutterstock; 248 (LO), AP Images/Nick Ut; 249, NASA/JHUAPL/SwRI/Gladstone et al./Science Source; 250, Mark Garlick/Science Source; 251, ZUMA Press/Alamy Stock Photo; 252, NASA/JPL; 253, ESO; 254–5, Ron Miller/Science Source; 256, NASA/Ames/JPL-Caltech; 257, NASA/JPL; 258–9, NASA/JPL-Caltech/STScI; 260 (UP), Kenneth Garrett/National Geographic Image Collection; 260 (LO), Beltrame, Achille (1871–1945)/Italian/Private Collection/© Look and Learn/Bridgeman Images; 261, Oselote/Shutterstock; 262, NASA, ESA, and the Hubble Heritage Team (STScI/AURA)-Hubble/Europe Collaboration, Acknowledgment: H. Bond (STScI and Penn State University); 263, Science Source; 264, ESA/Euclid/Euclid Consortium/NASA, image processing by J.-C. Cuillandre (CEA Paris-Saclay), G. Anselmi; 265, NASA, ESA/Hubble and the Hubble Heritage Team; 266–7, Ralf Kaehler and Tom Abel, Stanford University/KIPAC/SLAC; 268, ESA/Hubble & NASA, J. Kalirai, A. Milone; 269, ESA/Hubble & NASA, Acknowledgement: M. Novak; 270–1, NASA, ESA, CSA, STScI; 273, Babak Tafreshi/National Geographic Image Collection; 274, ESA/Hubble & NASA, Acknowledgement: Serge Meunier; 274–5, NASA, ESA, CSA, STScI, Brant Robertson (UC Santa Cruz), Ben Johnson (CfA), Sandro Tacchella (Cambridge), Phill Cargile (CfA); 276–7, NASA/JPL-Caltech/UCLA; 278, X-ray: NASA/CXC/Rutgers/G. Cassam-Chenai, J.Hughes et al.; Radio: NRAO/AUI/NSF/GBT/VLA/Dyer, Maddalena & Cornwell; Optical: Middlebury College/F.Winkler, NOAO/AURA/NSF/CTIO Schmidt & DSS; 279 (UP), CM Dixon/Print Collector/Getty Images; 279 (LO), Katie Dance, NGP; 280, Mark Garlick/

Science Source; 281, NASA, ESA, Caltech; 282–3, NASA, ESA, STScI; 285, John Bingham/Alamy Stock Photo; 286–7, NSF-DOE Vera C. Rubin Observatory; 289, Wally Pacholka/Astropics; 290, NASA/JPL-Caltech; 291, ESA/Hubble & NASA, M. Häberle (MPIA); 292–3, Betty-maya Foott; 294, Petr Horálek; 295, ESA/Hubble & NASA, C. Murray; 296, NASA Image Collection/Alamy Stock Photo; 298, Science Museum, London, UK/SSPL/UIG/Bridgeman Images; 299, NASA, ESA, S. Beckwith (STScI) and the Hubble Heritage Team (STScI/AURA); 300–1, Mikkel Juul Jensen/Science Source; 304, NASA, ESA, and R. Kirshner (Harvard-Smithsonian Center for Astrophysics and Gordon and Betty Moore Foundation) and P. Challis (Harvard-Smithsonian Center for Astrophysics); 305, ALMA: ESO/NAOJ/NRAO/A. Angelich. Hubble: NASA, ESA, R. Kirshner (Harvard-Smithsonian Center for Astrophysics and Gordon and Betty Moore Foundation) and P. Challis (Harvard-Smithsonian Center for Astrophysics). Chandra: NASA/CXC/Penn State/K. Frank et al.; 307, NASA, ESA, CSA, STScI; 308–9, X-ray: NASA/CXC/M.Markevitch et al.; Optical: NASA/STScI; Magellan/U. Arizona/D.Clowe et al.; Lensing Map: NASA/STScI; ESO WFI; Magellan/U.Arizona/D.Clowe et al.; 312–3, Richard Jones/Science Source; 314–5, Wladimir Bulgar/Science Source; 316–7, NSF-DOE Vera C. Rubin Observatory; 318–9, José Antonio Peñas/Science Source; 322, Victor de Schwanberg/Science Source; 323, NASA, ESA, and J. Lotz and the HFF Team (STScI); 324, Dorling Kindersley: Gary Ombler/HPS Museum of Leeds University/Alamy Stock Photo; 325, N.A. Sharp/NOIRLab/NSF/AURA; 326–7, Dabarti CGI/Shutterstock; 328, Berlanti Prod/DC ENT/Warner Bros.TV/Album/Alamy Stock Photo; 329, TCD/Prod. DB/Alamy Stock Photo; 330, CERN/Philippe Mouche; 331, Maximilien Brice, CERN/Science Source; 332, Mark Garlick/Science Source; 333, Lawrence Berkeley Laboratory/Science Source; 334–5, US Department of Energy; 336, The Yomiuri Shimbun via AP Images; 337, Martin Wolf, IceCube/NSF; 338, GL Archive/Alamy Stock Photo; 339, NASA/Christopher E. Henze/Science Source; 340–1, NASA/JPL-Caltech; 342–3, NASA/DDS/Science Source; 343, sjbooks/Alamy Stock Photo; 344 (UP), EHT Collaboration; 344 (LO) and 344–5, Dr. Lia Medeiros, University of Wisconsin, Milwaukee; 347, © Ron Miller/Novapix/Bridgeman Images; 348–9, NRAO/AUI/NSF; 350, FLHC 85/Alamy Stock Photo; 351, NASA/JPL-Caltech; 354–5, SSPL/Getty Images; 356, ESA — C. Carreau; 357, Creative Stock/Alamy Stock Photo; 358, Mark Godfrey/AIP Emilio Segre Visual Archives/Science Source; 359, Lawrence Berkeley National Laboratory/Marilyn Sargent © 2025 The Regents of the University of California, Lawrence Berkeley National Laboratory; 361, Xinhua/Shen Bohan/Alamy Stock Photo; 362–3, Jeff Gritchen/MediaNews Group/Orange County Register via Getty Images; 364, Mark Garlick/Science Source; 365, Pictorial Press/Alamy Stock Photo; 367, NOAA; 368, NASA/JPL-Caltech; 368–9, NASA/JPL-Caltech/SETI Institute; 370, AP Photo; 371, Rachid Dahnoun/Tandem-Stock; 372 and 373 (UP), Dr. Seth Shostak/Science Source; 373 (LO), Ozma of Oz, The Reilly & Britton Co, Chicago, 1907; 374, The U.S. National Archives and Records Administration; 375, courtesy of the Ohio History Connection, AL07146; 376–7, ESA/Webb, NASA & CSA, P. Zeidler, E. Sabbi, A. Nota, M. Zamani (ESA/Webb); 378, Collection Christophel/Alamy Stock Photo; 379, CBS Photo Archive via Getty Images; 380–1, X-ray: NASA/CXC/SAO/J.Drake et al., Optical: Univ. of Hertfordshire/INT/IPHAS, Infrared: NASA/JPL-Caltech; back cover (LE), NASA, ESA, CSA, STScI; (CT), NASA; (RT), NASA/JPL.

INDEX

Boldface indicates illustrations.

L

M

N

O

Since 1888, the National Geographic Society has funded more than 15,000 research, conservation, education, technology, and storytelling projects around the world. National Geographic Partners distributes a portion of the funds it receives from your purchase to National Geographic Society to support their mission to illuminate and protect the wonder of our world.

National Geographic Partners, LLC
1145 17th Street NW
Washington, DC 20036-4688 USA

Get closer to National Geographic Explorers and photographers, and connect with our global community. Join us today at nationalgeographic.org/joinus

For rights or permissions inquiries, please contact National Geographic Books Subsidiary Rights: bookrights@natgeo.com

Note: Some of the night sky portraits in this book implement a photo-stacking process that layers multiple images taken at different exposures. These are then combined into a single photograph, resulting in sharper details and reduced noise, allowing the image to convey more information with greater beauty than would otherwise be possible.

ISBN: 978-1-4262-2438-6

The authorized representative in the EU for product safety and compliance is Disney Trading B.V., Asterweg 15S, 1031 HL, Amsterdam, The Netherlands
email: DCP.DL-EU.bookscontact@disney.com

Printed in China

26/RRDH/1

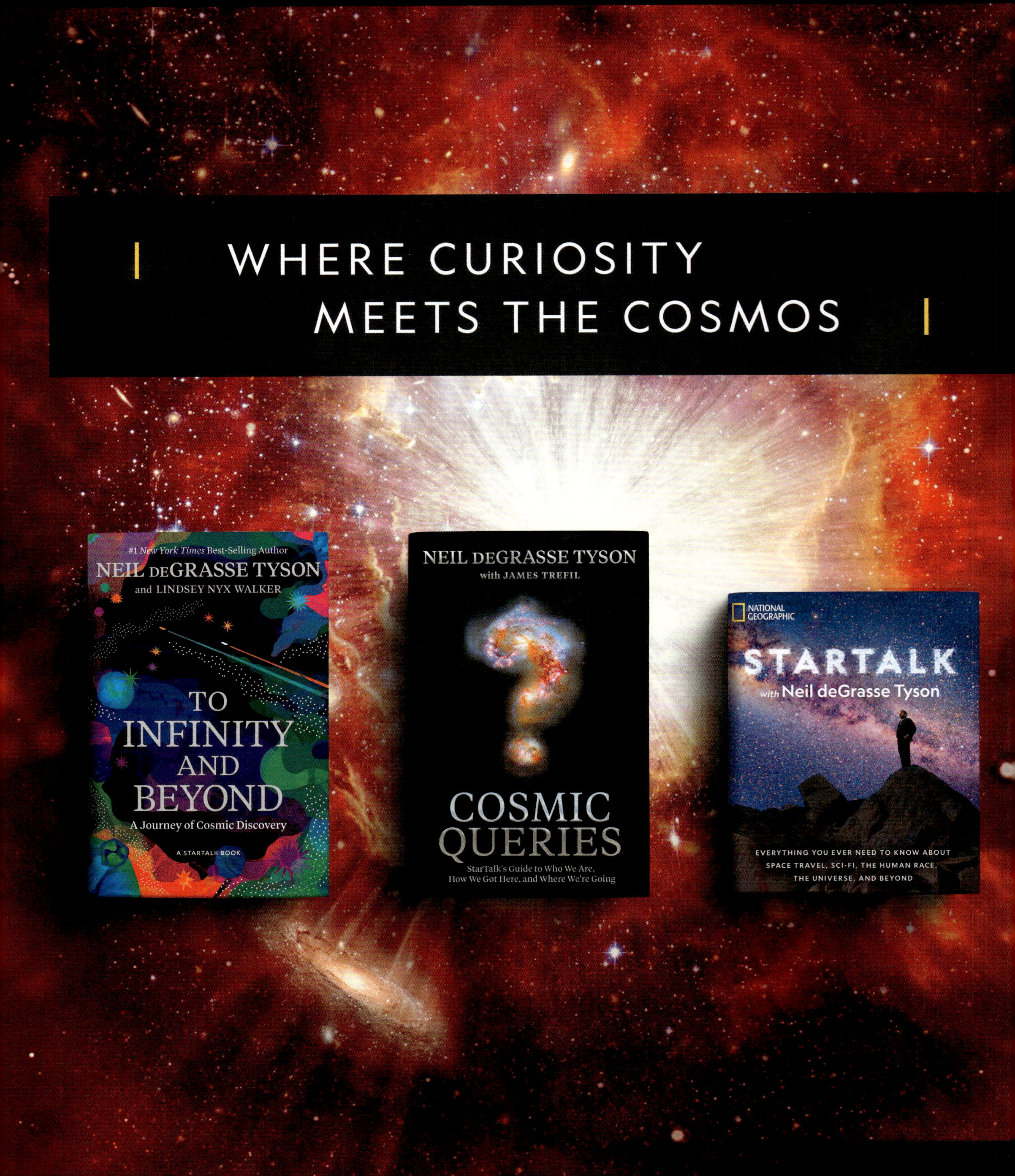
WHERE CURIOSITY
MEETS THE COSMOS
#1 New York Times Best-Selling Author
NEIL deGRASSE TYSON
and LINDSEY NYX WALKER
TO INFINITY AND BEYOND
A Journey of Cosmic Discovery
A STARTALK BOOK
NEIL deGRASSE TYSON
with JAMES TREFIL
COSMIC QUERIES
StarTalk's Guide to Who We Are, How We Got Here, and Where We're Going
NATIONAL GEOGRAPHIC
STARTALK
with Neil deGrasse Tyson
EVERYTHING YOU EVER NEED TO KNOW ABOUT SPACE TRAVEL, SCI-FI, THE HUMAN RACE, THE UNIVERSE, AND BEYOND